Digital Technology and Democratic Theory

Digital Technology and Democratic Theory

EDITED BY LUCY BERNHOLZ,
HÉLÈNE LANDEMORE, AND ROB REICH

The University of Chicago Press
Chicago and London

The University of Chicago Press, Chicago 60637
The University of Chicago Press, Ltd., London

Published 2021
Printed in the United States of America

30 29 28 27 26 25 24 23 22 21   1 2 3 4 5

ISBN-13: 978-0-226-74843-6 (cloth)
ISBN-13: 978-0-226-74857-3 (paper)
ISBN-13: 978-0-226-74860-3 (e-book)
DOI: https://doi.org/10.7208/chicago/9780226748603.001.0001

Library of Congress Cataloging-in-Publication Data

Names: Bernholz, Lucy, editor. | Landemore, Hélène, 1976– editor. | Reich, Rob, editor.
Title: Digital technology and democratic theory / edited by Lucy Bernholz, Hélène Landemore and Rob Reich.
Description: Chicago : University of Chicago Press, 2021. | Includes bibliographical references and index.
Identifiers: LCCN 2020024755 | ISBN 9780226748436 (cloth) | ISBN 9780226748573 (paperback) | ISBN 9780226748603 (ebook)
Subjects: LCSH: Democracy—Technological innovations—United States. | Computer networks—Political aspects—United States. | Information technology—Political aspects—United States. | Digital communications—Political aspects—United States.
Classification: LCC JC423 .D629 2020 | DDC 320.973—dc23
LC record available at https://lccn.loc.gov/2020024755

♾ This paper meets the requirements of ANSI/NISO Z39.48-1992 (Permanence of Paper).

Contents

Introduction

Lucy Bernholz, Hélène Landemore, and Rob Reich

This book explores the intersection of digital technology and democratic theory. We assembled scholars from multiple disciplines to wrestle with the question of how digital technologies shape, reshape, and affect fundamental questions about democracy practice and democratic theory. We asked our contributors to consider what democratic theory—broadly defined as scholars engaged in debate about the values and institutional design of democracy—can bring to the design, development, use, and governance of digital technologies. The simple goal: through a series of workshops that spanned nearly two years, we would examine how democratic ideals might provide us with a framework for understanding and, more importantly, shaping the digital transformation of our lives.

What kind of transformation? On the one hand, digital technologies allow for greater connectivity and transmission of information among individuals the world over. They empower people to produce and share content and to act collectively, all with access to more knowledge than ever before, across borders and distances previously considered insurmountable, and at virtually no financial cost. The Arab Spring, the Umbrella Revolution in Hong Kong, and many of the "color" revolutions around the world, as well as Black Lives Matter, the #MeToo movement, the Yellow Vests movement in France, and youth-led climate action in the United States and Europe, would not have been possible without social media.[1]

Digital technologies were similarly essential to the creation of the free, open-source, multilingual, and universal encyclopedia Wikipedia. They were central to the concept of "open government" first put in place by the Obama administration in 2009 and later expanded into the Open Government Initiative, an international network of countries seeking to fight government corruption via increased transparency and citizen participation. Digital technologies allowed Iceland to produce, in 2011, the first "crowdsourced" constitutional proposal in the world. In Taiwan, an innovative process for deliberative governance called vTaiwan has used digital technologies to facilitate successful negotiations among authorities, citizens, and companies like Uber and Airbnb. In France both the Great National Debate of January–March 2019 and the following Convention on Climate Change of October 2019–June 2020 relied heavily on digital technologies to experiment with deliberation at the scale of a large nation. Finally, as the COVID-19 pandemic ravages the world, digital technologies enable some people to telecommute, stay connected, exchange information, and serve as an essential element of public health surveillance.

On the other hand, the very same forces that connect individuals the world over and allow them to produce and share information also lead to a superabundance of content, and its global dissemination, which threatens the health of a functional public sphere and perhaps democracy itself. This flood of content has contributed to the rise of so-called post-truth politics,[2] in which citizens, lacking easy ways to sort good from bad information, are susceptible to misinformation, disinformation, and propaganda. Additionally, the business model of dominant tech firms—capturing and reselling users' attention—may reinforce or exacerbate political polarization by locking people in filter bubbles and echo chambers where the like-minded speak to the like-minded, ultimately increasing the distance between political discourse and reality.[3]

Additionally, network effects mean that a platform becomes more influential and valuable the more people who use it. This has enabled the domination of global platforms such as Google (including YouTube) and Facebook (including Instagram and WhatsApp), which, because they serve as the main gateways to information on the internet, confer on these companies extraordinary power. From the standpoint of any individual, or any nation, a tiny number of companies is responsible for algorithmically constituting our informational ecosystem. The policies of Facebook and Google are more consequential for permissible speech than is anything decided by Canada, France, or Germany. In Timothy Garton Ash's phrase, big tech firms are the new "private superpowers." And with such power naturally comes the possibility of abuse of power, manipulation, and mass surveillance.

It was once fashionable to regard technology through a starry-eyed utopian lens. The internet was an immaterial space beyond the reach of politics. The power of bits and bytes and microchips would topple authoritarian regimes, spread democracy, and connect distant peoples. Today conventional wisdom holds that technologists have brought the world addictive devices, an omnipresent surveillance panopticon, racist algorithms, and disinformation machines that exacerbate polarization, threatening to destroy democracies from within. Many suspect the companies are led by a band of ahistorical techno-libertarian merry pranksters who, after minting their billions by collecting our digital exhaust stream and selling personalized access to advertisers, are now either discovering the reputation-cleansing project of big philanthropy or readying to escape to Mars.[4]

This book went to press in the midst of the coronavirus pandemic and during a period of intensified activism and heightened media attention to endemic racism and police brutality in the US. The health crisis laid bare both the worst and the best of our digital dependencies, from an "infodemic" of false cures and conspiracy theories to heightened commitments to use and protect online tools for public health and for conducting the essential business of the democratic institution, such as the census, public deliberation, and court proceedings.[5] In an emergency, however, time horizons are shortened, and we look to any available means to cope.

What is needed is an avoidance of digital utopianism and dystopianism, and a more sober, long-term, assessment. How should we understand the powerful interactions between digital technologies and democratic ideals? We are overdue, in particular, for scholars of democratic theory to take stock of digital technologies and their promise and peril for reshaping democratic societies and institutions. In this collection of essays, scholars from across multiple disciplines examine enduring democratic commitments of equality and inclusion, participation, deliberation, a flourishing public sphere, civic and political trust, rights of expression and association, and voting through the lens of global digital networks. What we offer is a beginning—a research agenda for more cross-disciplinary scholarship. This volume makes clear that there is much work to be done to understand and, possibly with the aid of technology itself, to improve democracies in the digital age.

Whither Democratic Theory?

Despite the obvious ways in which digital technologies are having profound effects on democratic institutions and citizens, democratic theorists have yet to confront those effects. In recent years, democratic theorists have been more focused on the nontechnological aspects of the widely diagnosed crisis of democracy, whether the various democratic "deficits" built in existing institutions and particularly parties and electoral systems; the economic nexus of money in politics, neoliberalism, globalization, and increasing economic inequalities; political polarization and the decline of civil discourse and democratic norms; the rise of an age of distrust and shifting popular expectations in terms of political legitimacy; or the threat of populism or authoritarianism.[6] As a result, what scholarship exists on the intersection of digital technology and democracy has been done by scholars in media studies and communication theory, scholars of race and technology in African American studies and sociology, and by journalists, lawyers, and activists.[7]

Why have democratic theorists been so silent? The relative novelty of digital technology is one explanation. John Rawls observed that "the politician looks to the next election, the statesman to the next generation, and philosophy to the indefinite future."[8] The time horizons of philosophy are long, dealing more comfortably with perennial questions and slow-moving issues. But digital technologies and the phenomena they give rise to are new and puzzling objects. Facebook, for example, grew from a start-up in a Harvard dorm room to one of the most powerful companies on the planet in only a decade and a half.

We suspect as well that democratic theorists have been silent because familiar conceptual frameworks for thinking about politics and, specifically, democratic governance are maladapted to the task of analyzing the civic and political dimensions of the internet and the dominant tech companies. For instance, should we see giant platforms like Facebook and Twitter primarily as social networks or as media platforms? Are Facebook users best conceptualized as consumers, "produsers" (a portmanteau to refer to the dual identity of social media users as both users of a product and producers of content), subjects, or stakeholders? Is Google search best conceptualized as a service provided by a private corporation or as something akin to public or civic infrastructure? Do these platforms deliver products, services, public goods, or all of the above?

We haven't yet found agreement on whether technology companies are "only" technology companies or also publishers, news organizations, transportation providers, and even, in the vocabulary of philosopher Elizabeth Anderson, new forms of "private governments."[9] For example, there is robust debate about how to best to conceptualize just what a platform provider is.[10] Such decisions about how to think about technology companies are consequential: when Twitter in 2016 decided not to ban Donald Trump despite his repeated violations of its terms of service, the company appealed to a notion of public interest that shifted its role in the conversation from private company to civic platform. And in 2019 when an appeals court ruled that Trump could not block people on Twitter, it similarly interpreted Twitter as a public platform. At the same time, companies such as Facebook initially invited comparison to a digital public sphere but have more recently backed away from such analogies in favor of describing their services as hosting private communities.

Yet another difficulty lies in the fact that the big tech companies are global businesses, which raises questions at a supranational level that the conceptual toolbox of democratic theory has long had difficulty addressing. If we still cannot quite envision what a global democracy is or should look like, or even more minimally how to create a just global order in a world of nation-states, it is difficult to theorize about the regulation, control, and possible democratization of these already-global digital tools and services.

Beyond the conceptual difficulties, it is a familiar observation that new technologies race ahead of sensible public policy. Well-designed industry or public regulation arrives, if at all, only after the initial effects of a technology are widely felt and understood. For all of these reasons, political philosophy, including its more empirically sensitive branch of democratic theory, is playing catch-up with the fast-moving world of digital technologies.[11]

Finally, democratic theorists have not made serious attempts to work in tandem with engineers to understand the potential of digital technologies, let alone to design technologies to contribute to democratic practices and institutions with the aim of better fulfilling democratic ideals. There is enormous potential to harness, shape, and craft digital tools to serve various democratic aspirations. Consider experiments in online deliberation, crowdsourcing, e-voting, participatory budgeting platforms, and enhanced online communication between elected officials and constituents. Consider, finally, recent work on race and technology. Ruha Benjamin, for example, points out the many examples of "discriminatory design" built into our digital worlds and, in response, sketches out, together with the contributors to her edited volume *Captivating Technology*, a plurality of technological futures that aim to be anti-racist, liberatory, and more just overall. In an age of AI, big data, social media platforms, and justice (as well as discrimination) by algorithm, the lack of attention by democratic theorists to digital technologies is increasingly problematic and puzzling.[12] Consider also the creation of cryptocurrencies that are decoupled from state sovereignty, algorithmic determinations about how states distribute benefits or courts apply sentences, and the creation of a digital public sphere by social networks that shape what we get to see and do online. In an age of AI, big data, social media platforms, and (in)justice by algorithm, the lack of attention by democratic theorists to digital technologies is increasingly problematic and puzzling. This lack of attention to technology, interestingly, also affects theorists of race issues, as long as they operate from within the boundaries of political theory, rather than other fields. For example, Eddie Glaude's *Democracy in Black* (2016) and the otherwise excellent volume by Rogers and Turner (2020) on African American political thought address democracy and capitalism through the lens of race but without an explicit consideration of technologies).[13]

On the technical side, developers and engineers often overlook or ignore scholars in the humanities and social sciences, especially women of color, who have led the way in exploring the ethical and political dimensions of technology. Technologists and democratic theorists alike can draw from critical race and feminist theory on technological innovation and social systems. Many of the most powerful digital technologies have civic and political externalities, the foreseeable outcomes of shifting individual pieces of complex social and governing systems. Technologists, too, have been playing catch-up in dealing with the consequences of multipurpose technologies, from Twitter trolls and cyberbullying to algorithmic bias to a privacy-violating business model of surveillance capitalism to the simple fact that much of the internet seems dominated by white men with strong opinions. Indeed, the digital technologies that have changed our world were produced by a category of people that skews male, white, socially liberal, and libertarian (or at least antiregulation), and is geographically concentrated in Silicon Valley.[14]

Put most simply, the omnipresence of networked digital technologies requires us to reexamine several of the most fundamental questions of political theory: Who governs? What are the boundaries of the demos? What rights, freedoms, and obligations do individuals have?, and How do we track or generate something like a common good? Democratic values and aspirations also enable us to imagine—and perhaps design and implement—digital technologies that facilitate self-governance, expand the franchise, center justice, protect individual liberties, cultivate more informed civic and political participation, and nurture the pursuit of collective goods.

This volume arose out of a conviction that digital technologies are having a profound influence on democratic governance and that, reciprocally, the shape and content of democratic institutions can influence the design and evolution of digital technology. Our starting point was to move beyond the techno-utopian rhetoric that surrounded the emergence of the internet as well as the Chicken Little alarmism in the wake of the 2016 US election. The digital tools and platforms that are so ubiquitous in our lives are neither inherently liberatory nor destroyers of democracy.

Our goal was to bring a more rigorous approach to these questions by organizing an interdisciplinary conversation among scholars who are not frequent collaborators—political philosophers, social scientists, and technologists. Despite the civic and political dimensions of services such as Google search, Facebook, YouTube, and so on, engineers rarely work alongside social scientists and philosophers. And despite the increasing body of scholarship by social scientists on the effects of various technologies, we rarely see collaboration between them and the engineers who design and bring the technologies to market.[15] If we assembled scholars from a variety of disciplines, all with interests in democratic theory, to explore the intersection of digital technology and democracy, would they have a common conceptual vocabulary? Would there emerge new frameworks of analysis or novel approaches to the understanding of, and potential design of, technologies in support of democratic aims? Could interdisciplinary conversations contribute to disciplinary thinking as well as to ongoing public debate?

There is an abundance of opportunity for interdisciplinary scholarship. Political leaders and media analysts such as Stacey Abrams, Tiffany Cross, and Symone Sanders have written about the ways policy, race, and technology shape contemporary political campaigns. Race and gender scholarship on Black Twitter and hashtag activism examine questions of participation, identity, and exclusion—all core to the democratic project. In taking up digital technologies as a force within democracies, political theorists have considerable work to draw on from these adjacent domains.[16]

We did not imagine that we would come to any definitive conclusions. But we hoped to find ground for dialogue, to model constructive cross-disciplinary exchange, and to identify in our work a research agenda for the future. We view the contributions in this volume as the start rather than the conclusion of a scholarly conversation.

The Conceptual Landscape

Consider the basic terms of our title: digital technology and democratic theory. We did not impose definitions for either term, instead allowing our ongoing dialogue to reveal potentially shared conceptual territory.

Our conversations began with the idea that digital technology could be understood as the set of information and communication technologies that make use of the networked electronic generation, processing, and storage of data. "Digital technology" is thus an umbrella that covers computers, the internet, smart phones, and much more. Each chapter examines particular technologies of interest to the authors.

We characterized democratic theory as the field concerned with the study of democracy as both a theoretical and an empirical object, in both its descriptive and its normative dimensions. Central to democratic theory is the ideal of collective self-determination or self-rule, the ideal of the power of the people (*demos-kratos*), as rule of, by, and for the people.

But how to analyze the intersection of digital technology as such and democratic theory as such? For any useful analysis, we had to disaggregate various kinds of digital technologies and distinguish various institutional designs or components of a democratic society. Rather than attempting to analyze the effects on democracy of something as diffuse as "the internet," we had to consider the political consequences of different phenomena, like the online public sphere or the changing nature of trust in online interactions. And when thinking of how various aspects of democratic governance might be used to evaluate or design new technological objects and infrastructures, we had to think not in terms of "democratic governance" in the abstract, but from specific evaluative viewpoints, like inclusion, power, legitimacy, representation, and deliberation. On both the technological and political sides, productive analysis required a finer-grained set of definitions.

As a consequence, this book contains discussion of multiple digital technologies and different democratic theories. On the technological side, our contributors considered digitized data, the use of algorithms in processing digitized information, and decentralized technologies such as blockchain. Of course, we cannot seek to be exhaustive: absent from this volume are systematic discussions of important subjects such as artificial intelligence, quantum computing, and virtual or augmented reality. As a result, although at least two contributors (Ford and Landemore) venture into what could be considered science-fiction territory, the chapters tend to remain in the space of known or immediately foreseeable technological applications. None of the chapters explores the more radical and fantastical visions of an AI-based democracy, as imagined, for example, by Martin Hilbert or Cesar Hidalgo.[17]

On the political side, our contributors brought a variety of complementary theoretical perspectives on democracy. Many chapters are committed to a framework that emphasizes the role of deliberation in the informal or formal public sphere. Others operate in a (compatible but distinct) pragmatist lineage, emphasizing an experimentalist approach to solving social problems. Seen through this lens, democratic institutions are problem-solving tools, much like technology itself. Others are committed to a more classically Schumpeterian model of democracy, emphasizing electoral competition and vote aggregation. Finally, many put their emphasis on the rich ecology of offline or online associations that a thriving democracy depends on, inscribing themselves in the tradition of participatory democrats such as Robert Putnam, Carole Pateman, and Benjamin Barber. Meanwhile, some look to digital technologies as a means to move beyond representation and return to more direct forms of democracy at scale, whereas others yet see them as mechanisms to rethink democratic representation along nonelectoral lines.

Perhaps, more grandly still, some aspire to use digital technologies to move beyond the nation-state and globalize democratic governance. Cutting across these distinctions, some look to democracy mostly for procedural fairness; other mostly for the quality of outcomes, including in terms of knowledge aggregation.

Therefore, although our title is formulated in the singular, the book volume represents an exploration of the intersections of digital technologies and democratic theories. And these intersections remind us of the importance of asking old questions in democratic theory while being mindful of the promise and peril of new technologies.

An Emerging Research Agenda

The chapters in this volume lay the groundwork for an emerging research agenda in democratic theory. That agenda starts from a set of perennial questions in democratic theory, exploring them in light of the design and dissemination of various digital technologies. We believe that progress on this research agenda will be made through the collaboration of democratic theorists, social scientists, and engineers.

Among the perennial questions are: What are the informational conditions of a healthy public sphere and an informed and deliberative citizenry? How does the free speech ideal fit into a digital public sphere? Must we reestablish "gatekeepers" in order to build conditions of trust between citizens and experts, between citizens and politicians, or among citizens themselves? What are the geographic boundaries of the demos? Who counts as a citizen?

Democratic theory can also help formulate, question, and potentially address new issues as they are implicitly raised by digital technologies and the new categories those technologies create. How can a technology affect the inclusion or exclusion of citizens? How does the technological landscape affect the balance of power in a democratic society—between experts and others, corporations and the state, citizens and foreigners? What is and who owns digital "data"? What counts as online "consent," whether to terms of agreement or to the sharing of one's data? What, if anything, do companies owe to social media platform users? Could we imagine a "democratic," non-state-based, blockchain-enabled cryptocurrency, and what principles should it encode? What normative principles should shape the design of algorithms that curate information and news?

Digital technologies have also reopened questions seemingly closed by mainstream democratic theory before the advent of the Internet: Is direct democracy on a mass scale possible and, if so, desirable? Specifically, is deliberation on a large scale possible and, if so, desirable? Do we still need representation, and if so, of what kind? As online voting allows for vote delegation, bot voting, vote splitting, and vote budgeting, is "one person, one vote" still the best way to implement political equality?

This set of diverse questions—perennial, new, or freshly reopened—invites renewed inquiry in democracy theory and also new cross-disciplinary consideration. This volume models the practice of bringing different scholarly disciplines to shared questions.

Stimulated by the questions outlined here, the chapters in this volume revolve around several common themes: the informational conditions of a healthy democracy; the potential for, and nature of, expressive and associational spaces; the definition of a digital demos; the role of silence and exclusion in understanding participation; and the division in democratic societies between institutional and individual responsibilities. These are some of the core themes, we posit, at the heart of a new research agenda in democratic theory. Adjacent literatures important for theorists to engage include studies of race, power, artificial intelligence, justice, and algorithmic accountability.

THE INFORMATIONAL CONDITIONS OF A HEALTHY DEMOCRACY

Democracies rely on informed citizens. The contributors to this volume share some assumptions about what is necessary for a healthy digital public sphere: access to information, knowledge, and news; opportunities for deliberation; some form of minimal "truth tracking"; and trust (both vertical, between readers and journalists, citizens and government, and horizontal, among citizens themselves, between parties). An important conclusion emerges from reflecting on how digital technologies have transformed the public sphere: there exists a potential trade-off between quality and quantity (or diversity) of content. The internet reduces the cost of consuming and distributing information to near zero, enabling mass participation in the digital public sphere while also facilitating the amplification of misinformation or disinformation. How should we assess what some of the chapters refer to as the aperture and filter dimensions of digitally enabled communication when it comes to the overall quality of the public sphere?

The question of the quality of our existing public sphere is opened rather than closed by the chapters here. We still do not know, for example, if, despite popular opinion to the contrary, our current public sphere is less able to track the truth about various facts and the public good than the public sphere of the preinternet age. From a purely theoretical perspective, the literature on collective intelligence tells us that if digital technologies could cultivate inclusive deliberation in the new digital public sphere, its outcomes might be better than those produced by the more gated, less diverse model of the pre-internet era.[18] A key condition for this to be true, however, is that genuine deliberation takes place. At the moment, we are still groping for a way to make genuine online deliberation happen and are experiencing the turbulence of what could be a long transition phase out of an old order into a new one.

Our contributors fall, loosely speaking, into two camps on the question of free speech in the digital age: the Millians and the Rousseauvians. The Millians deplore the abuses of free speech online but see education and intersubjective monitoring of citizens as the appropriate answer, over and above legal or platform-based speech restrictions. The Rousseauvians, by contrast, favor more democratic or community-based self-rule and the active inclusion and protection of vulnerable minorities, even at the cost of some restrictions on free speech, whether imposed by governments or by platforms.

As a result, we see a research agenda for normative theorists (e.g., must we adapt our understanding of the free speech ideal in a digital era?), for social scientists (e.g., what is the effect of big tech companies on quality and quantity of information, on echo chambers and filter bubbles, on polarization?), and for engineers (e.g., how can digital tools and platforms be designed with democratic ideals in mind?). Several chapters here lay out a framework to understand the informational conditions of a healthy democratic society and try to take stock of the transformation wrought by digital technologies.

NEW EXPRESSIVE AND ASSOCIATIONAL SPACES

Liberal democracies seek to guarantee a slate of individual rights, including free expression and freedom of assembly. In addition, one of the core topics of debate in democratic theory is the role of information and an informed citizenry. Digital environments are the result of the interplay of, on the one hand, engineered designs and algorithmically mediated choices (which can in turn be shaped by commercial priorities), and, on the other hand, human desires for interaction, persuasion, knowledge, and entertainment. Digital technology opens up a wide variety of communication, including from person to person, from person to organization, from person or organization to government or state, from constituent to representative and back, and so on. These communications can be horizontal, taking place among individuals or groups positioned at a similar level of the governance or hierarchical structure, and vertical, between individuals or groups on the lower rungs and individuals and groups or entities (corporations, states, governments) on the higher rungs of this same chain structure.

These pairings open up new research questions concerning how digitally enabled expressive and associational spaces might aid the pursuit of democracy. In doing so we also must be alert to how digital spaces might be designed or used to suppress freedom or produce inequalities.

This presents us with two key tasks for new scholarly work. First, we must consider whether and how our networks of digital communications accommodate and protect individual rights to expression and association in ways that also protect privacy and resist corporate or state surveillance. If one research agenda is to analyze the digital public sphere, then a second and related research agenda is to delineate the infrastructure of digital civil society and to understand its promise and peril. Several chapters focus attention on the realm of expressive and associational spaces in a digital age.

REDEFINING THE DEMOS?

Digital technologies offer new opportunities to imagine and define the demos above and beyond the traditional boundaries set by geography and history. First, the ability to vote and partake in other political action online allows citizens to participate in their government from outside the physical borders of a democracy. Attempts to implement remote voting and participation through physical methods (like paper) have been cumbersome to the point of dissuading participation, and many countries give no political participation options at all to citizens living abroad. Digital technologies can massively lower the cost of enfranchisement for citizens who do not physically reside in their country.

Second, digital technologies enable virtual communities that crosscut or transcend traditional boundaries for community and to which people the world over can self-ascribe at will, for example, on the basis of shared interests. Digital technologies therefore allow for the creation of new, virtual *demoi*, completely independent of physical and territorial constraints. Finally, digital technologies could even empower something like a global democracy, a community that is physically located on planet Earth. Might technology enable a humanity-based demos, if only with respect to global issues such as climate change, North-South inequalities, transnational migrations, and terrorism? Scholars have yet to explore and assess the revolutionary potential of such new opportunities. Several chapters in this volume point to a research agenda enabling us to explore how technology can facilitate a reimagination of the demos.

Silence and Exclusion

For all their promise of bringing back together scattered *demoi*, creating new ones, or constituting a global demos, digital technologies also foster exclusionary group definitions or redefinitions.

If we take inclusion on equal terms as both a precondition of and an ideal for democratic society, then everyone's interests must count, and count equally, and everyone must have equal rights, and perhaps roughly equivalent means, to participate in civic and political life. Digital technologies promise greater access to the political sphere, but they also create new barriers, entrenching what is perhaps an ultimately unbridgeable digital divide. Some of these barriers are physical and economic. Many communities are sidelined because of insufficient built infrastructure, or because existing infrastructure is, in the absence of regulatory requirements that would ensure universal access, not accessible. In response, long-standing struggles for equal access expand their focus to digital inclusion.

Active participation takes many forms in the analog world, and so we must consider many possible parallels in the digital world. Protestors often make their point by removing themselves to the edge of the discussion. They exercise their voice and signal their disagreement by standing silently on the edge. Choosing to step out of digital spaces as a form of civil disobedience requires a new form of signaling. To understand—and design—digital spaces as democratic fora, we must account for those who are present, those who are excluded, and those who chose not to participate. This requires engineering for equity and inclusion. Two contributions here identify promising new research frontiers by foregrounding the democratic dimensions of silence and refusal in a digital age.

INDIVIDUAL AGENCY VERSUS INSTITUTIONAL DESIGN APPROACHES

Many of the chapters speak specifically to, or demand certain capacities of, individual actors in their roles as democratic participants. They point to specific digital literacies we need or must strive to develop. Others expect or call out for new individual responsibilities for our interactions in the digital public sphere, and some suggest that technology can be developed and deployed to enhance civic expertise and judgment.

Democratic theorists point out the basic tension underlying these options as determining what is the responsibility of individuals and what is the responsibility of the state. Where authors challenge the notion that individual agency is enough, and place their attention on regulatory, technological, and corporate changes, we have to also specify what may or may not be possible.

What is new here for democratic theory is the need to take full account of the "governing" roles that software code and digital systems can play by virtue of their default designs and the business models of their creators. The more our human systems of governance come to depend on software code and digital infrastructure, the greater the need to design democratic norms directly into these engineered systems. How to understand these dynamics between humans and digital systems, how to design digital systems specifically for democratic purposes, and how to hold such systems accountable is a new frontier for political theorists, digital technologists, and the public citizens. All the contributors to this volume explore in some fashion this new frontier in democratic theory.

Chapter Overview

We open the volume with two very different theoretical frameworks. The first chapter assumes a Habermasian paradigm of deliberative democracy, whereas the second chapter, also committed to deliberation, is centered on nonelectoral forms of democratic representation (specifically bodies constituted by random selection). The deliberate juxtaposition of these two visions is intended to engage readers immediately in a range of possibilities about the meaning and possible institutionalization of democracy as well as the role digital technologies can play in relation to these possibilities. From there we move on to more specific and empirical chapters that point out the democratic limits as well the possibilities and resources inherent to the deployment of digital technologies in the political space. Along this arc, we alternate disciplinary angles while preserving a thematic continuity from one contribution to the next. Because the chapters were written for a multiple-stage interdisciplinary workshop, they make frequent reference to one another. The chapters are in a loose dialogue.

Joshua Cohen and Archon Fung's opening chapter, "Democracy and the Digital Public Sphere," begins with the Habermasian two-track model of the public sphere and focuses on the informal sphere of "deliberation in the wild," the unstructured online public exchanges that are meant to inform and shape the debates in the formal decision-making sphere (e.g., Congress, Parliament, Courts). The authors also trace the path from "internet optimism" to "techno-pessimism." In so doing, they spotlight how digital fora raise anew questions of equality, access, identity, liberty, and deliberation central to democratic theory.

Hélène Landemore's "Open Democracy and Digital Technology" imagines how key institutional principles of her new paradigm of open democracy could be facilitated using modern technological tools. This chapter defends technologically empowered forms of nonelectoral yet democratic representation, centrally by means of networked "open minipublics" based on selection by lottery. Landemore argues that digital technologies, when properly harnessed, allow for the better realization of the democratic ideals of inclusion and equality.

The following three chapters look at questions of participation, representation, and institutional supports. Networked digital technologies and digitized data present opportunities for institutional reinvention. The key question that undergirds this opportunity—how to use private resources for public benefit—has been asked before. One answer from the past, when the resources at stake were rival and excludable, was nonprofit or nongovernmental organizations. Lucy Bernholz's "Purpose-Built Digital Associations" examines how efforts to protect and direct digital data toward public purposes finds us adapting old organizational forms to serve new associational purposes.

Seeta Gangadharan's "Digital Exclusion: A Politics of Refusal" draws our attention to a key question of democratic theory: who participates and how? By focusing on the physical and economic infrastructure that is required to provide broadband access, Gangadharan shows how entire communities and regions are excluded. Certain community responses to this exclusion are presented, however, as evidence of an assertion of political power rather than disadvantage. The chapter highlights the informational needs and associational options that are necessary for digital participation.

But the digital square, like its analog predecessor, is not defined solely by what is said or communicated. Those who abstain from participation as a form of protest are also part of the democratic process. Mike Ananny's chapter, "Presence of Absence: Exploring the Democratic Significance of Silence," argues that digital silence can be seen as a form of civil disobedience.

The grouping of chapters that follow considers questions of governance, trust, and legitimacy. In "The Artisan and the Decision Factory: The Organizational Dynamics of Private Speech Governance," Robyn Caplan examines the way that organizational structure influences how different platform companies curate content. Though governed by a common set of regulations, specifically Section 230 of the US Communications Decency Act and the General Data Protection Regulation of the European Union, Caplan finds that platform companies are effectively self-governing when it comes to content moderation and that organizational culture plays a large role in how Facebook, Twitter, and Google approach their roles as information gatekeepers.

Henry Farrell's and Melissa Schwartzberg's chapter, "The Democratic

Consequences of the New Public Sphere," asks how trust between citizens is shifting when news and information is produced and accessed online. With many more producers of information, and fewer gatekeepers, they find potentially troubling shifts in whom citizens find to be trustworthy providers of information.

How can we use digital technologies to cultivate human capital and prepare citizens of any age for a rapidly shifting economy? In "Democratic Societal Collaboration in a Whitewater World," David Lee, Margaret Levi and John Seely Brown examine the possibilities that microtasking and distributed participation present for addressing economic and educational inequality. In contrast to more familiar arguments that digital technologies will replace, or at best augment, human workers, their chapter examines how digital tools and platforms can be used to help people continuously learn new skills. Structuring educational processes in this way also involves redesigning nonprofit institutions, touching on new associational possibilities.

Julia Cagé's "From Philanthropy to Democracy: Rethinking Governance and Funding of High-Quality News in the Digital Age" considers the economics of digital technologies and how we might reimagine the provision of news and information. This chapter brings together historical insights from past financial structures, theory on the role of informed citizens in democracies, and contemporary analysis of digital-first business models.

In the final chapter, "Technologizing Democracy or Democratizing Technology? A Layered-Architecture Perspective on Potentials and Challenges," Bryan Ford brings his training as an engineer to propose that we use technologies to reinvent more than just new communication spaces, public spheres, voting systems, or general governance designs. Building on his own preferred model of "delegative" or "liquid" democracy, which allows for vote delegation—opening the status of democratic representatives beyond the list of usual (party-backed) suspects—Ford argues for a democratic digital currency that underwrites a universal basic income and the conditions of genuine political equality.

Toward a New Generation of Scholarship by Newly Trained Scholars

Digital technology changes and evolves so rapidly that, from the usual time horizon of scholarly inquiry and publication, it is impossible to remain fully informed about the latest technological trends. We acknowledge that we are writing and publishing this volume at a transitional moment, and that the ground on which we stand is ever shifting. The global pandemic of 2020 makes this dynamism ever more consequential.

This is why we have rooted our inquiry in enduring questions in democratic theory. Our aspiration is that this volume contributes to scholarship in two ways: to illuminate a broad research agenda on democratic theory in a digital age and to demonstrate the fruits of interdisciplinary modes of collaboration. One of the arguments for examining how digital technologies interact with the "familiar" questions of democratic theory is to help determine whether new questions emerge and where they may take us. Is there a need for "digital democratic theory," or do our existing frameworks fit current and future possibilities?

To meet the challenges of democracy in a digital era, we believe that universities need to train a new generation of scholars with the tools of engineers, social scientists, and humanists, and to stimulate cross-disciplinary conversations of the sort represented in this volume. More engagement with newer fields such as social and behavioral economics would also be welcome. We need to break down divides between schools of engineering and schools of humanities and sciences. We need "public interest technologists" and expanded intersections between humanistic and computational disciplines.

What should this new scholarship address? There are many questions at the intersections of digital technologies and the core concepts of democracy. The significant examinations of justice, access, and participation that exist within the literatures on race and technology offer inspiration.[19] It is in those questions—where both digital and democratic assumptions are examined together—that we find potentially lasting change and novel frameworks. The role of information and an informed citizenry is a central theme in democracy that ties directly to structural questions about inclusion and participation. What is it that matters—universal access to information, the quality of the information available, or whether people deliberate on the information and choices put before them? Viewed through the lens of democratic theory, both platform governance and algorithmic curating generate new questions about who bears the responsibility for an informed citizenry, the role of trust in information gatekeepers, and the many meanings of silence. What does democracy require from its participants and what, in turn, might they require from their information ecosystem? What responsibilities are born by information providers, by governments, and by individuals? How might we reimagine the pursuit of equality, participation, and self-governance given the tools we have at our disposal and those we might create?

Engineering frontiers such as artificial intelligence and biological enhancement have, and will continue to, create new opportunities and challenges to democracy. We hope to inspire scholarly practices and inquiry that speaks to the breadth and depth of these systemic developments.

Notes

1. Even as, for some of these movements, the fact that they were internet-fueled was also the main source of their vulnerability (Tufekci 2018).

2. MacMullen (2020).

3. A burgeoning social science literature assesses the effect of the internet on political polarization. For an overview, see Tucker (2018).

4. Harari (2018).

5. In February 2020 the World Health Organization called on technology companies to address the spread of false information about coronavirus, referring to it as an "infodemic" (see Thomas 2020). As national shelter-in-place orders were put into effect, government bodies as diverse as local city councils and the Supreme Court of the United States found technologically dependent ways to hear testimony and hold deliberations.

6. On parties and electoral systems, see Norris (2011, 2017); Papadopoulos (2013); Tormey (2015); Van Reybrouck (2016). On deliberation, see Parkinson and Mansbridge (2012). On the economic nexus of money, see Brown (2019); McLean (2017). On polarization and the decline of civil discourse, see Levitsky and Ziblatt (2018). On the age of distrust, see Rosanvallon (2008, 2011); Rosenfeld (2018). On populism and authoritarianism, see Müller (2017); Mounk (2018).

7. On journalists and activists, see Morozov (2013); Shirky (2008); Taylor (2014). On scholars from other fields, see Achen and Bartels (2016); Bucher (2018); Brown (2015); Coleman and Shane (2012); Crawford and Lumby (2013); Florini (2019); Gillespie (2010, 2017); Nakamura and Chow-White (2012); Napoli (2015); Nelson, Tu, and Hines (2001); Noble and Tynes (2016); Noble (2018); Noveck (2009, 2015); Pasquale (2015); Poblet and Plaza (2017); D. Roberts (2012); S. T. Roberts (2019); Taylor (2016); Tufecki (2018); van Dijck (2000); Webb (2020); Weltevrede, Helmond, and Gerlitz (2014). On African Americans and democracy theory, see Glaude (2016) and Rogers and Turner (2020).

8. Rawls (1971, 24).

9. Anderson (2017).

10. Read (2017); Gillespie (2018).

11. In fairness, there have been some attempts by democratic theorists to apply democratic theoretical frameworks to the new problems and opportunities raised by digital technologies or to update these frameworks in light of the empirical findings digital technologies generate. See, e.g., Coleman and Blumler (2012); Allen and Light (2016). To some extent, see also Neblo, Kevin, and Esterling (2018).

12. Benjamin (2019a, 2019b).

13. Costanza-Chock (2020).

14. For evidence of the libertarian preferences of technology entrepreneurs, see Broockman, Ferenstein, and Malhotra (2018).

15. Glaude (2016).

16. Abrams (2020); Cross (2020); Sanders (2020); Jackson, Bailey, and Foucault Welles (2020); Freelon, McIlwain, and Clark (2016).

17. Hilbert (2009); Hidalgo (2019).

18. E.g., Page (2007); Landemore and Elster (2012); Landemore (2013).

19. Kasy and Abebe (2020); Buolamwini and Gebru (2018).

References

Abrams, Stacey. 2020. *Our Time Is Now.* New York: Henry Holt.

Achen, Christopher, and Larry Bartels. 2016. *Democracy for Realists.* Princeton, NJ: Princeton University Press.

Allen, Danielle, and Jennifer Light, eds. 2015. *From Voice to Influence: Understanding Citizenship in a Digital Age.* Chicago: University of Chicago Press.

Anderson, Elizabeth. 2017. *Private Government: How Employers Rule our Lives (and Why We Don't Talk about It).* Princeton, NJ: Princeton University Press.

Benjamin, Ruha. 2019a. *Captivating Technology: Race, Carceral Technoscience, and Liberatory Imagination in Everyday Life.* Chapel Hill, NC: Duke University Press.

———. 2019b. *Race after Technology: Abolitionist Tools for the New Jim Code.* New York: Polity Press.

Broockman, David E., Gregory Ferenstein, and Neil A. Malhotra. 2018. "Predispositions and the Political Behavior of American Economic Elites: Evidence from Technology Entrepreneurs." Graduate School of Business Research Paper No. 17-61, Stanford University, Stanford, CA. http://dx.doi.org/10.2139/ssrn.3032688.

Brown, Simone. 2015. *Dark Matters: On the Surveillance of Blackness.* Chapel Hill, NC: Duke University Press.

Brown, Wendy. 2019. *In the Ruins of Neoliberalism: The Rise of Antidemocratic Politics in the West.* New York: Columbia University Press.

Bucher, Taina. 2018. *If . . . Then: Algorithmic Power and Politics.* Oxford: Oxford University Press.

Buolamwini, Joy, and Timnit Gebru. 2018. "Gender Shades: Intersectional Accuracy Disparities in Commercial Gender Classification." *Proceedings of Machine Learning Research* 81:1-15.

Coleman, Stephen, and Jay G. Blumler. 2012. *The Internet and Democratic Citizenship Theory, Practice and Policy.* Cambridge: Cambridge University Press.

Coleman, Stephen, and Peter M. Shane, eds. 2012. *Connecting Democracy.* Cambridge, MA: MIT Press.

Costanza-Chock, Sasha. 2020. *Design Justice: Community-Led Practices to Build the Worlds We Need.* Cambridge, MA: MIT Press.

Crawford, Kate, and Catherine Lumby. 2013. "Networks of Governance: Users, Platforms, and the Challenge of Networked Media Regulation." *International Journal of Technology Policy and Law* 1 (3): 270–82.

Cross, Tiffany D. 2020. *Say It Louder! Black Lives, White Narratives, and Saving Our Democracy.* New York: Amistad Press.

Florini, Sarah. 2019. *Beyond Hashtags: Racial Politics and Black Digital Networks.* New York: NYU Press.

Freelon, Deen, Charlton D. McIlwain, and Meredith D. Clark. 2016. *Beyond the Hashtag.* Washington, DC: Center for Media and Social Impact.

Gillespie, Tarleton. 2010. "The Politics of 'Platforms.'" *New Media & Society* 12 (3): 347–64.

———. 2017. "Governance of and by Platforms." In *SAGE Handbook of Social Media,* edited by J. Burgess, T. Poell, and A. Marwick, 254–78. London: Sage.

———. 2018. *Custodians of the Internet: Platforms, Content Moderation, and the Hidden Decisions That Shape Social Media.* New Haven, CT: Yale University Press.

Glaude, Eddie S. 2016. *Democracy in Black: How Race Still Enslaves the American Soul.* New York: Broadway Books.

Harari, Yuval Noah. 2018. "Why Technology Favors Tyranny." *The Atlantic*, October. https://www.theatlantic.com/magazine/archive/2018/10/yuval-noah-harari-technology-tyranny/568330/.

Hidalgo, Cesar. 2019. "A Bold Idea to Replace Politicians." *TED: Ideas Worth Spreading.* https://www.ted.com/talks/cesar_hidalgo_a_bold_idea_to_replace_politicians/transcript?language=en.

Hilbert, Martin. 2009. "The Maturing Concept of E-Democracy: From E-Voting and Online Consultations, to Democratic Value out of Jumbled Online Chatter." *Journal of Information Technology and Politics* 6 (2): 87–110.

Jackson, Sarah J., Moya Bailey, and Brooke Foucault Welles. 2020. *Hashtag Activism: Networks of Race and Gender Justice.* Cambridge, MA: MIT Press.

Kasy, Maximilian, and Rediet Abebe. 2020. *Fairness, Equality, and Power in Algorithmic Decision-Making*, preprint available https://www.cs.cornell.edu/~red/fairness_equality_power.pdf.

Landemore, Hélène. 2013. *Democratic Reason: Politics, Collective Intelligence, and the Rule of the Many.* Princeton, NJ: Princeton University Press.

Landemore, Hélène, and Jon Elster. 2012. *Collective Wisdom: Principles and Mechanisms.* Cambridge: Cambridge University Press.

Levitsky, Steven, and Daniel Ziblatt. 2018. *How Democracies Die.* New York: Random House.

MacLean, Nancy. 2017. *Democracy in Chains: The Deep History of the Radical Right's Stealth Plan for America.* New York: Viking.

MacMullen, Ian. 2020. "What Is Post-Factual Politics?" *Journal of Political Philosophy* 28 (1): 97–116.

Mounk, Yashka. 2018. *The People versus Democracy: Why Our Freedom Is in Danger and How to Save It.* Cambridge, MA: Harvard University Press.

Morozov, Evgeny. 2013. *To Save Everything, Click Here: The Folly of Technological Solutionism.* New York: Public Affairs.

Müller, Jan-Werner. 2016. *What Is Populism?* Philadelphia: University of Pennsylvania Press.

Nakamura, Lisa, and Peter A. Chow-White. 2012. *Race after the Internet.* New York: Routledge.

Napoli, P. M. 2015. "Social Media and the Public Interest: Governance of News Platforms in the Realm of Individual and Algorithmic Gatekeepers." *Telecommunications Policy* 39 (9): 751–60. https://doi.org/10.1016/j.telpol.2014.12.003.

Nelson, Alondra, Thuy Linh N. Tu, and Alicia Headlam Hines. 2001. *Technicolor: Race, Technology, and Everyday Life.* New York: NYU Press.

Noble, Safiya Umoja. 2018. *Algorithms of Oppression: How Search Engines Reinforce Racism.* New York: NYU Press.

Noble, Safiya Umoja, and Brendesha M. Tynes. 2016. *The Intersectional Internet: Race, Class, and Culture Online.* New York: Peter Lang.

Norris, Pippa. 2011. *Democratic Deficit: Critical Citizens Revisited.* Cambridge: Cambridge University Press.

———. 2017. *Why American Elections Are Flawed (and How to Fix Them).* Ithaca, NY: Cornell University Press.

Noveck, Beth. 2009. *Wiki Government: How Technology Can Make Government Better, Democracy Stronger, and Citizens More Powerful.* Washington, DC: Brookings Institution Press.

Page, Scott E. 2007. *The Difference.* Princeton, NJ: Princeton University Press.

Papadopoulos, Yannis. 2013. *Democracy in Crisis? Politics, Governance, and Policy.* London: Palgrave Macmillan.

Parkinson, John, and Jane Mansbridge, eds. 2012. *Deliberative Systems: Deliberative Democracy at the Large Scale.* Cambridge: Cambridge University Press.

Pasquale, Frank. 2015. *The Black Box Society: The Secret Algorithms That Control Money and Information.* Cambridge, MA: Harvard University Press.

Poblet, Marta, and Enric Plaza. 2017. "Democracy Models and Civic Technologies: Tensions, Trilemmas, and Trade-Offs." Paper presented at the workshop "Linked Democracy: AI for Democratic Innovation" (IJCAI2017), Melbourne, Australia, August 19. arXiv:1705.09015.

Rawls, John. 1999. *A Theory of Justice.* 2nd ed. Oxford: Oxford University Press.

Read, Max. 2017. "Does Mark Zuckerberg Even Know What Facebook Is?" *New York Magazine,* October. https://nymag.com/intelligencer/2017/10/does-even-mark-zuckerberg-know-what-facebook-is.html.

Roberts, Dorothy. 2012. *Fatal Invention: How Science, Politics, and Big Business Re-create Race in the Twenty-First Century.* New York: New Press.

Roberts, Sarah T. 2019. *Behind the Screen: Content Moderation in the Shadow of Social Media.* New Haven, CT: Yale University Press.

Rogers, Melvin L., and Jack Turner. 2020. *African American Political Thought: A Collected History.* Chicago: University of Chicago Press.

Rosanvallon, Pierre. 2008. *Counter-Democracy: Politics in an Age of Distrust.* Cambridge: Cambridge University Press.

———. 2011. *Democratic Legitimacy: Impartiality, Reflexivity, Proximity.* Princeton, NJ: Princeton University Press.

Rosenfeld, Sophia. 2018. *Truth and Democracy: A Short History.* Princeton, NJ: Princeton University Press.

Rule, Sheila. 1989. "Reagan Gets a Red Carpet from the British." *New York Times,* June 14.

Sanders, Symone. 2020. *No, You Shut Up: Speaking Truth to Power and Reclaiming America.* New York: HarperCollins.

Shirky, Clay. 2008. *Here Comes Everybody.* New York: Penguin Press.

Taylor, Astra. 2014. *The People's Platform. Taking Back Power and Culture in the Digital Age.* New York: Metropolitan Books.

Taylor, Keeanga-Yamahtta. 2016. *From #BlackLivesMatter to Black Liberation.* Chicago: Haymarket Books.

Thomas, Zoe. 2020. "WHO Says Fake Coronavirus Claims Causing 'Infodemic.'" BBC News, February 13. https://www.bbc.com/news/technology-51497800.

Tormey, Simon. 2015. *The End of Representative Politics.* Cambridge, UK: Polity Press.

Tucker, Joshua A., Andrew Guess, Pablo Barberá, Cristian Vaccari, Alexandra Siegel, Sergey Sanovich, Denis Stukal, and Brendan Nyhan. 2018. "Social Media, Political Polarization, and Political Disinformation: A Review of the Scientific Literature." Report prepared for the Hewlett Foundation, Menlo Park, CA. https://hewlett.org/wp-content/uploads/2018/03/Social-Media-Political-Polarization-and-Political-Disinformation-Literature-Review.pdf.

Tufekci, Zeynep. 2018. *Twitter and Tear Gas: The Power and Fragility of Network Protests.* New Haven, CT: Yale University Press.

van Dijck, J. 2000. *Models of Democracy and Concepts of Communication.* In *Digital Democracy: Issues of Theory and Practice,* edited by K. L. Hacker and J. van Dijck, 30–53. Thousand Oaks, CA: Sage Publications.

Van Reybrouck, David. 2016. *Against Elections: The Case for Democracy.* London: Bodley Head.

Webb, Maureen. 2020. *Coding Democracy: How Hackers Are Disrupting Power, Surveillance, and Authoritarianism*. Cambridge, MA: MIT Press.

Weltevrede, Esther, Anne Helmond, and Carolin Gerlitz. 2014. "The Politics of Real-Time: A Device Perspective on Social Media Platforms and Search Engines." *Theory, Culture & Society* 31 (6): 125–50. https://doi.org/10.1177/0263276414.

1

Democracy and the Digital Public Sphere

Joshua Cohen and Archon Fung

> The more the bonding force of communicative action wanes in private life spheres and the embers of communicative freedom die out, the easier it is for someone who monopolizes the public sphere to align the mutually estranged and isolated actors into a mass that can be directed and mobilized in a plebiscitarian manner. Basic constitutional guarantees alone, of course, cannot preserve the public sphere and civil society from deformations. The communication structures of the public sphere must rather be kept intact by an energetic civil society.
>
> JÜRGEN HABERMAS, *Between Facts and Norms*[1]

The "Topias"

The bloom is off the digital rose.[2]

Mobile technologies distract adults and depress kids. Twitter is "a honeypot for assholes."[3] Facebook and YouTube blend treacly cheeriness, insufferable self-display, and politically malignant deceit, delivered algorithmically to hyper-personalized, addictive filter bubbles that generate massive ad revenue by absorbing attention and amplifying extremism. The open web is a vast, soul-sapping wasteland of sexual depravity and medical malpractice, a "wormhole of darkness that eats itself,"[4] and the internet of things, an emerging panopticon. Behind all this destructive work lies a small band of

We are grateful for comments on earlier versions by Joseph Burgess, Brian Croll, Andrew Guess, Johannes Himmelreich, Larry Kramer, Brendan Nyhan, David Ruben, Charles Sabel, Noelle Stout, and Richard Tedlow; to audiences at the Princeton Program in Ethics and Public Affairs, the Brown University Political Philosophy workshop, Harvard's John F. Kennedy School of Government, and the September Group; and to the editors of *Philosophy and Public Affairs*. We particularly wish to thank Lucy Bernholz, Hélène Landemore, Rob Reich, and the other contributors to this book for their instructive comments. Because of the topic of the chapter, we need to draw attention to potential conflicts of interest for both authors. Joshua Cohen is a senior director at Apple Inc. Archon Fung teaches at the Harvard Kennedy School and has worked as an independent contractor for Apple, teaching at Apple University, since 2017.

predatory monopolies, animated by a toxic mix of swashbuckling ambition and self-satisfied moral preening. With brains hacked, communities fragmented, truth crushed, and public discussion subject to suffocating private control, "Social media is rotting democracy from within."[5]

Such currently fashionable techno-dystopianism is an inversion of the once-fashionable techno-utopian antecedents.

In 1999, Ira Magaziner, then senior adviser to President Clinton, confidently described the internet as "a force for the promotion of democracy, because dictatorship depends upon the control of the flow of information. *The Internet makes this control much more difficult in the short run and impossible in the long run*" (our emphasis).[6] Informed by his adviser's evident mastery of the new technologies, Bill Clinton amused his audience at the Nitze School for Advanced International Studies by comparing China's internet censorship efforts to "trying to nail Jello to the wall."[7] "The Internet," Nicholas Negroponte asserted, "cannot be regulated."[8] Moreover, he added (in a last-to-leave-the-party comment in 2010): "The real question is, 'Does the Internet overtly help causes like democracy, freedom, the elimination of poverty, and world peace?' My answer is: It does these things naturally and inherently."[9] George Bush lyrically urged us to "imagine if the Internet took hold in China. Imagine how freedom would spread." Rupert Murdoch asserted, more prosaically, that "advances in the technology of telecommunications have proved an unambiguous threat to totalitarian regimes everywhere."[10] "Censorship and content control are," Magaziner concluded—emphatically and yet with a "sense of humility"—"effectively impossible."[11]

The enthusiastic embrace of digital technology's democratic promise grew up at a singular political-economic moment: the end of the Cold War, with a history-concluding democracy seen by many as the only political game in town; American "democracy-promotion" linked to the "internet freedom agenda" and a dominant global role of American technology companies; and growing criticism of gatekeeping by dominant and concentrated corporate media.[12] Democratic benefits seemed inexorable because they appeared to flow from the very nature of digital technologies.[13] The argument for this optimistic conclusion, though flawed, was straightforward. Powered by Moore's law, digital technologies would drive the cost of bits to near zero, enabling virtually costless communication at a scale eluding central control. Low information and expression costs would reduce the need for gatekeepers; vastly expand opportunities for participation, informal communication, exploration, and persuasion; and increase the diversity of perspectives in circulation.[14] Easier horizontal communication would cut the costs of collective action and strengthen civil society. Citizens—sharing information, exposed to new ideas, and acting in newly nimble concert within broader networks—would have greater power to confront political and economic elites in democracies and gain the upper hand over sluggish autocrats.[15]

Not all early observers were so sanguine. Some were more impressed by proliferating pornography than promises of communicative empowerment.[16] Others, observing rapid downward spirals of communicative quality in unmoderated settings, saw how badly public discussion could go, even without technological barriers to entry, distracting advertising, or commercial pressures.[17] Worries arose, too, about social-political fragmentation resulting from ideological echo chambers or hyperpersonalized information, produced by self-segregating user choices or, in a later iteration, algorithms designed to capture user attention.[18] Nondemocratic governments turned out to be technologically adept and nimble.[19] Moore's law helped both racists and egalitarians. Concerns eventually emerged about confusing pop-up mobilizations with sustained collective action.[20] Platforms revealed vulnerabilities to manipulation and attack. And in due course, dominant platforms became the private, for-profit owners of functionally public spaces, with historically unprecedented curatorial and gatekeeping power.

Despite these accumulating challenges, digital information and communication technologies still seem to have considerable promise to enable movements (consider the Indignados, Yellow Vests, Five Star Movement, Black Lives Matter, Sunrise), mass protests in 2019 and 2020 (in Hong Kong, Chile, Bolivia, Iraq, and Lebanon), open-data initiatives in cities, and halting efforts at e-government.[21] This promise is especially salient to self-styled radical democrats, focused on the potential of digital technologies to support a public sphere in which free and equal persons are more fully enabled to use their common reason in public, political engagement.[22]

Many analysts have adopted a more defensive posture, focused (with good reason) on protecting public communication from lies, racism, misogyny, misanthropy, and misology. As complement, not substitute, we consider whether there is anything to say for a more hopeful picture of democracy and digital technology. Our short answer is yes. Less laconically: yes, but the democratic exploitation of technological affordances is vastly more contingent, more difficult, and more dependent on ethical conviction, political engagement, and good design choices than the technological determinists appreciated. In defending this affirmative answer, we aim to unsettle the current dystopian mood and replace dismal expectations with realistic hopes.

We begin by sketching an idealized democratic public sphere. The animating idea is to build a stronger marriage of broad participation and public discussion. To that end, we outline a structure of rights and opportunities that promises equal, substantive communicative freedom and a set of norms and dispositions concerning the use of those rights and opportunities.[23] We follow with a stylized contrast of two historical models of democratic public sphere—one dominated by mass media and the other by substantial digital communication—and we consider the strengths and weaknesses of the two models by reference to the conception of communicative freedom. Finally, using the same conception, we explore some ways to improve the digitally mediated, democratic public sphere. Companies need to invest more in designing their platforms to operate in ways that establish the rights and opportunities required by a democratic public sphere.[24] Governments, through legislation and the construction of regulatory capacity, need to press and monitor companies toward this end. And, because of the nature of digital technologies, citizens (aided by nongovernmental organizations) need to become more discerning consumers of information and more responsible creators and amplifiers of content.

We offer these improvements in an illustrative and tentative spirit. They are illustrative because our aim is not to provide definitive or complete solutions to very hard problems but to explore how an ambitious conception of democracy can help to guide public discussion about digital technologies. They are tentative because we want to steer clear of the overconfident assertions about the impact of fake news, filter bubbles, and echo chambers that characterize contemporary discussions of digital technologies and democracy. We are very impressed by how little is known, both because the terrain is changing and because careful investigation keeps upsetting conventional wisdom born of casual observation.[25]

Democracy and the Public Sphere

Our guiding conception of democracy is draws on three central ideas:

1. A *democratic society*, which means a society whose members are understood in the political culture to be free and equal persons. Such persons have a sense of justice, rightness, and reasonableness; an ability to bring their normative powers to bear on social and political issues, both in reflection and in discussion; and a capacity to act on the results of such reflection and discussion. Along with these common normative powers, persons hold divergent conceptions of the good and competing comprehensive doctrines; they have different interests, identities, capacities, social positions, and resources; and they stand in complex relations of cultural, social, and political power.

2. A *democratic political regime*, with regular elections, rights of participation, and the associative and expressive liberties essential to making participation informed and effective.
3. A *deliberative democracy*, in which political discussion about fundamentals of policy and politics appeals to reasons—including reasons of justice, fairness, and the common good—that are suited to cooperation among free and equal persons with deep disagreements. Moreover, the authorization to exercise collective power through the democratic political system traces to such argument.[26]

These three elements together describe the ideal of a political society whose free and equal members use their common reason to argue about the substance of public issues and in which the exercise of power is guided by that use. The animating idea is to marry broad participation by free and equal members with their engagement about the merits of different courses of public action: to combine mass democracy and public reasoning. This deliberative conception imposes more demanding expectations than either minimalist conceptions of democracy, which emphasize electoral competition (Joseph Schumpeter, William Riker, Adam Przeworski, Richard Posner), or fair aggregation conceptions, which emphasize an equal consideration of interests (Robert Dahl). Those conceptions dominate much current discussion about democracy and digital technology, which focuses on electoral threats.[27] While agreeing fully about the seriousness of these threats, we focus on what we regard as the best remedy: strengthening the communicative conditions of deeper democracy.

To achieve a marriage of participation and reasoning, political engagement cannot be confined to episodes of voting or lobbying, or even the activities of organized groups. Instead, democratic politics—as a discursive exercise of political autonomy—spills into informal, open-ended, fluid, dispersed public discussions of matters of common concern—discussions that are often created, focused, and expanded in scope by texts and other forms of representation, and that in turn shape public opinion, civic activism, and ultimately the exercise of formal political power.[28]

To bring these broad ideas about democracy closer to our subject, we distinguish two tracks in democratic decision making.[29] The first is the informal, dispersed, fluid, and unregulated exploration of issues in an unorganized, informal public sphere. Such exploration shapes public opinion(s) but does not produce authoritative collective decisions. The second is the formal political process, including elections and legislative decision making, as well as the processes and decisions of agencies and courts. In that formal process, ideally conceived, candidates and elected officials deliberate about issues, make authoritative decisions by translating the opinions formed in the informal sphere into legal regulations, and monitor the administrative execution of those decisions.

These two tracks are complementary. Informal communication in the public sphere provides—when it works well—a close-to-the-ground, locally informed, dispersed arena for detecting problems, exploring them and bringing them to public view, suggesting solutions, and debating whether the problems are important and worth addressing. The flow of information and communication enables problems to be identified more easily, and brought dramatically to common, public attention: think #MeToo or mobilization around gun regulation, the Occupy movement, Black Lives Matter, restrictions on abortion, color revolutions, anti-immigrant activism, or the Sunrise climate movement. To be sure, in all these cases, informal public discussion may be shaped by the mobilizing efforts of parties and public officials, thus qualifying the autonomy of the first track. But there also is, arguably, a significant role for more independent discussion and organized action that is neither prompted by nor organized by formal political organizations or agencies, and that also independently identifies needs, problems, and directions for solutions in nontechnical language.

Formal political processes—elections, legislatures, agencies, and courts—constitute the second track. When functional, they provide institutionally regulated ways to deliberate about proposals, evaluate solutions, and make authoritative decisions after due consideration, thus testing proposals that emerge from open-ended public discussion.[30] To focus our discussion, we put this second track aside. To be sure, democracy depends on integrating public discussion and opinion formation (track 1) with formal decision making and will or policy formation (track 2). And they are linked in many ways: what public officials say contributes to the shape of public debate and public opinion. But we focus here on the important challenge of creating public discussion among equals in the informal public sphere.

Think of a well-functioning, informal public sphere, then, as a space for a textually (or, more broadly, representationally) mediated, distributed public discussion in which participants are and are treated as free and equal persons. Because of the textual and representational mediators, members can think of themselves—despite their spatial separations and many differences and conflicts—as participants in common public discussions, which combine mass participation of equals with public reasoning. A well-functioning democratic public sphere, then, requires a set of rights and opportunities to ensure equal, substantive communicative freedom:

1. *Rights*: Each person has rights to basic liberties, including liberties of expression and association. The central meaning of expressive liberty is a strong presumption against viewpoint discrimination, which means a strong presumption against regulating speech for reasons having to do with its perspective. That presumption protects both the expressive interests of speakers and the deliberative interests of audiences and bystanders by enabling access to fundamentally different ideas.[31] It also secures the independence of public discussion from authoritative regulation.[32] The right to expressive liberty, thus understood, is not designed simply to afford protection against censorship of individual speakers; it is also democracy enabling. Protecting speech from viewpoint regulation helps establish the conditions that enable equal citizens to form and express their views and to monitor and hold accountable those who exercise of power. And it gives participants additional reason for judging the results to be legitimate. As an element in the constitution of the public sphere, the rights aims, as Meiklejohn says about the First Amendment, to secure "the freedom of those activities of thought and communication by which we 'govern.' It [the First Amendment] is concerned, not with a private right, but with a public power"—the power of citizens to make political judgments.[33]
2. *Expression*: Each person has good and equal chances to express views on issues of public concern to a public audience. While our rights condition requires the absence of viewpoint-discriminatory restrictions on expressive liberty, expression adds substance by requiring fair opportunities to participate in public discussion by communicating views on matters of common concern to audiences beyond friends and personal acquaintances. Expression requires a fair opportunity—dependent on motivation and ability, not on command of resources—to reach an audience given reasonable efforts. But the right to a fair opportunity for expression is not a right to have others listen or for one's views to be taken seriously.
3. *Access*: Each person has good and equal access to instructive information on matters of public concern that comes from reliable sources.[34] Access is not an entitlement to be informed, because becoming informed requires a measure of effort. Instead, access requires that those who make reasonable efforts can acquire information that comes from reliable sources and is instructive. Reliable sources are trustworthy and reasonable to trust, though of course not always accurate. Instructive information is relevant to the issues under discussion and understandable without specialized training. Like expression, access is a requirement for a substantive, fair opportunity: in this case, a fair opportunity—again, dependent on motivation and ability, not on command of resources—to acquire instructive information, as an essential requirement for having equal standing as a participant in free public discussion.[35]

4. *Diversity*: Each person has good and equal chances to hear a wide range of views on issues of public concern.[36] Unlike access, diversity is not simply about the opportunity to acquire factual information. It is about reasonable access to a range of competing views about public values—justice, fairness, equality, the common good—and the implications of those views for matters of public concern. Access to information about tax incidence and the implications of changes in incidence for growth and distribution is important, for example, but so are chances to hear different and conflicting views about the fairness of the tax incidence and distributional changes. Diversity is valuable both because exposure to disagreement is important for understanding the meaning and justification of one's own views, even if those views do not change, and because such exposure provides a good environment for forming reasonable and accurate beliefs.[37] Diversity thus confers individual benefits—on speakers, listeners, and bystanders—and arguably contributes to the quality of public deliberation.[38]
5. *Communicative Power*: Each person has good and equal chances to associate and explore interests and ideas together with others with an eye to arriving at common understandings and advancing common concerns.[39] Communicative power is a capacity for sustained joint (or collective) action, generated through such open-ended discussion, exploration, and mutual understanding. The communicative power condition thus helps to give substance to the equal rights of association contained in the rights requirement.

These five conditions together describe a structure of substantive communicative freedom among equals, essential to guiding our guiding conception of democracy. The freedom is communicative, not simply expressive, because the focus is not simply on speakers but also on listeners and bystanders; it is substantive because of the emphasis on fair opportunities as speaker, listener, and collective actor, not simply on rights against censorship. Equal, substantive communicative freedom is about—but not simply about—protecting people from state censorship, or the censorship of powerful private actors. It is also, more affirmatively, about creating conditions and affordances that enable broad participation in public discussion.

We have presented these five elements of a democratic public sphere very abstractly, but they have far-reaching political, social, economic implications. Equal standing in public reasoning requires favorable social background conditions, including limits on socioeconomic inequality and the dependencies associated with it.[40] Similarly, the conjunction of rights and expression has implications for concentrated private control of communicative opportunities. We return to these issues later. We note them here only to underscore that the features that define a well-functioning public sphere, though abstract, are not mere formalities.

Even if these rights and opportunities are in place, however, they are insufficient for the marriage of broad participation with public reasoning that defines a well-functioning democratic public sphere. The success of that marriage is doubly dependent on the norms and dispositions of participants in public discussion. Moreover, this dependence is especially strong—for reasons we explore later—in the digital public sphere.

It is dependent, first, because, those norms and dispositions shape the uses that people make of their fundamental rights and opportunities. Thus, participants might be indifferent to public concerns or to the truth of their utterances. They might disregard the essential rights and opportunities of others or be openly hostile to their equal standing. They might be so mistrusting that they lack confidence that others care about getting things right (especially others with whom they disagree). Or they might be so cynical that they deny any need to get things right or to defend their views with public reasons. Second, sustaining a stable structure of rights and opportunities depends on the norms and dispositions of participants.[41] Noxious behavior in the public sphere erodes the rights and opportunities that others enjoy. For example, online harassment reduces expressive opportunities for targets of that harassment. Thus the double dependence: as sources of substantive success in the exercise of communicative freedom and stability of the essential rights and opportunities.

In particular, three dispositions and norms are important both in constituting a well-functioning, democratic public sphere and in sustaining the enabling structure of rights and opportunities. We do not assume that these norms are legally binding (indeed, we assume that they are not legally binding). Rather, we think of them as parts of the political culture required for a well-functioning democratic public sphere:

1. *Truth*: First, participants in a well-functioning public sphere understand and are disposed to acknowledge the importance of truth, the norm associated with assertion.[42] That means not deliberately misrepresenting their beliefs, or showing reckless disregard for the truth or falsity of their assertions, or—in cases in which they know that others are relying on their representations, and in particular when the potential costs of that reliance may be large—showing negligence about the truth or falsity of their assertions.[43] Respecting a norm of truthfulness of course does not ensure getting things right all of the time, or even most of the time. Instead, it shows an effort to get things right, with a recognition that, on most important questions, it is difficult to get things right even when everyone

is aiming at the truth. Because uncertainty, error, and disagreement are normal features of public discussion, this norm requires a willingness to correct errors in assertion, particularly when one knows that others have relied on those assertions.

2. *Common Good*: Second, participants have a sense of and are concerned about the common good, on some reasonable understanding of the common good. "Reasonable understandings" respect the equal standing and equal importance of people entitled to participate in public discussion. A well-functioning public sphere does not depend on a shared view of justice or rightness or the common good. But it does depend on participants who are concerned that their own views on fundamental political questions are guided by a reasonable conception of the common good rather than a conception that rejects the equal standing of others as interlocutors or discounts their interests. Here, the value of equality is expressed not only in the rights and opportunities that define the structure of communicative freedom but also in the conceptions of justice, rightness, and reasonableness that participants bring to public discussion and that frame their contributions.
3. *Civility*: Third, participants recognize the obligation—founded on the equal standing of persons and a recognition of deep and unresolvable disagreements on fundamentals—to be prepared to justify views by reference to that conception. Thus, participants do not view political argument as simply serving the purpose of affirming group membership and group identity, much less as a rhetorical strategy for exercising power in the service of personal or group advantage. Following Rawls, we call this obligation to justify the *duty of civility*. Civility, thus understood, is not a matter of politeness or respect for conventional norms nor is it a legal duty. Instead, civility is a matter being prepared to explain to others why the laws and policies that we support can be supported by core, democratic values and principles—say, values of liberty, equality, and the general welfare—and being prepared to listen to others and be open to accommodating their reasonable views.[44] Civility, thus understood, is not about manners. Rather, it expresses a sense of accountability to others as equal participants in public discussion.

These conditions are demanding. We lay them out explicitly in order to consider how the existence of a digitally mediated public sphere—in which search, news aggregation, and social media provide important informational and communicative infrastructure—bears on these conditions of a well-functioning democratic public sphere.

In using them to evaluate the impact of digital technologies on democracy's public sphere, we make two kinds of assessments. The first—a baseline-regarding assessment—evaluates changes relative to historical baselines. The second—an ideal-regarding assessment—evaluates by reference to an idealized conception of a democratic public sphere such as we have just specified and asks which changes might improve matters relative to that ideal. We do some of both, but with a particular focus on the latter.

Two Public Spheres

We begin the baseline-regarding account by sketching highly stylized descriptions of two contrasting public spheres: the mass-media public sphere (we focus on the United States in the mid-twentieth century) and the digitally networked public sphere (we focus on the United States circa 2020).[45] Our stylized sketches are intended principally to serve normative rather than empirical or historical purposes. They have three dimensions: the industrial organization and technology of media, the dynamics of information production and flow, and the sociopolitical context in which media operate. This characterization sets the stage for exploring the democratic character of these different public spheres.

In its specifically American incarnation, then, the mass-media public sphere was dominated (though hardly exhausted) by a small number of private, for-profit organizations: three network broadcasters in the United States (supplemented especially in the 1990s by a growing number of cable news providers), a handful of national news magazines, and a small number of newspapers and news services dominating national and international news. These organizations delivered content generated by self-consciously professional journalists to large audiences. The 1947 *Hutchins Commission Report* powerfully defined the democratic challenges for the mass media and its liberal-democratic project. "Perhaps the most cogent and elegant report on media policy ever published in the English language,"[46] the report emphasized the power of the modern press, powerfully described its democratic challenges, and proposed a way to meet those challenges:

> The modern press itself is a new phenomenon. Its typical unit is the great agency of mass communication. These agencies can facilitate thought and discussion. They can stifle it. They can advance progress of civilization or they can thwart it. They can debase and vulgarize mankind. They can endanger the peace of the world; they can do so accidentally, in a fit of absence of mind. They can play up or down the news and its significance, foster and feed emotions, create complacent fictions and blind spots, misuse the great words, and uphold empty slogans. Their scope and power are increasing every day as new instruments become available to them. These instruments can spread lies faster and further than our forefathers dreamed when they established

> the freedom of the press in the First Amendment to our Constitution. . . . The press can be inflammatory, sensational, and irresponsible. . . . On the other hand, the press can do its duty by the new world that is struggling to be born. It can help create a new world community by giving men everywhere knowledge of the world and of one another, by promoting comprehension and appreciation of the goals of a free society that shall embrace all men.[47]

While wrestling with this duty, the Hutchins Commission accurately foretold the basic features of the mass-media public sphere. Major broadcast and major print media would be one-to-many communication technologies run by a few large, for-profit companies. The mass-media public sphere thus had a narrow aperture, with a few, industrially concentrated voices addressing large audiences, without much room for those in the audience to talk back, or develop new themes and topics among themselves. This was emphatically not a media system designed to foster communicative power, which is why C. Wright Mills called this the world of mass opinion, not public opinion.[48] Navigating between "corporate libertarian" and "social-democratic" conceptions of the media, the Hutchins Commission urged social responsibility through self-regulation.[49] While the media could go wrong in many ways, the commission argued, governmental regulation was not the right corrective.

In its broad outlines, American mass media grew to conform with the Hutchins Commission's vision and recommendations. Media corporations fulfilled their public responsibilities—to the extent that they did—through the work of professional journalists and editors, while acknowledging, at least in principle, the desirability of separating reporting, editorial, and commercial imperatives. Content was produced by professional journalists and editors who were guided by five norms: getting (and presenting) the truth, providing a representative picture of social groups, providing a forum for comment and criticism, clarifying public values, and offering comprehensive coverage. Professional journalists contributed to public discussion by bringing current events, political analysis, and investigative reporting to mass audiences. These journalists and media organizations were part of an epistemically coherent public sphere, professionalized and mutually correcting, with broadly shared standards of evidence, argument, and salience. This was a world in which Pizzagates and QAnons were not publicly visible. The relative coherence in the mass-media public sphere "fit" the relatively low levels of popular, activist, and elite polarization in which it emerged. Jonathan Ladd argues that relatively high public trust in the mass media, reflecting this fit, was produced by two structural factors: "First, low levels of economic competition enabled journalism to become highly professionalized. Journalists had the autonomy to enforce a professional norm of objectivity, greatly reducing salacious or explicitly partisan news coverage. Second, the lack of party polarization reduced political criticism of the institutional press."[50]

This narrow-aperture world did not always serve democratic aims well. Journalists relied heavily on official sources because of professional and social proximity and a desire to maintain access.[51] Successful journalists and political officials were racially homogeneous. They often shared cultural experiences and educational background, as well as perspectives and assumptions about America's global role, the benefits of globalization, and the virtues of market economies.

Reporting on Iraq at the end of the mass-media period provides an especially striking illustration. Hayes and Guardino analyzed evening network-news stories in the eight months before the 2003 US invasion of Iraq. They found that, on balance, the criticisms "of the media in the run-up to the Iraq War are justified: news coverage . . . was more favorable toward the Bush administration's rationale for war than its opponents' arguments against."[52] Network news stories did feature opponents of the war. But they were not domestic US voices: only 4 percent of the stories featured representatives of the Democratic Party and fewer than 1 percent of those were antiwar leaders. Moreover, the largest source of antiwar views on network news were Iraqi citizens and officials, and the largest share of the Iraqi sources quoted belonged to Saddam Hussein himself.[53]

This execrable performance was the general rule for the prior decades as well.[54] Coverage of the Vietnam War, especially before 1968, was decidedly prowar (i.e., progovernment policy), as was coverage of conflicts in Central America and elsewhere. In an aptly titled paper, "Government's Little Helper: US Press Coverage of Foreign Policy Crises," Zaller and Chiu argue that in foreign policy the content of news closely tracks the content of congressional speeches and testimony; mass-media news thus reflects the views of political officials rather than independently evaluating or criticizing those views.[55] "Reporters," Zaller and Chiu say, "report the story that political authorities want to have reported." They indexed heavily on official positions.

Domestic news was similar. Coverage of racial politics in major newspapers and broadcast media did not resemble what appeared in media in African American communities. Ronald Jacobs, for example, shows how the narratives of civil unrest in Watts and the police beating of Rodney King in mainstream outlets such as the *Los Angeles Times* and the *New York Times* differed dramatically from coverage of those very same events in African American media—what the Hutchins Commission called "Negro dailies and weeklies"—such as *Los Angeles Sentinel* and the *New York Amsterdam News.* Similarly for mass-media coverage of work and organized labor. As *New York Daily News* columnist Juan Gonzalez put it, there is a "class divide between those who produce news and information and those who receive it."[56]

Now contrast this with a comparably stylized picture of the digitally mediated public sphere. First, compared to the mass-media public sphere, the digital public sphere offers easier access to a vastly wider range of information, narrative, and political perspective. Second, the emergence of the digital public sphere has been accompanied by—some argue that it has caused, or perhaps amplified—political polarization.

Technologically, a fundamental difference between the mass media and the digital era is the shift from broadcast (one-to-many) to networked (many-to-many) communication, with effectively zero marginal costs of information and communication. The digital infrastructure of the public sphere is defined by this distinctive flow of information in which there are many more providers and distributers of content; people thus enjoy vastly greater choice among kinds and providers of information; and particular content can be directed (or targeted) by providers, advertisers, social media platform companies, or other actors to particular users or groups of users.

To be sure, a few firms dominate the infrastructure of the digital public sphere. Google dominates search (domestically and globally), and three sites—Facebook, YouTube, and Twitter—account for the vast share of users who get news on social media (44 percent, 10 percent, and 9 percent of US adults report getting news from these sources, respectively). Other social media platforms—such as Reddit, LinkedIn, Instagram, Snapchat, and Tumblr—account for a very small portion of news.[57] And while many Americans get their news from local or national television, online sources are already well ahead for eighteen- to twenty-nine-year-olds. And in 2017, fully 55 percent of Americans over 50 reported getting news on social media sites.[58]

But while the news organizations of the mass-media public sphere spoke with an editorial angle and voice, Facebook, YouTube, and Twitter (as well as other social media) are platforms that enable users to distribute content, consume it, and connect with one another. To be sure, social media platforms are not simply providing information pipes. They are in the businesses of curating, moderating, and amplifying—but that business is not the same as the editorial business of mass media.[59] And that is important. A digitally mediated public sphere expands opportunities to access a much wider array of perspectives. The shared sensibilities of journalists and editors, as well as professional norms, all narrowed the aperture of information and explanation that reached mass audiences. The digital public sphere is curated and moderated, but it widens the aperture, not least (though not only) because the sheer scale and scope mean that lots of moderation happens *ex post* and depends on user feedback.

What sources of news and information actually make it to audiences? A recent Pew Research Center study examined links to news about immigration that were shared on Twitter during the first month of the Trump administration. Of nearly ten million tweets, roughly 50 percent more were linked to legacy news and nonprofits than to digital native news and commentary. But the approximately four million tweets from digital native sources is a large number, and in part because the distribution mechanism is different, the range of voices is greater.

This widened aperture coincides with a sense of increasing political polarization. At the level of elected officials, Republican and Democratic members of Congress are more polarized than at any time in the past 140 years.[60] At the popular level, the dislike and fear that Democratic and Republican Americans have toward one another—so-called affective polarization—also registers at all-time highs.[61] But while there is widespread perception of deepening polarization (especially among the political engaged, who are pretty clearly polarized), the facts of mass polarization—as distinct from elite polarization and partisan sorting—are contested.[62] Moreover, even those who agree that there is mass polarization are uncertain about causal relationships with digital communication. Polarization has been growing most in the United States among older Americans, the group least likely to spend time online.[63] That may be because online diets are actually moderate in substance and broadly shared across partisan lines, in substantial measure because of popular reliance—especially for passive consumers of political information, as distinct from hyperactive partisans—on mainstream portals (Google, MSN, and AOL).[64] To the extent that there is polarization, the deepest drivers may not be wide aperture information and communication technologies but rather “long-term changes in American politics.”[65]

Still, the coincidence between the arrival of the digital public sphere and amplified polarization—especially affective polarization—is striking. And even skeptical arguments about causal effects of digital media on polarization suggest amplification. Social conditions of polarization and fragmentation combine with low barriers to content creation and high user choice to create an informational environment that is much more diverse but in which users find it easier to cluster into—or find themselves algorithmically shepherded into—homogeneous information spaces that share less epistemic common ground across different spaces than in the mass-media public sphere. To be sure, offline geographic, partisan, and associational sorting also produce information bubbles. Offline communication is hard to study, so it is difficult to know whether online is better or worse at reinforcing prior views. But in reality, we do not spend lots of offline time engaging people with whom we disagree,[66] and there is some reason to think that online exposure to those people is greater.

Perhaps what is most striking about online information sources, then, is neither the causal effects on offline polarization nor the fact that offline polarization is worse than it would have been in a counterfactual continuation of a mass-media public sphere, but that online information and communication are so much less than they could be. They fall far short of their democratic potential for fostering diverse and interconnected engagements.

These centrifugal forces are compounded by the role in the digital public sphere of sources of content that are shifting from professional journalists—with professional truth-seeking norms—to a wider range of other digital media, advocacy organizations, and political campaigns with other priorities. These other providers often have stronger incentives to develop content that appeals to specific audiences by reinforcing rather than challenging their perspectives. Advocacy organizations and political campaigns understandably seek to mobilize supporters by tailoring messages, arguments, and narratives that deepen their commitment and increase their activism. As media become more segmented and partisan, they affirmatively seek to address communities that share their assumptions and perspectives.

Assessing the Digital Public Sphere

How does the digital public sphere perform compared to the prior baseline of the mass-media public sphere with regard to the conditions of equal communicative freedom?

On expression and diversity, the digital public sphere appears at first blush to have clear advantages. Information in the mass-media public sphere was often generated by journalists and editors who worked for private companies and operated within relatively narrow ideological and social boundaries. The technological affordances of the digital public sphere enable many more people with many more perspectives to express themselves to larger public audiences.[67] That expanded expression in turn generates much greater diversity in the sources of information and views that are easily available—through both deliberate efforts and accidental exposure—in the public sphere. Now, we have not just the *Wall Street Journal* and Fox News, but also the *Drudge Report* and Breitbart. We have not just National Public Radio, but *Huffington Post* and *Chapo Trap House*. The enormous diversity of social media participants far exceeds even the expanded diversity of content from these web publishers.

It is similar regarding access. Digital technologies, including search, have enabled the creation of and access to vast amounts of the world's information (and explanations) about politics, policy, and society: aside from Google search, think of large portals (MSN, AOL, Yahoo), or Wikipedia, PubMed, publications from the Bureau of Labor Statistics, and Our World in Data. On the dimension of opportunities to acquire reliable and relevant information, it is simply much easier for more people to learn much more about most anything.

But these apparent gains in the availability of information supplied by diverse sources require careful qualification. The wider aperture also expands opportunities for expression and communication that violate norms of truth seeking, a common-good orientation, and civility.[68] A well-functioning, democratic public sphere offers relevant and reliable information and brings different arguments and views into connection and confrontation. But with a large mix of irrelevant noise, bullshit, and expressions of hatred, and the segregation of views from one another with each segment working to deepen its own views in opposition to the others, diversity, expression, and access may be limited, despite the apparent gains.

Doxxing, swarming, and threatening, for example, are familiar digital violations of the norms of common good and civility. Enhanced opportunities to engage in public discussion also enable people to impose costs on those who bear messages they don't like through these forms of online distraction and harassment. While an earlier generation of work explored the potentially silencing effects of hate speech and pornography, here, the "silencing" works not only by the content of speech but also by its sheer volume. And the burdens of the new forms of silencing are not equally shared. Moreover, by shrinking expression for some, the burdens limit the access of others to those points of view. At one limit, when online harassment spurs physical intimidation or violence, it threatens the right to expression and association.

Our hesitations about gains on expression, access, and diversity may seem puzzling. Earlier, we emphasized that these dimensions of communicative freedom are matters of opportunity. So why not say that the digitally mediated public sphere emphatically expands opportunities, thus improving on each of these dimensions, but that some people may choose—in the face of hostility and harassment, or from homophily, conflict avoidance, or deliberately selective exposure to confirming information—not to avail themselves of those opportunities? Our response—to simplify an issue of vast complexity—distinguishes homophily, conflict avoidance, and selective exposure from hostility and harassment. The former are not restrictions on opportunity, whereas the latter are. Any sensible account of fair opportunity needs to take into account the costs that a person can reasonably be expected to bear in order to gain access to a good. Consider an analogous point about fair employment opportunities: a person facing severe bullying at work lacks fair opportunity to advance, even if the company is hiring in that person's field. Similarly, the reduced costs of sending hostile and harassing messages increases burdens in ways that bear on the opportunities themselves, not simply on how people choose to exercise them.

Thus, as part of their professional commitment to norms of truth seeking and the common good, journalists and editors in the mass-media public sphere aimed to craft stories to direct audience attention to what they regarded as the most important public issues. By contrast, the digital public sphere amplifies noise and distraction, as well as targeted hostility. Most of this noise is a by-product of widely expanded avenues for expression—what many people regard as important for them to say simply isn't that important for others to hear (everything has been said, but not everyone has said it). Some of this distraction is intentional (consider the efforts of China's 50 Cent Party).[69] One challenge to the value of access to reliable and useful information, then, is that the "signal-to-noise" ratio in the digital public sphere may be lower than in the mass-media era. Obtaining reliable information in the digital public sphere may thus require considerable effort to distinguish reliable and instructive information from propaganda, screeds, and bullshit. With a wide aperture, obtaining relevant information from diverse perspectives may, moreover, require shouldering the burden of venturing into the web's nastiest corners and distinguishing who is aiming to make serious (even if fallible) contributions from those who seek merely to signal boost through sheer repetition.[70]

Misinformation, disinformation, and "fake news" pose an additional challenge for access to reliable and useful information in the digital public sphere. Consider violations of truthfulness, such as assertions of propositions believed to be untrue or displays of reckless disregard for or negligence about truth or falsity. The norms of professional journalism celebrated by the Hutchins Commission arguably made some difference in the mass-media public sphere. Although there was lots of incorrect reporting, burying of officially unwelcome stories, and sheer negligence, professional journalists took pride and care in making sure that the stories they filed were true; journalists and their organizations were embarrassed and apologetic when they got it wrong. They understood that they were supposed to get it right (the norm of truth seeking) and that trying to get it right served democratic values (the norm of a common-good orientation).

Norms, resources, and organization are now very different. With some notable exceptions, many content providers in the digital public sphere lack the resources to be vigilant about truthfulness. Others seek to mobilize support and so highlight partial truths that serve that end. And some have little regard for the truth at all and inject false information to mobilize and foster solidarity or hostility.[71] The low costs of supplying "information," the decentralization of its sources, and the absence of professional norms, together with powerful commercial and status incentives for capturing attention, create troubles that invite solution.

As a result of the fake-news scandals surrounding the 2016 election cycle, some of the major social media platforms have become somewhat more sensitive to accusations that they are spreading misinformation, and may be putting more resources into detecting and stopping fake news. While laudable—and better than disregarding the norm of truthfulness altogether—preventing fake news or retrospectively taking it down does not replace atrophied capacities to seek and publish truth. Many news organizations developed powerful capacities and commitments for common-good-oriented truth seeking. They retained journalists and opinion writers who were trained to advance a particular conception of the common good and to seek truth though investigation and observation. These news professionals produced most of the content of the mass-media public sphere. By contrast, the business model of social media platforms such as Facebook, YouTube, and Twitter focuses on capturing users' attention. Truth is at best a by-product, and the common good is thinned down to commitments to create connections, increase variety, and reduce costs of access.

Finally, consider communicative power—the capacity to act together with others to put forward information, arguments, narratives, values, and, especially importantly, normative considerations to a broader public. The mass-media public sphere provided cramped affordances for forming and exercising communicative power. Thus C. Wright Mills said, "The mass has no autonomy from institutions; on the contrary, agents of organized institutions penetrate this mass, reducing any autonomy it may have in the formation of opinion by discussion." Mills's assertion, which belongs to the mid-1950s, exaggerates the colonization of public discussion by commercial media and the transformation of the public into "mere media markets": consider the civil rights movement, the women's movement, and the antiwar movement.[72] Still, with media companies and journalists as gatekeepers and officials as privileged sources, prominent officials and powerful private actors dominated public discussion. Dependent on media gatekeepers for access to the public sphere, dissidents and activists struggled to generate communicative power by currying favor or making investments in making the kind of noise—through protests and other methods—that professional journalists and editors might deem newsworthy.

The digital public sphere features many more venues for publishing content. Indeed, anyone can potentially reach broad audiences by publishing on blogs and social media. Mere publication, of course, does not guarantee audience or attention. Still, the digital public sphere enhances communicative power by enabling many more citizens and activists to share stories, make arguments, and act together.

At the same time, the benefits for communicative power have been much smaller than early enthusiasts expected. To be sure, social media platforms such as Twitter and Facebook ease mobilization by reducing information costs. But information is not persuasion, mobilization is not organization, and knowing where people are gathering is not communicative power. Digital technologies enable activists to mount large protests more easily than in the mass-media public sphere, when that sort of collective action required efforts to build interpersonal trust and organizational infrastructure. But easy mobilization can be counterproductive for building communicative power.[73] Activists in the digitally mediated public sphere can achieve a short-term end (mass protest) without building the political, civic, and relational infrastructure that sustains collective action over the long term.

Moreover, it is not possible for us crookedly timbered creatures to sustain the public presence and vigilance that democracy requires in a world in which our lives have the kind of transparency that results when our private information is open to surveillance. Without better privacy protections, the costs of democratic presence and vigilance can be unbearable.

Toward a More Ideal Public Sphere

The bloom fell rapidly from the digital rose. But the digitally networked public sphere is relatively new, and its democratic effects—both positive and negative—will surely change as digital natives dominate, public policy evolves under the pressures of now-regnant digital skepticism, technologies change, and companies alter strategies in light of shifts in user populations, preferences, pressure from movements, and public policy. In this section, we draw on our earlier account of democracy and the digital public sphere to propose some guidance in thinking about how better to exploit the democratic possibilities of digital information and communication technologies.

Building a more democratic public sphere will require vigorous democratically oriented and concerted action by, among others, governments, private companies, nongovernmental organizations, and citizens themselves. All have been awakened to the damage to democracy caused by our current digital public sphere. But many of their responses so far have been reactive, defensive or simply indifferent. They have not yet risen to meet the intellectual, psychological, economic, or political challenges necessary to capture the democratizing potentials of the digital public sphere. Indeed, the guidance that we offer may seem unrealistic because of the motives and limitations on these actors. Governments may lack the institutional capacity or the independence from powerful technology companies to regulate them meaningfully. The attention-harvesting business models that have enabled a few powerful and commercially flourishing companies to dominate the infrastructure of the digital public sphere are at odds with democratic responsibilities.[74] And confirmation bias, out-group antipathy, and simple cognitive constraints may make the kind of digital citizenship that we suggest too burdensome.[75] These issues—about capacity and incentive alignment—constitute an important research agenda for social science. Our aim is primarily normative, although we have tried to be attentive to the nascent social science about the politics of the digital age. We are not persuaded that insurmountable constraints prevent these actors from acting in democratically responsible ways. So we offer some illustrative ideas about how to create a more democratic digital public sphere, focusing in particular on issues of regulating speech and powerful private corporations; the production of high-quality information; privacy and security; and the creation of a civic culture of responsible, democracy-reinforcing behavior.

SPEECH REGULATION

The digital public sphere reduces barriers to expression, which has been a mixed blessing. What is the appropriate response to the fouling of communicative air?

Some of what people say defames private persons or threatens serious imminent harm (including doxxing and swatting). Such expression has little democratic value but threatens expression, as well as diversity and communicative power, by imposing unreasonable burdens on potential contributors. Government (and private platforms) can and should act to address such noxious speech: with private defamation and threats of imminent harm, free expression concerns are at a minimum, even in the strongly speech-protective US setting.[76] Platform companies are better positioned to prevent the spread of content that defames or threatens violence; they should invest in improved methods for doing so. Those investments protect individuals from discrete harms and also respect the values of equal communicative freedom.

We are much more skeptical about protecting communicative freedom through hate speech rules, especially when those rules are designed by dominant platform companies as supplements to public hate-speech rules (which exist in most jurisdictions outside the United States). Private hate-speech rules tend to be too open ended and not tied to concerns about imminent harms, thus sweeping up much legitimate public discussion.[77] When designed by companies, moreover, they threaten to substitute an unaccountable exercise of private power for heightened public vigilance.

We support, more cautiously, some efforts to address concerns about so-called fake news. Internet platform companies might be able to take action to detect and demote clearly false news content in order to advance access and diversity while encouraging concerns about truth.[78] In doing so, however, they should avoid the viewpoint discrimination that is at the heart of rights. Independent, journalistic fact-checking websites (including PolitiFact, Snopes, and FactCheck.org) can help provide inputs for platform judgments (in some geographies).

The responsibility for vigilance in policing the supply of news is easiest to see for platforms like Apple News that are both formally and functionally in the news business. Formally, because they are in the business of curating content, thus certifying to users that they are concerned with veracity in the sources they provide access to. Functionally, because they drive very substantial traffic to professional journalism. But similar reasoning applies to platforms that are formally in the search, blogging, or social network business, but that function effectively as news providers—for example, Facebook, Twitter, and YouTube, as well as Google and other large portals.[79] Because they provide important informational infrastructure for the informal public sphere (whatever their Standard Industrial Classification in filings with the Securities and Exchange Commission), they have a public responsibility to police—and not to promote—reckless and perhaps negligent misrepresentations, even though their judgments are bound to be contentious (or at least contested) and even if the exercise of their responsibility may come at a cost of user engagement.[80]

Why only cautiously supportive? Although there are now several independent sources that identify fake news, the categories are fluid, their judgments are contested, and the disagreements provoke legitimate concerns about viewpoint regulation. Moreover, platform regulation, especially by large, market-dominating players, may jeopardize expression, diversity, and communicative power, reproducing or even worsening the deficiencies of the mass media's narrow-aperture world. Some views and some people—typically, assuming past as precedent, people with less power—are likely to bear the burden of the regulations. Moreover, the sheer scale of searching and messaging is so great that private censorship may be overwhelmed; the rapid execution demanded by enormous volume may generate arbitrariness. But there may be effective measures short of private censorship that foster greater regard for civility and promote it. Firms have begun experimenting with labeling sources as more or less dependable. And they can promote stories and posts from sources that they judge to be more truthful and self-correcting and demote those that have less reliability.

Out of abundant free-expression caution, governments should act only indirectly—and then only with circumspection—to discourage fake news. Some have considered revisions to Section 230 of the US Communications Decency Act, which shields platforms from liability claims based on content posted by third parties (as well as takedowns of their content) as a way to address fake news.[81] However, citizens can and should do much more to stem fake news. Out of regard for truth, they—with assistance from platform affordances—should try to detect fake news, avoid it, refrain from amplifying it, and call it out when they see it. Recent research has argued that the problem may be exaggerated. To be sure, some people—in particular those who are older, conservative, and consume large amounts of political content—have been more likely to be exposed to and share fake news than others.[82] But it is not clear that the problem is endemic to the low-cost-information environment. Perhaps we can learn how citizens can become more vigilant by understanding the causes of these variations in fake-news engagement.

CONCENTRATED POWER

Some have argued that efforts to encourage more democracy-enhancing behavior by companies are doomed by the political economy of the digital public sphere. Private companies, their rhetoric aside, are often in the business of advertising and attention, not providing reliable, relevant information and enabling public communication. Moreover, some are simply too big to impose on. Bigness not only drives markups, profitability, and inequality; it also creates outsized power over the infrastructure of public discussion while enabling firms to resist regulatory measures that would require them to help build a more democratic public sphere.[83] Addressing these concerns might, then, involve efforts to revive competition policy, including greater efforts to police (prospectively and retrospectively) vertical integration.[84] In the face of increased market concentration and market power, such a revival makes good sense for a variety of reasons, including concerns about political influence and economic inequality, which would improve the background conditions for a democratic public sphere.

But would making the large social media enterprises smaller—and invigorating competition—help more specifically to democratize the digitally mediated public sphere? Not so clear. On the positive side, large platforms have extraordinary gatekeeping powers. In the United States, they operate under, and have defended, a regulatory regime that gives them a pretty free editorial hand (putting aside digital copyright) in part by protecting them from intermediary liability.[85] That gives a small number of powerful private actors enormous power to define, interpret, and enforce terms of service that users need to comply with on pain of exclusion. The problem is not merely the abuse of that power but also the very existence of that power. Current concerns about the dominance of private curating and gatekeeping would be significantly reduced were the power not so highly concentrated. A larger number and variety of social media companies might improve rights (by reducing concerns about viewpoint regulation) and increase diversity. Looking forward, reducing the political influence of these large media companies might reduce the resource-backed lobbying and interest-group hurdles to a broader regulatory project aimed at democratizing the digital public sphere.

But maybe not. Larger firms may be better able to pay the costs of stronger privacy rules (as under the General Data Protection Regulation, or GDPR).[86] Because of their visibility, they also provide more easily identifiable targets for activists aiming to hold platforms accountable for bad behavior on the platform, and they may be better able to bear the costs of monitoring conduct on their sites. Moreover, they create a common place for public communication rather than a set of small, potentially fragmented communities. In addition, while a single major attack surface may be a more attractive target for adversaries, it also may be easier for owners of large surfaces to defend against hostile information attacks. Contra Brandeis, bigness is not always a curse.[87]

Imagine, for example, the successful application of retrospective merger review to Facebook's 2014 acquisition of WhatsApp. Would we see an advance in equal, substantive communicative freedom with a dozen baby WhatsApps? Neo-Brandeisians might reasonably respond that each baby WhatsApp would have had many fewer users, which would reduce incentives for creating epistemic pollution in the first place. But the results are not so clear.

On balance, we see promise in revived antitrust activity, mostly for its impact on the background conditions that create equal standing in public reasoning, but we think the jury is still out on more direct effects on the digitally mediated public sphere.

PRODUCING HIGH-QUALITY INFORMATION

Perhaps the best cure for fake news is not to suppress it but rather to increase access to better information that is both informative and reliable: raise the floor instead of imposing ceilings. In the mass-media era, good news was produced (however imperfectly) by professional journalists. Many have shown how the shift in revenue from commercial advertising in local and national news to digital advertising on the largest platforms such as Facebook and Google decimated the news business. From a 1990 baseline, employment decline in the newspaper business has been greater than in steel or coal mining.[88] The search for successful business models for news—to date an unsolved problem—continues, and we hope that these efforts on the behalf of news start-ups, local news organizations, and philanthropists find success.

But platform companies, governments, and citizens can do more to finance good news. The past offers some models. C-SPAN, created as a source of public-interest news at the dawn of the cable television era, is funded by cable and satellite operators that charge each subscriber six cents on the cable bill. At the dawn of radio, the British Broadcasting Company was initially financed with "receiving licenses" in which radio owners paid a licensing fee to support content production and operations. Today, the BBC is financed by television licensing fees amounting to approximately £150 per household.[89] Marci Harris has suggested levying a tax on digital advertising revenue in order to provide a sustainable source of financing for professional journalists to produce high-quality, reliable news. A variant on this idea would be to fund the work of independent, public journalism from revenue generated by a "data tax" on firms that use large amounts of data that are drawn from uncompensated human activities.[90] These and other novel combinations of corporate, government, and citizen efforts could help finance production of good news: information that is important for citizens's democratic deliberations about local and national affairs.

Access to important information might also be increased by public regulation's creating of an internet-appropriate analog to must-carry provisions that require cable operators to transmit local television programming. The rationale behind "must carry" is that there is a citizen interest and a public interest in some providers and some kinds of content that companies will not transmit if driven by business considerations alone.[91] The rise of the digital public sphere has been accompanied by the collapse of local news not just in rural areas of the United States but also in many cities.[92] Digital must-carry provisions—in the form of regulations or industry conventions—could, for example, require companies to feature local news and information in order to bring democratically significant information to citizens' attention and foster demand for such content.

Even with more high-quality information and broadly improved access, the digital public sphere will never generate convergence on the truth or broadly agreed-on facts, evidentiary standards, and reasonable perspectives. But convergence is an unreasonable expectation. And the epistemic convergence that is sometimes associated with mass-media gatekeeping is overrated and overstated. It was overrated because professional journalists sharing an elite culture and dependent on official sources sometimes converged on critically erroneous positions and excluded views outside of conventional "wisdom." It may have been overstated because the diversity of views in many publics—"subaltern counterpublics" such as racially and ethnically minority communities, political groups beyond the center left and center right, religious and culturally distinct subcultures—likely far exceeded the convergence on facts and perspectives that was manifest in network news and mainstream press.[93] The digital public sphere should not attempt to reproduce—even if it could—the informational conditions of the twentieth-century mass media by employing gatekeepers who narrow the informational aperture to exclude content that is "not safe for democracy." Such restrictions would set back the values of expression, access, diversity, and communicative power. Rather, our hope is that taking measures to increase the quality of information in the digital public sphere and encouraging participants to seek truth with attention to the common good and to one another will enable successful democratic governance without narrowing the range of available perspectives and interpretations.

PRIVACY AND SECURITY

Increasing privacy and security is important for many reasons, among them that it enables citizens to participate more effectively in the public sphere. Better security would foster expression by helping to protect speakers from personal attacks—potential and actual, in real life and also online (e.g., doxxing)—for holding unpopular views. Enhanced privacy could increase access and diversity by reducing the power of data-driven algorithms to exacerbate polarization and reinforce filter bubbles by micro-targeting audiences. As we indicated earlier, increasing user privacy and data security can also protect the public vigilance needed for building communicative power.

Both firms and regulators could help to enhance privacy and security. On the side of private companies, Apple sets a rather high standard that social media companies could aspire to, with its policies of data minimization, on-device processing, differential privacy, end-to-end encryption (for iMessage and FaceTime), and information siloing.[94] But privacy protection is fundamental to a democratic public sphere: it should not simply be a company policy but should be taken out of competition. GDPR-type data-privacy regulations may help, but some caution is needed. Privacy regulations can cut pretty deeply into the availability of information relevant for public discussion.[95] Moreover, some GDPR-type regulations will arguably consolidate the power of large private firms due to the costs of compliance.

INDIVIDUAL BURDENS

Looking forward, we think—not from ethical conviction but because of the nature of digital technologies—that users themselves, both individually and through coordinated action, will have a significant role to play in democratizing the digital public sphere.[96] Although we are concerned about overtaxing individuals, we think that the democratic success of a digitally networked public sphere will place a larger burden on the efforts of individuals than did the mass-media public sphere: that strikes us as the kernel of truth in technological determinism. By dramatically reducing the costs of transmitting messages, the digital public sphere has increased the aperture of information and communication. With flatter communication at an unprecedentedly large scale, a one-in-a-million event happens every few minutes. If we wish to avoid reducing opportunities for expression and imposing significant hurdles on access to information,[97] the digital public sphere will impose greater burdens on individuals and groups to distinguish information from manipulation, exercise greater restraint in deciding what to communicate, and sanction others who abuse these norms. Platforms and governments can help to enable that active, affirmative role.

Many of the democratic benefits of the mass-media public sphere depended on journalists who genuinely embraced democratic norms and responsibilities.[98] Education and professional socialization encouraged a commitment to those norms as central to the success of journalism, thus to democracy itself. The organizations that employed them—as well as professional organizations—sometimes rewarded those democratic norms. Although the analogy of citizens and groups to professional journalists is very imperfect, internet companies should help users behave as citizens by designing their platforms to foster participants' democratic orientation. Platforms themselves can take responsibility for enhancing digital literacy by more explicitly recognizing that some sources are negligent about truth, by spreading habits of checking, and by encouraging users to encounter diverse perspectives.

But design is not enough; we will also need bottom-up efforts that elicit the right kind of engagement and content generation from users. This apparently is a widespread view: a 2017 Gallup and Knight Foundation Survey found that "the public divides evenly on the question of who is primarily responsible for ensuring people have an accurate and politically balanced understanding of the news—48 percent say the news media and 48 percent say individuals themselves."[99]

Meeting this challenge, if it can be done at all, will be hard. To be sure, the digital public sphere is new, and part of what we are observing is not intrinsic to the new technologies, but a normal lag of social change (including norm change) behind technological change. We have yet to clarify, articulate, and adopt the intellectual and behavioral norms needed to engage constructively in this new, rapidly mutating medium, which was designed to generate advertising revenue and foster expression of all manner of direct communication, not as a forum for accurate news and political discussion to enable citizens to govern themselves democratically.

We need a better understanding of the kinds of communicative behavior that are appropriate for the digitally mediated public sphere, both to give it real substance and to sustain it: a *Hutchins Commission Report* for the digital public sphere. Focusing on truth and the common good, what do these norms imply for communicative responsibilities?

In both spheres of mass media and the digital public, truth requires citizens—who are not just content consumers but also producers and amplifiers—to be media literate in the sense that they can distinguish information from propaganda.[100] The proliferation of sources and content—many of dubious quality and pedigree—make that task more challenging and also more important. Aiding others' opportunities for access to reliable information requires making some effort to check on the veracity of stories before liking or forwarding them. Platforms can help by promoting content from reliable sources, offering tools that enable users to assess the veracity and trustworthiness of specific content and of their information diets in general, by rewarding—and enabling users to reward—truth-seeking behavior on their platforms and also instructing in the importance of epistemic humility. Teachers, as well as students of information and communication, could help by developing updated methods (e.g., attention to the quality of sources, cross-checking assertions, and awareness of confirmation bias) that enable citizens to better discern reliability and relevance in the digital environment.

Or consider the commitment to the common good. Although citizens disagree about what justice requires or the ends that society ought to pursue, a commitment to the common good means that citizens acknowledge that they have reasons to resolve these differences on a basis that respects the equal importance of others. This commitment entails that citizens support perspectives and policies that they believe will advance the common good of all, not just of a particular religion, interest group, class, profession, or political party.

Projecting this broad commitment into the digital public sphere, a common-good orientation will often require citizens to avoid narrow news diets.[101] It also requires citizens to be attentive to a broader array of information in order to be able to form views and make appropriate judgments. Doing that, in turn, requires learning about the interests, perspectives, and pain of others. For example, as Richard Rorty argued, we should seek out information that reveals the cruelty that both social practices and our own behaviors cause to others.[102] A commitment to the common good thus requires citizens to inhabit parts of the digital public sphere that are common in the sense that they encounter information from diverse perspectives. Citizens should help create these common spaces by putting forth views and perspectives that appeal to others rather than to narrow interests and commitments.

From the platform side, filter bubbles, homophily in social networks, and confirmation bias's reinforcing of newsfeed algorithms all arguably exacerbate in-group loyalties and out-group hostilities. Platform architects should seek to expose users to ideas that lie outside their familiar territory and to content that is visibly common (like Apple News's Top Stories). Broad adherence to such common-good-oriented behaviors would foster greater access, expression, and perhaps diversity in the digital public sphere.

A range of other actors, including nongovernmental organizations, can contribute as well. Formal and popular education could help spread those methods widely. Second, such normative and prescriptive accounts ought to be part of civic education and socialization. Just as many people learn how to treat one another decently and engage politically in families, clubs, and classrooms, so they should learn how to engage democratically in the digital public sphere. Democratic norms should be part of digital literacy. Clear rules of thumb and expectations would help individuals direct their own attention and offer them an important ethical dimension of judgment and guide their expressions of approval or disapproval. Moreover, third-party organizations—analogous to independent organizations that have emerged around fake news and open educational resources—might call out content or users that violate norms or demote the priority of such posts on news feeds.

Final Thoughts

We have examined the implications of digital information and communication technologies for a demanding conception of a democratic public sphere. We have been exploring whether and how that conception might guide thinking about how to make those technologies deliver better on their erstwhile democratic promise.

In service of that inquiry, we have tried to clear away some of the gauzy romanticism about the mass-media public sphere. But baseline-relative comparison is not where the real action is. And on the ideal-regarding assessment, our thinking is unsettled. Resisting techno-dystopian defeatism, we have explored some ways to foster the rights, opportunities, norms, and dispositions of a democratic public sphere. Much more needs to be said and done. But when all is said and done, what is needed may not be possible, at least not now, in the face of vast inequality, deep social fragmentation, highly concentrated digital media organization, and the dysfunctions of representative democracies. So perhaps we have not landed in an especially hopeful place, not for now, but in Kafka's place:

> "I remember," Max Brod writes, "a conversation with Kafka which began with present-day Europe and the decline of the human race. 'We are nihilistic thoughts, suicidal thoughts that come into God's head,' Kafka said. This reminded me at first of the Gnostic view of life: God as the evil demiurge, the world as his Fall. 'Oh no,' said Kafka, 'our world is only a bad mood of God, a bad day of his.' 'Then there is hope outside this manifestation of the world that we know.' He smiled. 'Oh, plenty of hope, an infinite amount of hope—but not for us.'"[103]

Maybe the bad mood will end. And maybe we can hasten that end by enlarging our ambitions.

Notes

1. Jürgen Habermas, *Between Facts and Norms*, trans. William Rehg (Cambridge, MA: MIT Press, 1996), 369.

2. We completed this contribution before the COVID-19 pandemic. Although definitive assessments of the quality of public information and discussion around COVID-19 will be some time coming, both the dangers and the promise of the digital public sphere have been manifest in discussions to date. There is no shortage of misinformation—from both official and surreptitious sources—about the causes of the disease (e.g., Chinese or American biological labs), its dynamics (e.g., the face-mask controversy, minimized or maximized projections about severity), and effective therapies and vaccines. In contrast, we have been impressed by the quality of information that is available—for those who look for it—and the democratic accountability it has fostered. A crowdsourced project called the COVID Tracking Project (covidtracking.com) became the authoritative information provider about the pace and dearth of COVID-19 testing in the United States and perhaps brought pressure to expand testing. Social media gave researchers at the University of Washington (@UWVirology on Twitter) a way to publicize their early testing successes in the face of lethargy at the national level and make their success available to others. Social media provided space to question (successfully) the official positions that COVID-19 did not spread asymptomatically (World Health Organization) and that individuals should not wear masks to protect themselves and others (Centers for Disease Control and Prevention). Moreover, efforts under way to create decentralized, private-preserving, proximity tracing (DP3T) apps may play an important role in creating capacity for contact tracing on top of expanded testing. In the midst of the vast uncertainty, ignorance, and expert fallibility that the pandemic brings, the diversity of perspectives and capacity for concerted action created by the digital public sphere have been providing valuable correctives to official, conventional, and individual errors.

3. Charlie Warzel, "'A Honeypot for Assholes': Inside Twitter's 10-Year Failure to Stop Harassment" *Buzzfeed News*, August 11, 2016, https://www.buzzfeednews.com/article/charliewarzel/a-honeypot-for-assholes-inside-twitters-10-year-failure-to-s.

4. "Scarlett Johansson on Fake AI-Generated Sex Videos: 'Nothing Can Stop Someone from Cutting and Pasting My Image,'" *Washington Post*, December 31, 2018.

5. Zach Beauchamp, "Social Media Is Rotting Democracy from Within," *Vox*, January 22, 2019, https://www.vox.com/policy-and-politics/2019/1/22/18177076/social-media-facebook-far-right-authoritarian-populism.

6. Ira Magaziner, "Creating a Framework for Global Electronic Commerce," *Progress and Freedom Foundation*, July 1999, http://www.pff.org/issues-pubs/futureinsights/fi6.1globaleconomiccommerce.html.

7. Quoted in Jack Goldsmith and Tim Wu, *Who Controls the Internet? Illusions of a Borderless World* (Oxford: Oxford University Press, 2006), 90.

8. Goldsmith and Wu, 3.

9. Nicholas Negroponte, "Letter," *Foreign Policy*, 15 June 2010, https://foreignpolicy.com/2010/06/15/wiring-democracy/.

10. Stephen Kotkin, "How Murdoch Got Lost in China," *New York Times*, 4 May 2008, https://www.nytimes.com/2008/05/04/business/media/04shelf.html.

11. Magaziner, "Creating a Framework."

12. Jack Goldsmith, "The Failure of Internet Freedom," Knight First Amendment Institute https://knightcolumbia.org/content/failure-internet-freedom; Ben Bagdikian, *The Media Monopoly* (Boston: Beacon Press, 1983); Edward S. Herman and Noam Chomsky, *Manufacturing Consent* (New York: Pantheon Books, 1988).

13. Consider, for example, Ithiel de Sola Pool's claim that "electronic technology is conducive to freedom," and that "the easy access, low cost, and distributed intelligence of modern

means of communication are a prime reason for hope." Ithiel de Sola Pool, *Technologies of Freedom* (Cambridge, MA: Harvard University Press, 1983), 231, 251. For critical discussion of some less prosaic versions of this line of thought, see Franklin Foer, *World without Mind* (New York: Penguin Books, 2017); Goldsmith and Wu, "Visions of a Post Territorial Order," in *Who Controls the Internet?*, chap. 2.

14. See Eugene Volokh, "Cheap Speech and What It Will Do," *Yale Law Journal* 104 (1994–1995): 1805–50: exploring the results of a "shift [in] control from intermediaries—record labels, radio and TV station owners, newspaper, magazine, and book publishers—to speakers and listeners" (1834).

15. Techno-hopefulness was not always technologically deterministic. In the internet's early days, Mitch Kapor (founder of Lotus and cofounder of the Electronic Frontier Foundation) argued that its democratic benefits depended on making the right architectural choices that would enable users not only to acquire information but also to express ideas at low cost. Mitchell Kapor, "Democracy and the New Information Highway," *Boston Review* (September–October 1993), http://bostonreview.net/archives/BR18.5/democracyinfo.html.

16. See the Communications Decency Act of 1996. Title V of the Telecommunications Act of 1996, Pub. L. No. 104-104, 110 Stat. 56 (1996), codified at 47 U.S.C. §§223, 230; Benjamin Barber, "Democracy and Cyber-Space: Response to Ira Magaziner," http://web.mit.edu/comm-forum/legacy/papers/barberresponse.html.

17. Steven Schneider, "Expanding the Public Sphere through Computer-Mediated Communication: Political Discussion about Abortion in a Usenet Newsgroup," PhD diss., Massachusetts Institute of Technology, 1997, https://dspace.mit.edu/handle/1721.1/10388.

18. On the former, see Cass Sunstein, *Republic.com* (Princeton, NJ: Princeton University Press, 2001). On the latter, see Eli Pariser, *The Filter Bubble: How the New Personalized Web Is Changing What We Read and How We Think* (New York: Penguin, 2012). Fragmentation, Bill Gates, recently observed, "turned out to be more of a problem than I, or many others, would have expected." See Kevin Delaney, "Filter Bubbles Are a Serious Problem with News, Says Bill Gates," *Quartz*, February 21, 2017, https://qz.com/913114/bill-gates-says-filter-bubbles-are-a-serious-problem-with-news/.

19. In *Who Controls the Internet?*, Jack Goldsmith and Tim Wu identified this issue very clearly in 2006, especially in their chapter 6. For more recent discussion, see the analysis of porous and customized censorship through "friction and flooding" in Margaret Roberts, *Censored: Distraction and Diversion Inside China's Great Firewall* (Princeton, NJ: Princeton University Press, 2018), esp. chaps. 5 and 6.

20. Zeynep Tufekci, *Twitter and Tear Gas* (New Haven, CT: Yale University Press, 2018); Yochai Benkler, Robert Farris, and Hal Roberts, *Network Propaganda* (Oxford: Oxford University Press, 2018), 343 ("first failure mode").

21. Jorn Berends, Wendy Carrara, and Heleen Vollers, "Analytical Report 6: Open Data in Cities 2," European Data Portal, June 2017: "Open data . . . refers to the information collected, produced or paid for by public bodies that can be freely used, modified, and shared by anyone for any purpose."

22. See Joshua Cohen and Archon Fung, "Radical Democracy," in *Swiss Journal of Political Science* 10, no. 4 (2004): 169–80.

23. We borrow the term *communicative freedom* from Habermas. See the epigraph to this chapter.

24. We assume that those efforts will not be spontaneously undertaken but will require political pressure from legislators, regulators, movements, and nongovernmental organizations.

25. For instructive and chastened reviews of the social-scientific literature, see Joshua A. Tucker, Andrew Guess, Pablo Barberá, Christian Vaccari, Alexandra Siegel, Sergei Sanovich, Denis Stukal, and Brendan Nyhan, "Social Media, Political Polarization, and Political Disinformation," Hewlett Foundation, March 2018, https://hewlett.org/wp-content/uploads/2018/03/Social-Media-Political-Polarization-and-Political-Disinformation-Literature-Review.pdf; Kris-Stella Trump, Jason Rhody, Cole Edick, and Penny Weber, "Social Media and Democracy: Assessing the State of the Field and Identifying Unexplored Questions," Social Sciences Research Council, April 2018, https://www.ssrc.org/publications/view/AE0A0077-C168-E811-80CB-005056AB0BD9/. For a striking example of the conflict between conventional wisdom and the results of more systematic investigation, see the critique of echo chambers in Andrew Guess, Benjamin Lyons, Brendan Nyhan, and Jason Reifler, "Avoiding the Echo Chamber about Echo Chambers: Why Selective Exposure to Like-Minded Political News Is Less Prevalent Than You Think," Knight Foundation, 2018, https://kf-site-production.s3.amazonaws.com/media_elements/files/000/000/133/original/Topos_KF_White-Paper_Nyhan_V1.pdf. The conventional wisdom often overlooks that most people are not active seekers of political information and that the relationship between exposure and subsequent behavior is not at all straightforward.

26. Joshua Cohen, "Democracy and Liberty," in *Deliberative Democracy*, ed. Jon Elster (Cambridge: Cambridge University Press, 1998), 185–231.

27. See, e.g., Joseph Nye, "Protecting Democracy in an Era of Cyber Information War," Belfer Center, January 2019, https://www.belfercenter.org/publication/protecting-democracy-era-cyber-information-war.

28. The idea that reduced costs of communication have democratic potential does not require an aspiration for an e-democracy, with virtually assembled citizens directly deciding on the substance of policy.

29. Here, we follow Habermas, *Between Facts and Norms*, chaps. 7 and 8. Although we draw on Habermas's distinction between two tracks, our account of the democratic public sphere does not depend on the details of his formulation. Thus, we find large areas of agreement with the account of democratic diarchy, with its distinction between opinion formation and will formation, in Nadia Urbinati's account of democratic proceduralism: *Democracy Disfigured* (Cambridge, MA: Harvard University Press, 2014).

30. Communication between citizens and elected officials is an intermediate case. Thus, the use of digital technology in building "directly representative democracy" through online deliberative town halls falls outside our scope. For a creative exploration of this idea, see Michael Neblo, Kevin Esterling, and David Lazer, *Politics with the People* (Cambridge: Cambridge University Press, 2018).

31. On the enabling role of rights to expressive liberty, see Alexander Meiklejohn, *Free Speech and Its Relation to Self-Government* (New York: Harper, 1948); Owen Fiss, *The Irony of Free Speech* (Cambridge, MA: Harvard University Press, 1996). On speaker, audience, and bystander interests, see T. M. Scanlon, "Freedom of Expression and Categories of Expression," *University of Pittsburgh Law Review* 40 (1979): 519–47.

32. For helpful reflections on the nature and foundations of the presumption against viewpoint discrimination, set in the context of current debates about hate speech regulation, see Vincent Blasi, "Hate Speech, Public Assurance, and the Civic Standing of Speakers and Victims," *Constitutional Commentary* 32 (2017): 585–97.

33. Alexander Meiklejohn, "The First Amendment Is an Absolute," *Supreme Court Review* (1961): 245–66.

34. Dahl's condition of "gaining enlightened understanding," a defining feature of democ-

racy, is about opportunities (or capabilities), not achievement. See Robert Dahl, *On Political Equality* (New Haven, CT: Yale University Press, 2006), 9.

35. Deciding whether access is satisfied requires substantive and contestable judgments about whether sources are reliable and instructive. These judgments cannot be made simply by considering whether all sources and all views are presented in the name of balance. We are grateful to Bryan Ford for suggesting this clarification.

36. For discussion, see Joshua Cohen, "Freedom of Expression," *Philosophy and Public Affairs* 22, no. 3 (Summer 1993): 228–29, on the deliberative interest; Scanlon, "Freedom of Expression," on audience interests; Michael Ananny, *Networked Press Freedom: Creating Infrastructures for a Public Right to Hear*, esp. chap. 2. On audience interests in diverse content as a value in communications policy, see the discussion of must-carry rules for cable, in Philip Napoli, *Foundations of Communications Policy* (Cresskill, NJ: Hampton Press, 2001), 58–62. Napoli misleadingly describes conditions analogous to access and diversity as based on a "collectivist" conception of the First Amendment and speech rights. But they can equally well (perhaps better) be understood as based on an account of the social preconditions for protecting individual interests in favorable conditions for forming beliefs.

37. The empirical literature on the psychological impact of exposure to opposing views is mixed. Some research suggests "backlash" or "backfire" effects, where exposure increases polarization. See Christopher A. Bail, Lisa P. Argyle, Taylor W. Brown, John P. Bumpus, Haohan Chen, M. B. Fallin Hunzaker, Jaemin Lee, Marcus Mann, Friedolin Merhout, and Alexander Volfovsky, "Exposure to Opposing Views on Social Media Can Increase Political Polarization," *Proceedings of the National Academy of Sciences*, 115, no. (2018): 9216–21. Other literature sees no such effect. See Andrew Guess and Alexander Coppock, "Does Counter-Attitudinal Information Cause Backlash? Results from Three Large Survey Experiments," *British Journal of Political Science* (November 2018): 1–19.

38. Mill, *On Liberty*, chap. 2, in *"On Liberty" and Other Writings*, by John Stuart Mill, edited by Stefan Collini, Cambridge Texts in the History of Political Thought (Cambridge: Cambridge University Press, 1989). These two considerations are independent from the more familiar Millian arguments about the benefits of disagreement in correcting errors and arriving at true beliefs.

39. On communicative power, see Hannah Arendt, *On Violence* (New York: Harcourt Brace, 1970); Habermas, *Between Facts and Norms*, chap. 4. On Habermas, see Jeffrey Flynn, "Communicative Power in Habermas's Theory of Democracy," *European Journal of Political Theory* 3, no. 4: 433–54. We agree with Flynn's "wide reading" of the idea of communicative power, which emphasizes dispersed public discussion in the informal public sphere rather than exclusively in formal institutions. C. Wright Mills distinguishes "public" communication from "mass" communication. The features we have noted here broadly fit with his conception of a public, which he defined in broadly process terms, not in terms of a common set of convictions or goals. See his *The Power Elite* (Oxford: Oxford University Press, 1959), 303–4.

40. Nancy Fraser, "Rethinking the Public Sphere," *Social Text* 25–26 (1990): 56–80; Robert Dahl, *Democracy and Its Critics* (New Haven, CT: Yale University Press, 1989), on the "essential prerequisites" for democracy; John Rawls, *A Theory of Justice* (Cambridge, MA: Harvard University Press, 1971) on the background conditions required for fair equality of opportunity.

41. "Actors who support the public sphere are distinguished by the dual orientation of their political engagement: with their programs, they directly influence the political system, but at the same time they are also reflexively concerned with revitalizing and enlarging civil society and

the public sphere as well as with confirming their own identities and capacities to act" (Habermas, *Between Facts and Norms*, 370).

42. We are touching here on areas of vast complexity: In precisely what sense is truth a norm for assertion? What is the role of assertion—presenting a proposition as true—as distinct from other forms of rhetorical engagement in public discussion? The surface form of assertion may mask deeper rhetorical purposes, perhaps evident from context. We cannot explore these issues here.

43. On the notion of truth and its role in public reasoning, see Joshua Cohen, "Truth and Public Reason," *Philosophy and Public Affairs* 37, no. 1 (Winter 2009): 2–42. The distinction between reckless disregard and negligence comes from the law of libel.

44. John Rawls, *Political Liberalism*, expanded ed. (New York: Columbia University Press, 2005), 217; Joshua Cohen, "Reflections on Civility," in *Civility and American Democracy*, ed. Cornell Clayton, ed. (Pullman: Washington State University Press, 2013). The conception of civility we embrace here has some affinity with the "mere civility" attributed to Roger Williams in Teresa M. Bejan, *Mere Civility* (Cambridge, MA: Harvard University Press, 2017), chap. 2. For a remarkable exploration of the implications of a failure of the duty to listen, see Douglas Ahler and Gaurav Sood, "The Parties in Our Heads: Misperceptions about Party Composition and Their Consequences," *Journal of Politics* 80, no. 3 (July 2018): 964–81: "For instance, people think that 32 percent of Democrats are LGBT (vs. 6 percent in reality) and 38 percent of Republicans earn over $250,000 per year (vs. 2 percent in reality). . . . When provided information about the out-party's actual composition, partisans come to see its supporters as less extreme and feel less socially distant from them."

45. We use the mass-media public sphere as a point of contrast because it looms large in contemporary criticisms of the digital public sphere. But historically speaking, it was very unusual, associated with professionalization, scale economies in news production, relatively unpolarized politics, and deliberate efforts (through the Hutchins Commission) to preserve a relatively unregulated media through an emphasis on professional norms. On the economics of news production, see James T. Hamilton, *Democracy's Detectives: The Economics of Investigative Journalism* (Cambridge, MA: Harvard University Press, 2016), and Julia Cagé's contribution to this volume.

46. James Curran, *Media and Democracy* (New York: Taylor & Francis, 2011), 9. Baker calls it "the most important, semi-official, policy-oriented study of the mass media in US history," in C. Edwin Baker, *Media Concentration and Democracy: Why Ownership Matters* (Cambridge: Cambridge University Press, 2007), 2.

47. Commission on Freedom of the Press and Robert Devore Leigh, *A Free and Responsible Press: A General Report on Mass Communication: Newspapers, Radio, Motion Pictures, Magazines, and Books* (Chicago: University of Chicago Press, 1947), 3–4.

48. Mills, *Power Elite*, chap. 13.

49. The terms come from Victor Pickard, *America's Battle for Media Democracy: The Triumph of Corporate Libertarianism and the Future of Media Reform* (Cambridge: Cambridge University Press, 2015), chaps. 5, 6.

50. Jonathan M. Ladd, *Why Americans Hate the Media and How It Matters* (Princeton, NJ: Princeton University Press, 2011), 6.

51. Herman and Chomsky, *Manufacturing Consent.*

52. Danny Hayes and Matt Guardino, "Whose Views Made the News? Media Coverage and the March to War in Iraq," *Political Communication* 27, no. 1 (2010): 59–87.

53. Hayes and Guardino, p. 75. Just as a reminder: 23 senators and 133 members of the House voted against the October 2002 Authorization for Use of Military Force against Iraq, see Deborah White, "Congress Members Who Voted Against the 2002 Iraq War," *ThoughtCo*, August 17, 2019, https://www.thoughtco.com/2002-iraq-war-vote-3325446.

54. Large exception: Knight Ridder reporting.

55. John Zaller and Dennis Chiu, "Government's Little Helper: Us Press Coverage of Foreign Policy Crisis, 1946–1999," in *Decisionmaking in a Glass House: Mass Media, Public Opinion, and American and European Foreign Policy in the 21st Century*, ed. Brigitte Nacos, Robert Y. Shapiro, and Pierangelo Isernia (New York: Rowman & Littlefield Publishers, 2000), 61–84.

56. Robert Bruno, "Evidence of Bias in the *Chicago Tribune* Coverage of Organized Labor: A Quantitative Study From 1991 to 2001," *Labor Studies Journal* 34, no. 3 (2009): 385–407.

57. Pew Research Center, "News Use Across Social Media Platforms 2016," https://www.journalism.org/2016/05/26/news-use-across-social-media-platforms-2016/.

58. Amy Mitchell, Jeffrey Gottfried, Michael Barthel, and Elisa Shearer, "The Modern News Consumer: News Attitudes and Practices in the Digital Era," Pew Research Center, 2016, https://www.journalism.org/2016/07/07/the-modern-news-consumer/; Jeffrey Gottfried and Elisa Shearer, "Americans' Online News Use Is Closing In on TV News Use," Pew Research Center, September 17, 2017, http://www.pewresearch.org/fact-tank/2017/09/07/americans-online-news-use-vs-tv-news-use/.

59. Kate Klonick, "Facebook v. Sullivan," Knight Foundation (October 2018), discusses Facebook's use of judgments of newsworthiness. This is an attenuated case of being in the editorial business: attenuated because the judgments of newsworthiness are used in the special case of deciding whether to delete posted messages that facially violate terms of service and thus would be deleted but for their newsworthiness.

60. Nolan McCarty, *Polarization: What Everyone Needs to Know* (Oxford: Oxford University Press), 30.

61. See Marc J. Hetherington and Thomas J. Rudolph, *Why Washington Won't Work: Polarization, Political Trust, and the Governing Crisis* (Chicago: University of Chicago Press, 2015); Pew Research Center, "Partisanship and Political Animosity," 2016, https://www.people-press.org/2016/06/22/partisanship-and-political-animosity-in-2016/; Pew Research Center, *The Partisan Divide on Political Values Grow Even Wider*, 2017, http://www.people-press.org/2017/10/05/the-partisan-divide-on-political-values-grows-even-wider/; Shanto Iyengar and Sean J. Westwood, "Fear and Loathing across Party Lines: New Evidence on Group Polarization," *American Journal of Political Science* 59, no. 3 (2015): 690–707; McCarty, *Polarization*, 61–66.

62. For skepticism, see Morris P. Fiorina, *Unstable Majorities* (Stanford, CA: Hoover Institution Press, 2017).

63. Levi Boxell, Matthew Gentzkow, and Jesse M. Shapiro, "Greater Internet Use Is Not Associated with Faster Growth in Political Polarization among US Demographic Groups," *Proceedings of the National Academy of Sciences* 114 (2017): 10612–17.

64. Andrew M. Guess, "(Almost) Everything in Moderation: New Evidence on Americans' Online Media Diets" (unpublished paper, on file with the authors), 2 July 2018.

65. Drawing on data about recent engagement with media on Facebook, Twitter, and the open web, Benkler, Faris, and Roberts, in *Network Propaganda*, argue that media polarization, like elite political polarization, is asymmetric, that professional norms and concerns about getting things right remain important outside of the right (e.g., Breitbart, Fox, Zero Hedge, Free Beacon, InfoWars).

66. Diana Mutz, *Hearing the Other Side* (Cambridge: Cambridge University Press, 2006).

67. Sarah Jackson, Moya Bailey, and Brooke Foucault Welles, *#HashtagActivism: Networks of Race and Gender Justice* (Cambridge, MA: MIT Press, 2020).

68. According to a 2017 Gallup and Knight Foundation Survey, *American Views: Trust, Media and Democracy*, "By 58% to 38%, Americans say it is harder rather than easier to be informed today due to the plethora of information and news sources available." And, "Half of US adults feel confident there are enough sources to allow people to cut through bias to sort out the facts in the news—down from 66% a generation ago." https://kf-site-production.s3.amazonaws.com/media_elements/files/000/000/130/original/Knight-Gallup_Survey_Topline_FINAL.pdf.

69. See Roberts, *Censored.*

70. See gab.ai, for example.

71. Benkler, Faros, and Roberts, in chapter 3 of *Network Propaganda*, claim that the "propaganda feedback loop" that characterizes right-wing media is founded on competition for "identity-confirmation" and polices deviations from such confirmation, which means that there are not incentives to correct misstatements of fact.

72. Mills, *Power Elite*, 304.

73. Tufekci, *Twitter and Tear Gas.*

74. See, e.g., Tim Wu, *The Attention Merchants: The Epic Scramble to Get Inside Our Heads* (New York: Vintage Books, 2017); Shoshana Zuboff, *The Age of Surveillance Capitalism: The Fight for a Human Future at the New Frontier of Power* (New York: PublicAffairs, 2018).

75. For influential reasoning along these lines, see Christopher Achen and Larry M. Bartels. *Democracy for Realists: Why Elections Do Not Produce Responsive Government* (Princeton, NJ: Princeton University Press, 2016).

76. Expression that is directed to inciting imminent violence and is likely to incite such violence can be legally regulated in the United States, consistent with current understandings of the First Amendment. See *Brandenburg v. Ohio*, 395 U.S. 444 (1969). Cases of defamation directed against people who are not public figures are similar.

77. The literature on the subject is vast. For recent discussion, see Jeremy Waldron, *The Harm in Hate Speech* (Cambridge, MA: Harvard University Press, 2012); Sarah Cleveland, "Hate Speech at Home and Abroad," in *The Free Speech Century*, ed. Lee C. Bollinger and Geoffrey R. Stone (Oxford: Oxford University Press, 2019), 210–31.

78. For skepticism about the importance of fake news in persuading and driving behavior, see Andrew Little, "Fake News, Propaganda, and Lies Can Be Pervasive Even If They Aren't Persuasive," *Comparative Politics Newsletter* 28, no. 2 (2018): 49–55, http://andrewtlittle.com/papers/little_fakenews_cp.pdf.

79. On the importance to the public sphere of being functionally in the news business, even if not formally in the business of editing, see Robert Post, "Data Privacy and Dignitary Privacy: Google Spain, the Right to be Forgotten, and the Construction of the Public Sphere," *Duke Law Journal* 67 (2018): 981–1072. For an effort to provide First Amendment protections for search engines for their editorial role, see *Jian Zhang v. Baidu.com, Inc.*, 10 F. Supp. 3d 433 (S.D.N.Y. 2014, 1989).

80. Alcott, Gentzkow, and Yu argue that Facebook cut engagements with fake news by 50 percent after the 2016 election. See Hunt Alcott, Matthew Gentzkow, and Chuan Yu, "Trends in the Diffusion of Misinformation on Social Media" (Stanford Institute for Economic Policy Research Working Paper 18-029), September 2018, https://siepr.stanford.edu/research/publications/trends-diffusion-misinformation-social-media. A recent analysis of Twitter by

Hindman and Barash argues that "most fake news on Twitter links to a few established conspiracy and propaganda sites," so that deliberate action directed to a relatively small number of sites could "greatly reduce the amount of fake news people see." Matthew Hindman and Vlad Barash, "Disinformation, 'Fake News,' and Influence Campaigns on Twitter," Knight Foundation, October 2018, https://kf-site-production.s3.amazonaws.com/media_elements/files/000/000/238/original/KF-DisinformationReport-final2.pdf.

81. Tim Hwang, "Dealing with Disinformation: Evaluating the Case for CDA 230 Amendment," 2017, https://papers.ssrn.com/sol3/papers.cfm?abstract_id=3089442.

82. Nir Grinberg, Kenneth Joseph, Lisa Friedland, Briony Swire-Thompson, and David Lazer, "Fake News on Twitter during the 2016 US Presidential Election," *Science* 363, no. 6425 (2019): 374–78.

83. Jan De Loecker and Jan Eeckhout, "The Rise of Market Power" (Working Paper No. 23687), National Bureau of Economic Research, Cambridge, MA, August 24, 2017, https://www.nber.org/papers/w23687; Thomas Philippon, *The Great Reversal: How America Gave Up on Free Markets* (Cambridge, MA: Harvard University Press, 2019); Federico J. Díez, Daniel Leigh, and Suchanan Tambunlertchai, "Global Market Power and Its Macroeconomic Implications" (Working Paper No. WP/18/137), International Monetary Fund, https://www.imf.org/en/Publications/WP/Issues/2018/06/15/Global-Market-Power-and-its-Macroeconomic-Implications-45975/.

84. For discussion, see Lina Khan, "Amazon's Antitrust Paradox," *Yale Law Journal* 126, no. 3 (January 2017): 710–805; Tim Wu, *The Curse of Bigness: Antitrust in the New Gilded Age* (New York: Columbia Global Reports, 2018); Jonathan Tepper, with Denise Hearn, *The Myth of Capitalism* (Hoboken, NJ: Wiley, 2019).

85. For an argument in favor of modifying the blanket immunity provided by current interpretations of Section 230, see Danielle Keats Citron and Benjamin Wittes, "The Internet Will Not Break: Denying Bad Samaritans Section 230 Immunity," July 2017, https://papers.ssrn.com/sol3/papers.cfm?abstract_id=3007720.

86. For discussion of large and small firm compliance, see Craig McAllister, "What about Small Businesses: The GDPR and Its Consequences for Small US-Based Companies," *Brooklyn Journal of Corporate, Financial and Commercial Law* 12, no. 1 (December 2017): 187–211.

87. Louis Brandeis, *The Curse of Bigness* (New York: Viking, 1934).

88. Evan Horowitz, "Even Fishing and Coal Mining Are Not Losing Jobs as Fast as the Newspaper Industry," *Boston Globe*, 3 July 2018), https://www.bostonglobe.com/business/2018/07/03/even-fishermen-and-coal-miners-are-faring-better-than-newspaper-employees/snK5o6ritw8UxvD51O336L/story.html.

89. "License Fee and Funding," https://www.bbc.com/aboutthebbc/governance/licencefee.

90. Imanol Arrieta-Ibarra, Leonard Goff, Diego Jiménez-Hernández, Jaron Lanier, and E. Glen Weyl, "Should We Treat Data as Labor? Moving beyond 'Free,'" *AEA Papers and Proceedings* 108 (May 2018): 38–42.

91. See *Turner Broadcasting System, Inc. v. FCC*, 512 U.S. 622 (1994), on the important governmental interest in the "widespread dissemination of information from a multiplicity of sources."

92. Yemile Bucay, Vittoria Elliott, Jennie Kamin, and Andrea Park, "America's Growing News Deserts," *Columbia Journalism Review* (Spring 2017): https://www.cjr.org/local_news/american-news-deserts-donuts-local.php.

93. See Fraser, "Rethinking the Public Sphere."

94. Data minimization is also a GDPR requirement. See Commission Regulation 2016/679,

2016 O.J. (L 119), at para. 156: "Safeguards should ensure that technical and organisational measures are in place in order to ensure, in particular, the principle of data minimisation." For an informal statement of Apple's differential privacy approach, see Apple, "Apple Differential Privacy Technical Overview" (undated), https://www.apple.com/privacy/docs/Differential_Privacy_Overview.pdf. For a more formal version, see Apple Differential Privacy Team, "Learning with Privacy at Scale," https://machinelearning.apple.com/docs/learning-with-privacy-at-scale/appledifferentialprivacysystem.pdf. For critical reflections, see Jun Tang, Aleksandra Korolova, Xiaolong Bai, Xueqiang Wang, and Xiaofeng Wang, "Privacy Loss in Apple's Implementation of Differential Privacy on MacOS 10.12," arXiv:1709.02753[cs.CR], https://arxiv.org/pdf/1709.02753.pdf. Finally, arguments about how ad-driven business models put such efforts out of reach for some companies seem quite weak. Businesses based on advertising revenue can manage reasonably well, commercially speaking, while exercising more restraint.

95. Post, "Data Privacy and Dignitary Privacy."

96. See generally Habermas, *Between Facts and Norms.*

97. On frictions, see Roberts, *Censored.*

98. Hamilton, *Democracy's Detectives.*

99. *American Views: Trust, Media, and Democracy* (A Gallup/Knight Foundation Survey 2017), https://knightfoundation.org/reports/american-views-trust-media-and-democracy.

100. On media literacy, see Monica Bulger and Patrick Davison, "The Promises, Challenges, and Futures of Media Literacy," *Data and Society*, February 2018, https://datasociety.net/library/the-promises-challenges-and-futures-of-media-literacy/; for thoughtful hesitations, see danah boyd, "You Think You Want Media Literacy . . . Do You?," March 2018, *Points: Data & Society*, https://points.datasociety.net/you-think-you-want-media-literacy-do-you-7cad6af18ec2. For a more encouraging assessment of potentials for digital literacy, see Andrew M. Guess, Michael Lerner, Benjamin Lyons, Jacob M. Montgomery, Brendan Nyhan, Jason Reifler, and Neelanjan Sircar, "A Digital Media Literacy Intervention Increases Discernment between Mainstream and False News in the United States and India" (draft on file with the authors). One recent study, underscoring the challenge, found that students find it difficult to distinguish news from advocacy online: Sam Wineburg, Sarah McGrew, Joel Breakstone, and Teresa Ortega, "Evaluating Information: The Cornerstone of Civic Online Reasoning," *Stanford Digital Repository* 8 (2016): 2018, https://purl.stanford.edu/fv751yt5934.

101. Many people already do not have narrow diets. Guess, "(Almost) Everything in Moderation."

102. Richard Rorty, *Contingency, Irony, and Solidarity* (Cambridge: Cambridge University Press, 1989).

103. Walter Benjamin, *Illuminations* (New York: Random House, 2015), 115.

2

Open Democracy and Digital Technologies

Hélène Landemore

Meet Angeliki Papadopoulos. She is a twenty-eight-year-old woman who lives on the outskirts of Athens in the year 2036. It's almost 8:30 on a beautiful April Sunday morning, and she is just about to pour herself a cup of coffee before reading the news on her laptop. Angeliki logs into her Citizenbook account, a platform initially called Facebook, which was renamed by the public in the year 2025, when its founder, Mark Zuckerberg, decided to withdraw from the company and donate all his shares to a nonprofit foundation run along democratic lines. The company was then repurposed as a deliberative platform for democracy. As Angeliki opens the news tab, her attention is drawn to a flashing sign alerting her that she is overdue for two votes, one on a somewhat complicated issue of environmental law and the other on the choice of a delegate to represent Europe at the next international trade summit in Rio de Janeiro. She decides to ignore the flashing sign and contemplates instead whether she shouldn't delegate the first vote to her uncle, who is a marine biologist and would know better, and just abstain on the second vote. She just does not have time this weekend to read up on the relevant literature.

It's now 8:35 a.m. and Angeliki goes to her inbox, which seems quite full. Her attention is drawn to an email with a bolded title from the Office of the House of the European People, which she opens with a mix of curiosity and excitement. Yes, this is what she thought, the golden ticket! An invitation to join the House of the European People for the next four years in Brussels. She has been randomly selected to join a group of 499 other citizens and set the agenda for the European Parliament over the coming three years. The invitation comes with a generous stipend and financial aid for her and her immediate family to relocate to Brussels. Angeliki is both excited and overwhelmed. She knows how lucky she is to have received this honor. The likelihood of being selected during one's lifetime is only a tiny fraction of a percent. It is much more likely, however, to be selected for the thousands of popular juries gathered for punctual decision making and one-off issues at the European, national, regional, and municipal levels every year. Indeed, Angeliki has already participated in one of those, at the municipal level in Greece. She was a member of the municipal lottocratic body of forty-nine citizens setting the agenda for the city of Thessaloniki ten years earlier. She developed deep friendships with several of the members, many of whom she would have never met through her regular life, and is still in touch with many of them. She also fondly remembers being able to work with elected officials to develop a pilot project for a universal basic income. However, her frustration grew at the time because she realized that she wasn't able to influence the decisions that really mattered, most of which took place at the European level. Becoming an official, full-time lottocratic representative at the European level is another ball game altogether—and will come with a lot more power.

Angeliki herself grew up in an economically precarious family. Her parents' savings were wiped out in the 2008 crisis—the year she was born. After studying philosophy in Greece she became the manager of a bed-and-breakfast (unlike many of her friends at the time, she decided not to flee the country in search of elusive economic opportunities elsewhere). Her husband is a kindergarten teacher. This invitation from the European Union feels like an opportunity to learn about the world, develop new skills, travel, meet interesting people, network in the capital of Europe, and make a concrete difference, possibly, to the world's future as well as to her own. It feels a bit like the chance to go back to college. As to her job, she will have to take a civic leave of absence from her employer.[1] But better opportunities might arise for her after her three years in Brussels, for example, in the booming industry of "democratic jobs" that consist of helping organize and facilitate the minipublics now structuring much of public life and, more and more, the private sector as well. Her husband would also have to take a leave of absence from his teaching job. They could try leaving on her salary as a lottocratic representative (which seems incredibly generous compared to what she is now making), at least for a while, so he can finally work on his novel.

One of the first thing she does after learning the news is to post to her network the announcement that she has been chosen to be part of the European random sample. Maybe hearing other people's reactions will make this more real and help her process how she feels about it. The congratulations immediately start pouring in—everyone seemingly assuming she will accept the position (and indeed, although participation is nonmandatory, only three out of ten chosen people decline the position, usually for family-related reasons). She also notices a hint of envy in some of the comments.

Angeliki smiles. While she's still online, she posts on the marketplace corner of the platform a picture of her sofa and a bunch of clothes that she had been meaning to get rid of anyway. She might as well start preparing for the big move. Then she checks the global politics tab of her page and reads the news for twenty minutes. The e-voting call flashes at the top of her screen once again. She swiftly delegates her vote on environmental policy to her uncle but, reinspired by her new political future, decides to figure out which candidate to endorse as a European trade delegate. She thoroughly browses the profiles of the five candidates (out of 523) that an algorithm has selected for her as most closely matching her values. The summaries have been written by Citizens' Reviews—small, jurylike groups of citizens selected by stratified random sampling—and are fair, clear, and balanced assessments of each candidate's views. Because she has a lingering doubt about the meaning of a particular proposition, Angeliki logs in quickly to her assigned "chat room," a secure virtual space where a few hundred randomly selected citizens like her regularly check in to deliberate about issues. Using her usual avatar (a purple unicorn), she posts her question and, while waiting for an answer, browses through some of the previous exchanges. Within minutes, she has her answer and has herself settled a few misunderstandings by other people. She can now log off from the chat room and cast her vote with one click of her mouse.

After voting, she clicks on an icon that lets her show to the rest of her friends that she has voted, hopefully nudging them to do so as well. She then goes to the earnings tab and checks her earnings. She made fifty coin units since last week, from all the clicking and posting, and even the voting (which comes attached to "mining rights").

Time to walk the dog.

As she is about to cross the street with her mutt to go to the park, an elderly woman is nearly run over by a speeding car. Angeliki pulls out her phone and snaps a picture of the offender disappearing on the horizon before rushing to the shaken elderly lady. Even if she did not manage to get the plate number, there is something else Angeliki has been meaning to do for a long time. She logs in to her Citizenbook platform and goes to the tab SeeClickFix_My Neighborhood, where she files a request for a pedestrian crossing and a slow-down sign on this problematic section of the avenue. She also sends an invitation to the old lady to go and sign the online petition. Then she shares the post on Citizenbook with all her friends and neighbors. She also tags on the issue the lottocratic group representing her neighborhood so they can follow up on the issue for her.

The melody of the ice-cream truck makes her look up. Perfect timing. She selects pistachio and vanilla and then flashes her phone to the seller, who deducts the couple of coin units she just earned from her last hour of online activity on Citizenbook. She then ambles back to the park, finds a convenient spot under an oak tree, and sits down. Time to put down the phone, let the dog run free, and enjoy nature.

*

This chapter looks at the connection between democratic theory and technological constraints, and argues for renovating our paradigm of democracy to make the most of the technological opportunities offered by the digital revolution. The most attractive normative theory of democracy currently available—Habermas's model of a two-track deliberative sphere—is, for all its merits, a self-avowed rationalization of representative democracy, a system born in the eighteenth century under different epistemological, conceptual, and technological constraints. In this chapter I show the limits of this model and defend instead an alternative paradigm of democracy that I call "open democracy," in which digital technologies are assumed to make possible the transcending of a number of dichotomies, including that between ordinary citizens and democratic representatives.

Rather than just imagining a digitized version or extension of existing institutions and practices, I thus take the opportunities offered by the digital revolution (its technological "affordances," in the jargon) to envision new democratic institutions and means of democratic empowerment, some of which are illustrated in the vignette with which this chapter opens. In other words, rather that start from what is—our representative democracies—I start from what democracy could mean if we reinvented it more or less from scratch today with the help of digital technologies.

To do so, however, I first need to lay out, in the first section, the problems with and limits of our current practice and theory of democracy and trace these problems, in the second section, to conceptual design flaws partially induced by eighteenth-century conceptual, epistemological, and technological constraints. The third section then lays out an alternative theory of democracy I call "open democracy," which avoids some of these design flaws, and introduces the institutional features of this new paradigm that are specifically enabled by digital technologies: deliberation and democratic representation. Once this radical normative and institutional framework is in place, I turn to speculation about the ways in which digital technologies could be mobilized further to render open democracy possible, first at the nation-state and ultimately perhaps at the global scale.

This chapter most closely relates to two other chapters in this volume. Like Joshua Cohen and Archon Fung, I am interested in the ways digital technologies can empower a form of deliberative democracy. Deliberation is indeed as central to open democracy as it is to the Habermasian two-track model to which Cohen and Fung subscribe. However, Habermas's model is an avowed idealization of the conditions of possibility of our existing practices. In other words, Habermas's normative ideal is derived from the sociological reality and, crucially, the technologies, of the eighteenth-century public sphere and the iterations of that reality to our day. In contrast, I start from institutional principles derived from the abstract concept of democracy itself, defined as popular rule in which all are equally empowered. Like Cohen and Fung, I seek to set an ideal benchmark to evaluate the flaws and potential of the current status quo (including our current twenty-first-century digital public sphere), albeit one that is unconstrained by the past and thus theoretically maximally ambitious.

Second, whereas Cohen and Fung are mostly concerned with the first track of Habermas's deliberative democracy—the informal public sphere in which deliberation is supposed to take place "in the wild"—I resist this dichotomy of the formal and informal deliberative tracks and strive to imagine a democratic system in which there is a much more fluid and integrated relationship between the deliberations of ordinary citizens and those of political decision makers. This integration is, incidentally, not meant to blur the distinction between public and private—very much preserved here—but to allow for meaningful deliberation among private citizens in a way that is currently not available either in our analog informal public sphere or our digital marketplace of ideas.

This chapter also shares a lot of similarities with Bryan Ford's, "Technologizing Democracy or Democratizing Technology?" As Ford does, I start from a set of abstract principles (though a somewhat different list from his) and build toward a technologically empowered version of such principles. However, whereas Ford focuses on voting systems (liquid democracy) and a democratic currency (an inflationary version of Bitcoin), my focus is on deliberation, representation, and a reinvented articulation between ordinary citizens and democratic decision making.

The Limits of Representative Democracy, as Practice and as Reconstructed Ideal

Democracy has historically been associated with various ideals, such as popular sovereignty, self-rule (or autonomy), and equality (Kloppenberg 2016,

16). These various ideals, though abstract and vague taken individually, together point to a relatively defined political system in which all members of the demos are equally entitled to participate in decision making about laws and policies. Concretely this political system has usually been translated into a system of free and fair elections that is based on the idea of "one person, one vote" and universal suffrage.

An enriched definition, however, has been defended in the past forty years by so-called deliberative democrats, focusing on the step before voting, namely, deliberation. Deliberative democracy is thus a theory of democratic legitimacy that traces the authority of laws and policies to the public exchange of arguments among free and equal citizens, rather than, strictly speaking, the moment of voting. This theory was developed in the late 1980s and 1990s as an alternative to the then-dominant theory of aggregative democracy, according to which democratic legitimacy stems simply from the proper aggregation of votes in free and fair elections pitting various elites against one another.[2] Because I share deliberative democrats' belief in the centrality of deliberation to democratic legitimacy, I propose as my working definition of *democracy* the following: a political system in which all the members are equally entitled to participate in the association's decisions about its laws and policies, including in the prevoting deliberative stage.

By this demanding definition, our current democratic systems appear deeply flawed. Deliberation on matters of public policy and laws in which all have a genuinely equal opportunity to participate almost never happens, for obvious reasons. It is not possible to gather millions of citizens in a common space and give them equal opportunities to participate in deliberation. Even the 2019 recent French Great National Debate, a large-scale exercise in public deliberation that President Emmanuel Macron tried out as a response to the Yellow Vests' revolt, only managed to involve a tiny percentage of the population in roughly ten thousand town-hall meetings of anywhere between twelve and three hundred people and twenty-one randomly selected regional assemblies gathering a grand total of about 1,400 participants.[3] In any case, the imperfect, second-best solution to mass deliberation has always been to delegate the actual deliberation preceding a formal decision to representatives, who conduct it on our behalf with our consent (or rather, the consent of a majority among us).

Deliberative democrats in the Habermasian vein have inventively rationalized this de facto division of labor between representative elites and the public and tried to carve out a central role for the public by distinguishing, at a normative level, two deliberative "tracks" (as per the analysis of chapters 7 and 8 in Habermas's [1996] *Between Facts and Norms*). The first track is the space of formal decision making (e.g., parliament, the courts, the administrative agencies), and the other is the space of informal public deliberation, where public opinion is formed. The first deliberative track is meant to shape decision making per se, whereas the informal one, taking place in unstructured and even anarchic ways in the wilderness of the larger public sphere, feeds content to the formal deliberative track in substantive but not directly binding ways.

Another central Habermasian metaphor, the sluice, illustrates the connection between these two tracks (Habermas 1996, 556, as borrowed from Peters 1993; see also Peters 2008).[4] According to Habermas, "binding decisions [made in the first track], to be legitimate, must be steered by communication flows that start at the periphery [in the second track] and *pass through the sluices of democratic and constitutional procedures situated at the entrance of the parliamentarian complex or the courts*" (Habermas 1996, 356, my emphasis). Sluices are systems of water channels controlled at their heads by a gate. Their image is used to characterize the intermediary bodies and procedures (e.g., parties, elections) between official decision makers and the public, intermediary bodies that ensure transmission of information from the outer periphery of diffuse public opinion to the center.

This picture of representative democracy is enormously attractive, essentially because of the central place it gives to deliberation and the circular and reciprocal relation between the two tracks. In theory, thanks to the sluices, the relationship between the two tracks is meant to approximate the ideal of equal entitlement to participate in the deliberation of the polity about its laws and policies. Yet not only is the reality far from resembling the idealized Habermasian reconstruction of it.[5] The picture itself is problematic.

First, even on Habermas's model democracy suffers from the problem of the separation between a ruling group of elected officials, appointed courts, and administrative bodies on the one hand and the mass of ordinary citizens on the other, with the first group firmly positioned at the center of power and the other relegated to the periphery (Habermas 1996, 354).

The metaphor of the sluices, though meant to capture the ways in which the two tracks are connected in constructive and even dialogical ways, thus emphasizes that the public and its representatives are meant to be kept separate from each other. Political elites and institutions thus form a necessary intermediary between ordinary citizens and actual decision making and also the bottleneck where popular ideas come to cluster and, for many of them (perhaps most), die. It appears as if ordinary citizens and their contributions are structurally marginalized in what is the idealized version of the system.

Furthermore, it is worth noting that the sluice connotes mechanical, hierarchical, rigid, and slow processes. The two-track Habermasian metaphor thus appears out of sync with modern democratic expectations of more immediate participation and voice, particularly as enabled by the digital revolution in almost any other sphere of life. Think, by contrast, of the ease and speed with which individuals can access and generate information on services and products in the marketplace with applications like Rotten Tomatoes, Yelp, Zillow, and so on.

Last but not least, in the Habermasian model, political elites are supposed to be engaged in a circular, reflexive, and dialogical exchange with the public via intermediaries such as the media, political parties, and the pressure of an informal public opinion formed in civic associations of all kinds.[6] Yet it is not entirely clear what power deliberators in the wild really have over the formal deliberative track. The idea that the decentralized deliberations of citizens in the wild add up to a meaningful way of setting the agenda for the formal decision-making sphere is not entirely plausible, as Habermas himself occasionally seems to acknowledge.[7] There are many reasons to think that the larger public sphere is shaped by the formal deliberative track in a way that is not fully reciprocal. The collective action problems faced by the public are enormous compared to those faced by the smaller number of agents at the center of formal decision structures. Moreover, even in the best-case scenario of a functional public sphere, it is hard to imagine how a series of haphazard, unregulated, and decentralized deliberations among groups of different sizes and compositions, which are not intentionally oriented toward this outcome, could be the proper way of setting the agenda for the formal deliberative track. Does such deliberation "in the wild," one might ask, even amount to deliberation at all?[8]

Habermas's sociologically rooted model of a two-track public sphere is to date the most attractive and powerful rationalization of our existing systems at their best. But it is too wedded to the dichotomies of representative democracy (represented ordinary citizens versus representing elected elites) and too constrained by the technologies of yesteryear. It also leaves deliberation in the wild to the vagaries of self-organizing communities in ways that may or may not facilitate the crosscutting exposure required for deliberation.

The Problem with Elections

The main design flaw of representative democracy, at its inception in the eighteenth century and now, is that it is centered on the principle of periodic elections (Manin 1996). As per Manin's diagnostic, this principle is ambiguously democratic at best. While it is generally combined with the egalitarian principle of "one person, one vote" it does not treat candidates to elections equally, giving more chances to those who can stand out in the eyes of others, on the basis of properties unevenly distributed in the population (typically social and economic advantages). Not only are elections an ambiguously democratic selection mechanism; their use is also arguably premised on the wrong picture of what it takes to create a representative assembly with good deliberative and thus problem-solving capabilities. As I have argued in previous work (Landemore 2013, building on Page 2007), the problem-solving capabilities of deliberative assemblies are likely not a mere linear function of the individual competence of their members. Instead, they are likely to be more a function of a group property, cognitive diversity, that characterizes the diversity of views and ways of thinking present among the members.[9] If the goal is to compose an all-purpose assembly of democratic representatives, for which there is ex ante uncertainty as to what the relevant diversity should be, and assuming that on average citizens are at least competent enough to address most political questions, a good strategy is to take a random sample of the larger population and form a statistically representative minipublic (Landemore 2012). In contrast, recruiting members of a deliberative assembly by elections will naturally entail a loss of cognitive diversity that will come at an epistemic cost (and is likely responsible for the many blind spots of democratic decision making diagnosed today).

Finally, another negative implication of elections as a selection mechanism for democratic representatives is that they give rise to a partisan logic that ultimately runs against the open-mindedness required to conduct proper deliberation. Electoral democracies are today systems in which the public debate is structured as a competition between policy platforms backed by partisan justifications. Parties are essential intermediary bodies between individual citizens and the institutions of the state, in that they aggregate views, perspectives, solutions, and information into a cognitively manageable amount of bullet points, value statements, and other ideological shortcuts. To the extent that parties are necessary, so is the virtue of partisanship that sustains them in existence.

Yet parties and partisanship come at a deliberative cost. Diana Mutz's empirical work on the relation between participation and deliberation strongly suggests that we cannot have it both ways: either people will be willing to engage with dissenting others and enjoy the benefits of exposure to diverse, or even conflicting, views, or they will be willing to vote, campaign for candidates, and generally be engaged as partisans in the political arena.[10] But they cannot be open-minded and politically engaged at the same time. This is so, she argues, because most people, when faced with even minimal disagreement in the political realm—what she calls "cross-cutting perspectives"—recoil from engaging and prefer to retreat to the sphere of their like-minded peers and political friends.[11] In other words, Mutz finds that partisan political participation and the kind of deliberative mindset assumed by deliberative democrats do not go together. To the extent that exposure to diversity and disagreement through political discourse threatens interpersonal harmony, people will tend to avoid entering into political territory at all. They will apply the etiquette of the polite guest—let's not talk about politics—or they will seek the company of like-minded people.[12]

Representative government, from its early elitist beginnings to today's partisan version, is thus the contingent product of eighteenth-century ideological, technological, and epistemological constraints. Today, however, we have better social-scientific tools, a better understanding of what makes groups smart, and digital technologies to help us achieve what eighteenth-century institutional designers could only dream of. One way forward in reimagining our institutions could thus be, instead of rationalizing away, with Habermas, the representative democracy we have inherited from the eighteenth century, to start imagining different institutions. In what follows I go back to the drawing board to sketch a vision of "open democracy," which in a way returns us to earlier versions of democracy (specifically, classical Athens) but that digital technologies arguably render feasible at scale and allow us to tweak in innovative ways.

Open Democracy

In this section I lay out a normative paradigm of democracy that I call "open democracy" (for a sketch, see Landemore 2017; for fuller development of this paradigm, see Landemore 2020). Open democracy is meant to be not just an improved, more participatory, or differently representative version of representative democracy but a different paradigm altogether. Its core ideal is to put ordinary citizens at the center of the political system rather than at the periphery, emphasizing accessibility and equality of access to power over mere consent to power and delegation of power to elected elites.

In Landemore (2020) I defend open democracy as constituted by a series of five institutional principles: participatory rights, deliberation, the majoritarian principle, democratic representation, and transparency.[13] This is not the space to go over all of these principles. Instead, I want to zoom in on two of the principles uniquely enabled by digital technologies: deliberation and democratic representation. Before I do, though, let me add a word about the concept of openness.

Openness is, first, the opposite of closure, in both a spatial and temporal sense. In a spatial sense, openness can mean various things, depending on the context, from degrees of porosity to accessibility, participation, and inclusion. Openness is to both voice and gaze. This openness is inclusive and receptive—of people and ideas. This openness characterizes a system that lets ordinary citizens in, whether the spatial openness is facilitated by architectural design or by technological tools. In a temporal sense, openness means open-endedness, and thus adaptability and revisability. It means, concretely, that democratic institutions must change as the people they are meant to serve change. An open-ended system is more likely to adjust to rapid changes in complex, large-scale, connected societies. Openness, finally, is a property of the type of minds a democracy should cultivate in its citizens, as opposed to narrow-mindedness (or its close cousin, partisanship).[14]

In many ways the concept of openness is already pervasive in the vocabulary of activists, grassroots associations, and even the jargon of government officials.[15] The concept of openness also owes a lot to the world of coders and advocates for self-organization and freedom on the internet. The open-source movement promotes so-called open-source software, which is software with source code that anyone can inspect, modify, and enhance.[16] The image of open-source software is applicable and relevant to democracy because if, as some have argued, "code is law" (Lessig 2000), then one can argue, conversely, that democratic law should be more like code, or at least code of the kind made available in Linux or other open-source communities. In other words, instead of being something created and guarded by small groups of insiders or experts, in a democracy the law should be something to which all have access and on which all can make an impact. Everyone should be able to write and claim authorship over the law. This is what Icelanders tried to do with their 2010 revolutionary constitutional process, whereby they let a national forum of 950 randomly selected citizens set constitutional values and priorities and a twenty-five-person council of nonprofessional politicians write the new constitution in collaboration with online crowds (see, e.g., Landemore 2015). This is also the idea behind experiments in participatory budgeting, crowdsourced law reform, or the most recently forged all-encompassing concept of crowdlaw.[17]

The concept of openness in open democracy, finally, is also indebted, and a nod, to the liberal Popperian tradition of the "open society" (Popper [1945] 2013). Building on a contrast between closed and static traditional societies and modern open ones, Popper defined the open society as a dynamic society in which government is expected to be responsive, tolerant, and relatively transparent, and citizens are free to use their critical skills to criticize laws and traditions. Open democracy can be interpreted as a subset category of an open society, in which the government is not just liberal but genuinely democratic and, furthermore, democratic in an "open" manner that facilitates participation of ordinary citizens. Open democracy is the democratic answer, and in many ways a complement, to the essentially liberal concept of the open society. Unlike in the liberal tradition, the subject of openness is the space of political power itself, the place from where power is exercised, not just the society ruled or structured by it.

Let me now turn to two principles of open democracy that could be uniquely facilitated or enhanced by digital technologies: deliberation and democratic representation.[18] Deliberation is explicitly borrowed from the recommendations, over the past forty years, of deliberative democracy theorists, for whom, as already mentioned, political decisions and policies can be legitimate if and only if they could be (Cohen, Rawls) or de facto are (Habermas, assuming ideal conditions) the product of a deliberative exchange of reasons and arguments among free and equal citizens. Open democracy thus consciously embraces deliberation as a key institutional principle.[19] The problem with deliberation, as we saw earlier, is that we do not know how to render it genuinely possible at scale, for millions of people. Digital technologies, however, have rendered the promise of deliberation at scale considerably more plausible, by offering the possibility of replacing face-to-face, necessarily small meetings taking place in the here and now, and always fraught with the danger of power dynamics tracking visible physical differences, with much larger meetings of disembodied or reembodied (using pseudonyms or avatars) individuals, in which quality deliberative exchanges are facilitated by augmented reality tools, the argument-centric organization of the contributions by the participants themselves (Klein 2006; Spada et al. 2016), and even artificial intelligence tools (Hilbert 2017). Ultimately one can imagine these online deliberative platforms being facilitated and aided by natural-language analysis performed by artificial intelligence algorithms.

Democratic representation is another central principle of open democracy and its subtlest point of departure from representative democracy, which is centrally characterized, in contrast, by electoral representation. Representation is defined, here, minimally and descriptively, as the act of standing for someone in a way that a relevant audience recognizes, as per Rehfeld's (2006) definition. By contrast, normative representation can be thought of as representation authorized by the relevant constituency and meeting some other normative, possibly substantial criteria of justice (e.g., good representation, representation in the interests of the represented, as in Pitkin [1967]).

Democratic representation I define more specifically as an act of standing for others that is the result of an inclusive and equal selection process. By contrast, oligarchic representation is an act of standing for others that is the result of an exclusionary and unequal selection process. Note that electoral representation arguably awkwardly sits between the two categories, if we believe that it is indeed a hybrid form of representation, with a face turned toward democracy and another toward oligarchy (as per Manin's account of elections as Janus-faced).

Democratic representation includes both what I call "lottocratic" and "self-selected" representation. Lottocratic representation is representation performed by citizens selected at random or, as a close second best in theory and often an improvement in practice, stratified random sampling (which allows the targeting of minorities at risk of being underrepresented in a true random sample). Lottocratic representation is on display in the many variations of so-called minipublics that gather a (more or less) random sample of the entire demos. These assemblies are not equally accessible to all in the here and now, because one needs to be selected to enter them, but they can be characterized as "temporally open" assemblies since over time (provided enough rotation and a sufficient number of those assemblies) all citizens should have equal access to them. Randomly selected assemblies thus produce a type of democratic representation whereby each citizen has the same equal chance of playing the part of a democratic representative.[20]

Self-selected representation, by contrast with lottocratic representation, translates into what I characterize as "spatially open" assemblies, namely assemblies that are accessible to all those willing to participate (though sometimes only up to capacity when the meetings are physical rather than online). Examples of self-selected representative assemblies are the People's Assembly of classical Athens, Swiss *Landsgemeinden*, participatory budgeting meetings, town-hall meetings, Occupy assemblies, the meetings of Yellow Vests on traffic circles, or even online crowdsourcing platforms allowing deliberative exchanges. These open assemblies allow for the willing fraction of the population that typically shows up for such events to stand for the whole and make decisions on their behalf in ways that are recognized as representative by the larger system. I thus distinguish open assemblies from direct democracy moments, like referenda, in that in direct democracy moments or processes everyone, or at least, a majority, is expected to participate. In open assemblies, by contrast, only a small fraction is expected to turn out.

Contrary to elected assemblies, which are at best accessible to the willing and ambitious, and contrary to lottocratic bodies, which are only accessible to all over time (in the best-case scenario and with sufficiently frequent rotation)—in theory, everyone is able to participate in self-selected assemblies. There is no qualification needed to be included, whether social salience and ambition or luck. All it takes, in theory, is the will to participate. Similarly, in the Athenian People's Assembly, in theory, every citizen had the same right to participate and, once there, to say something and to be heard. These generalizations are of course true only at a high level of idealization, which brackets the substantive conditions for participation in general, such as time and social, educational, and economic resources.[21] Whether this idealization is tolerable depends in large part on the empirical question of whether the substantive conditions for equality of opportunity to participate can be plausibly achieved. If they cannot, then self-selection may turn out to reinforce existing inequalities. Additionally, in nonideal contexts, silence and exclusion (including digital exclusion) could also be read as active refusal and a form of civil disobedience (see chapters by Ananny and by Gangadharan in this volume).

Both lottocratic and self-selected representation are "open" by contrast with electoral representation, which is only accessible to those who stand out in the eyes of their fellow citizens, as per the "distinction principle" noted by Manin (1996). In electoral representation, access to the status of representative is neither fully spatially open (the slate of candidates is usually restricted by the hierarchies of parties and other organizations and then the "aristocratic" principle of election only selects the most salient people among those). Nor is it fully open over time (electoral elites tend to reproduce in ways that are exclusionary for the rest of us).

There are, arguably, ways to reimagine electoral representation via schemes of so-called delegative or liquid democracy (Ford 2012; Blum and Zuber 2016), based on vote delegation (or vote recommendation) to allow for what one might consider a greater democratization of the status of elected representative.[22] Delegative or liquid democracy is a system in which people can give their votes to anyone they like, either for a given term, or just on certain issues, with the option of recall at any time and the possibility to retain one's right to direct input throughout.[23] This type of democracy crucially differs from electoral democracy "in the principle that each voter should have free, *individual* choice of their delegate—not just a choice among a restricted set of career politicians" (Ford 2014, 1). Conversely, delegative democracy aims to lower the barrier to participation for would-be delegates. While delegative or liquid democracy schemes typically claim to want to get rid of representation altogether or strike a middle-ground between direct and representative democracy,[24] they can also be described as aiming to strike down the barriers to entry to the status of elected representative, thus rendering electoral representation more inclusive. This arguably constitutes such a radical break from electoral representation that I propose conceptualizing the representation at stake in liquid democracy schemes as the distinct notion of delegative or liquid representation.[25]

The point of rethinking democratic representation in these nonelectoral ways is that we could imagine a democracy that need not translate into elections alone or even elections at all. Thus, no mention is made at the level of the fundamental institutional principles of the principle of elections because elections, far from being *a*, let alone *the*, ultimate democratic principle, are merely one selection mechanism among others capable of translating the representative principle in a democratic fashion. Whereas periodic elections are a defining institutional feature of representative democracy, open democracy is not committed to elections per se. Instead, it embraces a rich ecology of various forms of democratic representation.

How do these two central principles—deliberation and democratic representation—ultimately create a different type of democracy from representative democracy? Considering that there is no such thing, yet, as an open democracy, I am forced to render the difference at the model level, where the contrast is with the Habermasian ideal.

In contrast to the Habermasian metaphor of a two-track deliberative sphere, the central metaphor for open democracy is, I propose, that of the "open minipublic," that is, an all-purpose, periodically renewed, randomly selected body of citizens made entirely porous to the direct input of the larger public and permanently connected to subordinated single-issue minipublics, all of which are also open and porous to the larger public's input. Instead of the dichotomy between ordinary citizens deliberating in the informal sphere of opinion formation and elected elites making binding decisions in the formal sphere of will formation, open democracy pictures a constant rotating of ordinary citizens in and out of the variously nested and networked decision-making loci, all the while maintaining constant communication flows between the temporarily represented and the temporary representatives in ways that bypass the classic bottlenecks or "sluices" of elections and other party structures. This prevents the ossification of permanent difference between ruling elites and ruled. With the rotation principle built into lottocratic representation, we come closer to the Greek ideal of "ruling and being ruled in turn." You might call it the modern principle of "representing and being represented in turn."[26]

Additionally, whereas in Habermas's model the larger public sphere is left to self-organize, open democracy aims to structure large-scale public deliberations as much as it structures the small-scale deliberations of decision makers by bringing randomly selected citizens into contact with one another at all levels of the polity to create as seamless a continuum as possible between the open minipublics at the heart of political decision making and the multiplicity of deliberative minipublics taking place around them in a networked fashion. As one can easily imagine from this description, such a metaphor is not an idealized reconstruction of anything already in existence. In fact, what it describes is conceivable only with the help of modern digital technologies and, at best, in some nearby future.

As a side note, the 2019 French Great National Debate was, in my view, an effort to structure the informal public sphere in a way that might indeed render it more effective at setting the agenda for the public sphere. One might think of what could be done on that model, if institutionalized and made truly permanent, as the creation of a third track in the public sphere, which would be more structured than deliberation in the wild per se but still limited to opinion formation as opposed to will formation. In the French case, at least the goal was to generate propositions that the formal sphere (essentially the executive power) would still be the only one to decide on, but with a lot more pressure than usual to follow the agenda of the larger public. Although this model of a three-track public sphere is certainly an improvement in many ways, it is not the radical break I have in mind. In open democracy we would no longer need this clear-cut separation between tracks and between ordinary citizens on the one hand and professional politicians on the other.

How Digital Technologies Can Help

How can digital technologies help us bring about open democracy, or at least some of its core principles, deliberation and democratic representation? What would that look like? Let us first turn to the possibility of deliberation, that is, deliberation involving all on an equal basis, and how to use digital technologies to organize it in a rational way throughout, as opposed to leaving it either to the structured but adversarial exchanges of party leaderships and career politicians in the first deliberative track or the unstructured and currently rather dysfunctional space of the larger public sphere (including the digital sphere).

To my mind the deliberative ideal should be, ultimately, "many connected brains" seamlessly and almost simultaneously exchanging information and arguments in ways that are costless and frictionless, resulting in enlightened individuals and enhanced collective intelligence. Given the physical limitations of human beings and long-standing technological constraints, we have so far used delegated deliberation to a group of elected men and women physically gathered in a parliament over long periods of time (in theory, years, although the actual physical time spent in parliamentary chambers is surprisingly short). These men and women are very loosely and imperfectly connected to their constituencies via the feedback mechanism of periodic elections and direct contact with them during regular "office hours" in their constituencies, most intensely during campaigning periods. Digital technologies have rendered access to elected representatives much easier, more immediate, and efficient, allowing constituents to tweet at their representatives, for example, or engage them on their Facebook pages. Michael Neblo, Kevin Esterling, and David Lazer (2018) see the future as "direct representation"—classic electoral representation enhanced with direct participation by citizens. By direct representation these authors envision a better version of our current system, in which technologies (phones in their case) are mobilized to enhance the flow of communication between constituents and elected representatives. For all its merits, in particular its feasibility, this defense of "direct representation" optimistically assumes that just because the elected representatives will be more directly exposed to a greater variety of constituents' views, they will be able to process the input correctly and reflect it in the deliberations they are then part of in the US Congress and in the resulting decision-making process. But every elected representative's mind in this approach still plays the role of a bottleneck and a funnel, which risks leaving out too many aspects of the original content. It would be better, it seems to me, to open up the formal deliberation itself to a variety of minds and contributions.

What technologies could enable, in this respect, is a dematerialization of face-to-face deliberations and a vast expansion of the number of people involved in them, allowing us to bypass the bottleneck of elected representatives' minds having to synthesize and transfer this massive input from the constituents. Could technologies even get us to "direct democracy on a mass scale" or "mass online deliberation" (Velikanov 2012)? This is the hope cautiously entertained by cyberdemocrats or e-democrats, some political and communication theorists, and visionary engineers (see, e.g., Bohman 2004; Dahlgren 2005; Hindman 2008; Velikanov 2012). Online deliberation permits the recording and archiving of all people's thoughts, comments, and ideas while economizing on the necessity of being present all at once. As long as everyone can have access to the same virtual "room" (the platform they sign into), individuals are able to read the same content when it is convenient for them and at their own pace. Assuming a sufficiently long window of time before the decision is to be taken, deliberation among all can thus be distributed over time in a way that fosters great inclusivity. Digital, text-based deliberation thus potentially takes care of some of the constraining time and space aspects of analog, face-to-face deliberation.

The reality, however, is that the closest we can get to mass deliberation in the physical world is via a multiplicity of minipublics, themselves operating on the basis of small groups (e.g., tables of twelve to fifteen people in the typical setup of Jim Fishkin's [2018] deliberative polls; or six to seven only in the methodology used in the 2019 French regional assemblies during the Great National Debate and more recently during the Citizen Convention on Climate Change). These micro–deliberative groups merge into temporary plenary assemblies to dissolve again into differently constituted groups again (during the so-called pollination phase), until many people have touched on and contributed to many subjects.

Moving such deliberations online holds the promise of expanding the number of people who can meaningfully deliberate all at once. According to some, social media and other applications of the Web 2.0 in particular have the potential "to fulfil the promise of breaking with the longstanding democratic trade-off between group size (direct mass voting on predefined issues) and depth of argument (deliberation and discourse in a small group)" (Hilbert 2017, 2). At the moment, however, even the most promising existing platforms succeed in expanding the number of people who deliberate directly with one another in this way to a few hundred people (Spada et al. 2016). Enabling a few hundred people to deliberate with one another directly is a clear improvement over the limits of face-to-face deliberation. This is nonetheless a far cry from the millions that would need to be included for direct democracy to be possible in existing polities. Current promises of true "mass online deliberation" are at best conceptual prototypes at this point.

Direct democracy, to the extent that it involves a deliberative phase, is probably feasible only for small groups, even in the digital age. If this is true, then the possibility of direct democracy breaks down as soon as the group expands beyond a few hundred people.

Another argument against the meaningfulness and possibility of direct e-democracy has to do with the nature of politics rather than technological or human limitations per se. The claim is that representation is necessary and desirable, in and of itself, as a way to constitute interests and preferences. It is sometimes expressed as the view that "representation is always constitutive of democracy and democratic practices" (see, e.g., Plotke 1997, 10; Urbinati and Warren 2008). Or, as Plotke (1997, 19) puts it even more explicitly: "Representation is not an unfortunate compromise between an ideal of direct democracy and messy modern realities. Representation is crucial in constituting democratic practices. 'Direct' democracy is not precluded by the scale of modern politics. It is unfeasible because of core features of politics and democracy as such." What Plotke and other proponents of the so-called constructive turn in representation theory argue is that interest and preferences, unlike say, a taste for vanilla or chocolate ice cream, are not given. Only on a very crude (or economistic) understanding of politics can one expect individuals to be able to speak their interests and preferences (let alone judgments) off the top of their heads and without prior elaboration. This elaboration will usually require the creation of interest groups, associations, or parties, which can then enter the deliberation, negotiations, and bargaining taking place at the collective level in a meaningful and informed manner. Figuring out, clarifying, and articulating interests is, in other words, a prerequisite to deliberation. If this is so, representation is fundamentally unavoidable and would remain so even if deliberation could be scaled to millions of people. In other words, except, perhaps, for very small groups whose interests can be identified in the course of a direct deliberation, direct (deliberative) democracy is never really an option.

If this is true, then representation is unavoidable and deliberation must take place in relatively small units compared to the size of any modern polity. Yet nothing says that deliberation among democratic representatives must be confined to the familiar groups of elected politicians. Instead of replicating an elected chamber in digital format, and trying to connect it to a larger, unstructured public sphere left to its own devices (which we would want to do if we were to simply digitally enable the Habermasian model), digital technologies could be used to implement something different, better integrating the deliberations of the decision makers and the citizenry.

Lottocratic representation, or representation based on random selection, is perhaps easiest to imagine implemented in a digital format, because we have only to picture an online version of classical Athens's large juries (between five hundred and one thousand citizens) and the kind of online version of the minipublics now practiced in various guises (e.g., deliberative polls, citizens' assemblies) around the world.

Self-selected representation is also uniquely enabled by digital technologies, which allow at little cost the gathering of input from online "crowds" on any issue of relevance. A great pilot for what this could look like was the 2011 Icelandic crowdsourcing consultation of the public on twelve successive constitutional drafts (Landemore 2015).

"Liquid representation," finally, is premised on a core concept—vote delegation—perhaps even more uniquely dependent on digital technologies. It would indeed be difficult to envision something like liquid democracy on a mass scale using regular mail (although corporations have long used somewhat similar systems, proxy voting). Examples of liquid representatives can be found in the Demoex party in Sweden, which first used a liquid democracy system between 2002 and 2016 (Norbäck 2012). Around 2006, software platforms were created to facilitate not just comment functions and vote delegation but delegation-based online discussion and deliberation as well, under the names "LiquidFeedback" and "Adhocracy." LiquidFeedback was adopted and used for the past several years by the German Pirate Party (Swierczek 2014).

Now, assuming that the legitimacy of such new forms of democratic representation is accepted, what form should deliberation take? Given what we saw earlier about the impossibility (for now) and perhaps even the undesirability of "mass online deliberation," a second-best alternative is this: having myriad randomly appointed small groups deliberating independently, with their inputs aggregated up to a final level of decision making, or simply fed to a central decision-making body with ultimate sovereign power. The number of these minipublics would have to be large enough to ensure that any member of the demos could join one if she so chose. The deliberations of these minipublics could be made public and their exchanges open to the input of external crowds via crowdsourcing platforms. I call such a structure an open minipublic, a deliberative unit that is uniquely possible in a digital world.

To capture this idea, imagine first a version of Facebook devoted to the task (perhaps among other democratic ends) of growing and curating a deliberative platform for any given democracy. (I set aside here the possibility of a global deliberation, which assumes the existence of a global demos, or virtual, cloud-based demoi). Let us call this fictional, utopian version of Facebook—something so different indeed that it is worth changing the name—"the Citizens' Book," or Citizenbook for short (as per the earlier vignette). Assume that every citizen would be automatically electronically registered on it at birth (and let us set aside for now the dystopian possibilities that something like this will also inevitably raise).

Imagine if Citizenbook could be used as a safe and secure space for online deliberation among all citizens to talk about collective issues, such as climate change, economic inequalities, immigration, or gun control (in the United States) except that instead of letting us talk only to our chosen circles of friends and acquaintances (as in the current version of most existing social networks), the system would match us to randomly selected others and invite us to join deliberative chat rooms with them for a certain amount of time (as per the opening vignette).

Imagine—and this is probably the least demanding leap of imagination—if all members registered on Citizenbook could then securely vote on this platform, easily accessible from their smartphones, in online referenda and other popular vote processes, including those allowing "liquid democracy" schemes where people can delegate their votes to whomever they want (as per the initial vignette as well).

The voting part seems conceptually easy, although the devil is in the details and security issues would need to be worked out. The deliberative part, however, deserves more thought. The deliberative chat rooms could be virtually augmented to render them attractive and fun to participate in, along the now-well-established principle of gamification. The chat rooms could be made public or private, depending on the choice of the participants. If made public, they would be instantly connected to a crowdsourcing platform allowing individuals outside of it to directly submit input or comments on the deliberations. All the information in the world, properly processed and synthesized, would be at the fingertips of such minipublics, available in user-friendly format after proper vetting by, for example, randomly selected online juries of other citizens collaborating with professional journalists and experts the world over. A good model here are Oregon Citizens' Initiative Reviews (CIRs), citizen panels of about two dozen people selected to form a cross section of the larger population. These CIRs are tasked with deliberating for several days about upcoming ballot initiatives or referendums. At the end of their deliberations they produce a citizens' statement, which offers a balanced assessment of the ballot initiative or referendum that is distributed to all registered voters in the hope of promoting more informed voting.[27] In addition to such measures, facilitators and even "political translators"—people able to help disempowered individuals and groups find their voice in multilingual and multicultural settings (Doerr 2018)—would be available to structure deliberations and ensure protection for vulnerable minorities against the usual power dynamics induced by differences in languages, linguistic skills more generally, social and economic status, and so on. All these actors would be the practitioners of these new "democratic jobs" mentioned in the opening vignette (only some aspects of which, like basic facilitation, can be reasonably expected to be automated). Finally, to maximize participation, one could imagine financially compensating or incentivizing participants by crediting them real money instantly on their digital account. One could imagine Citizenbook also financially compensating citizens for the data they generated in virtue of their online activities, including their political activities, as these activities would serve as essential training material for the artificial intelligence tools and functionalities on which Citizenbook, as well as the constellations of properly regulated (perhaps at the global level) private corporations thriving in that ecosystem, would depend (for similar ideas, see, e.g., Lanier 2013; Posner and Weyl 2018). One could also imagine each citizen being automatically credited, at birth, and perhaps at regular intervals after this, with a basic universal income, above and beyond online earnings.

Another solution would be not to give final say to these online minipublics but to feed their aggregated input (synthesized with the help of natural language analysis software) to a central, open minipublic.

Regardless of what turns out to be the better format, what these solutions have in common is that they structure and curate public deliberation from beginning to end, instead of leaving the larger public sphere completely to its own anarchic ways. This is not to say that something like Habermas's deliberation in the wild would not subsist—we would still have political conversations at the kitchen table or with taxi drivers—but these would no longer be the only or even primary sources of information and the loci of deliberation for citizens.

In other words, this model preserves the differentiation of deliberation for opinion formation and deliberation for decision making but without necessarily creating two (or even three) separate deliberative tracks with two different logics (one structured, one unstructured). Indeed, in both cases the deliberations are structured. This presents several advantages. First, the structuration of the informal public sphere through online deliberations in open minipublics (whether advisory or binding) would break the silos, filter-bubbles, and echo-chambers in which individuals currently prefer and are in fact encouraged to segregate themselves by platform designs created to maximize ad revenue rather than quality deliberation. Open minipublics would facilitate, in other words, crosscutting exposures in ways that minimize unpleasantness and may in fact prove rewarding, empowering, and educational. Third, there would be much less of a loss of information between the deliberations of citizens and the decision-making moment, whether because citizens ultimately vote themselves after being part of a minipublic or because the deliberations of the sovereign minipublic would be so tightly connected and porous to the conclusions of satellite minipublics. Fourth, the spillover effects of such online deliberations, whether they end up in voting or not, would presumably affect the world of offline and online conversations with families, friends, and peers (Habermas's second track), thereby introducing a greater diversity of perspectives to these often too same-minded environments and spreading a spirit of moderation and open-mindedness to counterbalance the polarizing effects of groupthink.

Now, there are, of course, a number of questions that the idea of a Citizenbook as deliberative platform may raise.[28] Should Citizenbook be a not-for-profit company? If so, how would it be funded, by whom? Would this really suffice to remove all the problems currently faced by for-profit social networks? What about newspapers and journalists in this new environment? Do we still have the problem of noise, fake news, and filter bubbles? What about the potential for mass surveillance? What about people who refuse to register, or can't? What about racial biases? What about inequality of resources or, assuming this economic dimension can be taken care of (via a universal basic income, for example), the problem of scarcity of attention? Considering that the feasibility of open democracy heavily depends on the use of algorithms, who would be in charge of curating those algorithms? I do not have the space here to develop any of these thoughts, nor am I yet able to picture how, exactly, all these interlocking parts should work together. I trust that some of the other contributions to this volume can help shed some light and provide guidance on some of these questions and that the volume as a whole will trigger further fruitful conversations allowing us to refine the ideal of a technologically empowered open democracy.

Conclusion

It is tempting, in today's climate and given recent events, to imagine the worst dystopian version of an online deliberative platform, especially if it were to grow global and was enabled by a tentacular, privacy-shattering, and totalitarian corporation (in the vein of, say, the novel-turned-movie *The Circle*). In this chapter I have chosen to pursue a more utopian and optimistic approach, one in which digital technologies are deployed to support and deploy a new and better kind of democracy. I have argued, specifically, that scientific developments and new technologies now allow us to think and invent institutions beyond the dichotomy of the voter and the elected representative, including at scale. They might even one day allow us to reimagine the possibility of true mass deliberation. If so, then there is no reason to stick to representative (electoral) democracy or even to the Habermasian two-track model that rationalizes it as the most desirable normative framework (at least as currently formulated). The model of open democracy sketched here, centered on the model of the open minipublic, offers what I hope is a new and more democratic framework meant to guide future institutional reforms and technological innovations.

Notes

1. A right that appeared with the first institutionalized House of the People in Europe, created in France in 2019 during what is affectionately referred to as the "second French Revolution." The model quickly spread to other countries and a few years later to the European Union itself.

2. Deliberation is valued by deliberative democrats for a number of reasons, among which are that it allows laws and policies resulting from it to be supported by public reasons and justifications (rather than mere numbers); gives all citizens a chance to exercise their voice (including via their legitimate democratic representatives); has beneficial consequences, such as educating citizens, building a sense of community, and promoting civic engagement; generalizes interests (Habermas); and increases the chance of the group successfully solving various collective problems (a dimension more specifically emphasized by so-called epistemic democrats). I embrace all of these reasons to want to put deliberation front and center in a theory of democracy.

3. For greater detail, see my analysis of the exercise in the *Washington Post*, at https://www.washingtonpost.com/politics/2019/04/24/can-macron-quiet-yellow-vests-protests-with-his-great-debate-tune-tomorrow/. Up to two million people may have additionally contributed to the online governmental platform, but, ironically, the platform did not include any deliberative feature and so their participation does not obviously count as "deliberation."

4. See Peters 2008 for an exploitation of the sluice metaphor into a full-blown model pointing out the double meaning of a sluice and corresponding dual functionality (gate and filter).

5. In practice parliaments mostly operate as bargaining chambers and the public sphere as a cacophony of polarized enclaves. Additionally, the circulation of ideas and preferences from the wider public sphere to the formal one and back is far from smooth, as the gap between what majorities want and what they get would seem to indicate. In the United States, empirical studies (e.g., Gilens and Page 2014) point to a worrying lack of causal efficacy of majorities on public policies, in contrast to business and economic interests. The low approval rates of most representative institutions in advanced Western democracies (since such polls were first conducted in the 1970s) speak to the same problem. The system can thus be diagnosed as rather dysfunctional when it comes to agenda setting from the informal to the public sphere.

6. This circularity between the sphere of opinion formation and will-formation is also theorized in the model of Nadia Urbinati (2008).

7. E.g., Habermas (1996, 358), where he recognizes that the "problematic assumption" in the model of power circulation he borrows from Peters (1993) is the assumption that "the periphery has . . . a specific set of capabilities," allowing it "to ferret out, identify, and effectively thematize latent problems of social integration (which require political solutions)." Habermas unfortunately does not explain what happens to the model if this assumption proves indeed too "problematic," nor does he provide an account of the exact mechanisms by which the second track could generate or be endowed with those capabilities.

8. Habermas (1996, 307) himself acknowledges the limitations of such an "anarchic structure," which renders "the general public sphere . . . more vulnerable to the repressive and exclusionary effects of unequally distributed social power, structural violence, and systematically distorted communication than are the institutionalized public spheres of parliamentary bodies." Habermas goes on to note that "on the other hand, it has the advantage of a medium of unrestricted communication" (307). Somehow, however, the unrestrictedness of communication does not seem to be nearly worth the trade-off of immense power asymmetries inherent to an anarchical system.

9. For a discussion of the Hong and Page's diversity-trumps-ability theorem behind this argument, see critics such as Quirk (2014, 129); Thompson (2014); Brennan (2016). For a critic of the critics, see Landemore (2014); Page (2015); Singer (2018); Kuehn (2017).

10. See Mutz (2006).

11. Mutz (2006).

12. See Landemore (2015, 25), providing detailed analysis of Mutz's argument.

13. I consider additional principles in the concluding chapters to take into account the need to expand the definition of the demos in an interconnected, globalized world where affected interests transcend national boundaries, as well as the need to expand democracy to the economic sphere. The two additional principles considered are dynamic inclusiveness and substantive equality.

14. One of the probably controversial claims I make is that to the extent that classic electoral democracy thrives or even just depends on partisanship, this is one more reason to want to move past it.

15. On activists, see, for example, the influential Open Democracy media platform, at https://www.opendemocracy.net/en/. President Obama's administration famously launched an Open Government Initiative whose motivation, according to a 2009 White House memorandum, was "transparency, public participation, and collaboration." See https://www.whitehouse.gov/the_press_office/TransparencyandOpenGovernment. See also O'Reilly (2011).

16. In other words, it is software that is accessible to all at all times, not just in term of being visible but in terms of being usable, shareable, manipulable, and modifiable by all. By contrast, so-called closed source or proprietary software is software that only one person, team, or organization has control over and can modify. Open-source software is best known for some of the cocreated public goods it has generated, for example, the operating system Linux and the generalist online encyclopedia Wikipedia.

17. See "Crowdlaw," https://crowd.law/.

18. Transparency can of course also be helped by the use of digital technologies, but because the debate about the benefits of open data and open government is already well established, I prefer to focus on the more central and in some ways original principles of open democracy.

19. By contrast, it is worth emphasizing that representative democracy is not essentially committed to deliberation, in that it can be and has been implemented in purely aggregative and Schumpeterian versions that emphasize elite competition and voting procedures over deliberation.

20. Of course, one needs to assume here the equivalent of universal franchise in terms of the pool from which lottocratic representatives are chosen (an assumption that was not verified in classical Athens, as the Greeks both required volunteering for participation to certain lottocratic functions and put age restrictions on who was allowed to volunteer in the first place).

21. See Rose (2016) and Cohen (2018) on the political value of citizens' time.

22. I won't draw a hard distinction here between vote delegation and vote recommendation, although some see in it a reason to distinguish delegative from liquid democracy as two distinct projects (see, e.g., Davies-Coates 2013). The current consensus seems to be that they are roughly the same.

23. In the words of Bryan Ford (2002, 1), one of its first theorists, who called it "delegative" democracy, although the name didn't stick as much, liquid democracy is thus "a new paradigm for democratic organization which emphasizes individually chosen vote transfers ('delegation') over mass election" and replaces "artificially imposed representation structures with an adaptive structure founded on real personal and group trust relationship."

24. In the earliest documented use of the term, a wiki by "Sayke" (a pseudonym whose real owner is unknown), liquid democracy is described as "probably best thought of as a voting system that migrates along the line between direct and representative democracy" and "combines the advantages of both, while avoiding their flaws." See Sayke, "Liquid Democracy,"

June 16, 2004, https://web.archive.org/web/20040616144517/http://www.twistedmatrix.com/wiki/python/LiquidDemocracy.

25. A worry with this form of representation is that it is still premised on a "distinction" principle that may leave many people—the shy, inarticulate, or socially invisible—out of the pool of democratic representatives. To the extent that "star voting" can be resisted via social norms or technological solutions, however, delegative or liquid representation offers a much more "open" form of votation-based representation.

26. I owe this neat reformulation of my position to a participant in a two-day seminar in London at the New College for the Humanities: "How to Improve Public Debate?," March 29, 2019.

27. Citizen Initiative Reviews have been a legislatively authorized part of Oregon general elections since 2010. The review gathers a representative cross section of two dozen voters for five days of deliberation on a single ballot measure. The process culminates in the citizen panelists writing a citizens' statement that the secretary of state inserts into the official *Voters' Pamphlet* sent to each registered voter (see, e.g., Gastil et al. 2017).

28. For larger questions about the open-democracy paradigm, including the question of accountability of nonelectoral bodies, see Landemore (2020), esp. chap. 7.

References

Aitamurto, Tanja, and Hélène Landemore. 2016. "Crowdsourced Deliberation: The Case of an Off-Traffic Law Reform in Finland." *Policy & Internet* 8 (2): 174–96.

Blum, Christian and Christina Isabel Zuber. 2016. "Liquid Democracy: Potentials, Problems, and Perspectives: Liquid Democracy." *Journal of Political Philosophy* 24 (2): 162–82.

Bohman, James. 2004. "Expanding Dialogue: The Internet, the Public Sphere and Prospects for Transnational Democracy." *Sociological Review* 52 (1): 131–55.

Brennan, Jason. 2016. *Against Democracy*. Princeton, NJ: Princeton University Press.

Cohen, Elizabeth. 2018. *The Political Value of Time: Citizenship, Duration, and Democratic Justice*. Cambridge: Cambridge University Press.

Dahlgren, Peter. 2005. "The Internet, Public Spheres, and Political Communication: Dispersion and Deliberation." *Political Communication* 22: 147–62.

Davies-Coates, Josef. 2013. "Liquid Democracy Is Not Delegative Democracy." United Diversity. https://uniteddiversity.coop/2013/07/19/liquid-democracy-is-not-delegative-democracy/.

Doerr, Nicole. 2018. *Political Translation*. Cambridge: Cambridge University Press.

Fishkin, James. 2018. *Democracy When The People Are Thinking*. Oxford: Oxford University Press.

Ford, Bryan. 2002. "Delegative Democracy." http://www.brynosaurus.com/deleg/deleg.pdf.

———. 2014. "Delegative Democracy Revisited." *Bryan Ford* (blog), November 14. https://bford.github.io/2014/11/16/deleg.html.

Gastil, John, and Erik Olin Wright. 2018. *Legislature by Lot: Transformative Designs for Deliberative Governance*. London: Verso.

Gastil, John, Katherine R. Knobloch, Justin Reedy, Mark Henkels, and Katherine Cramer. 2018. "Assessing the Electoral Impact of the 2010 Oregon Citizens' Initiative Review." *American Politics Research* 46 (3): 534–63.

Gilens, Martin, and Benjamin Page. 2014. "Testing Theories of American Politics: Elites, Interest Groups, and Average Citizens." *Perspectives on Politics* 12 (3): 564–81.

Guerrero, Alex. 2014. "Against Elections: The Lottocratic Alternative." *Philosophy and Public Affairs* 42 (2): 135–78.

Habermas, Jürgen. 1996. *Between Facts and Norms*. Translated by William Rehg. Cambridge, MA: MIT Press.

Hansen, Mogen Hermans. 1999. *The Athenian Democracy in the Age of Demosthenes: Structure, Principles and Ideology*. Norman: University of Oklahoma Press.

Hilbert, Martin. 2009. "The Maturing Concept of E-Democracy: From E-Voting and Online Consultations, to Democratic Value out of Jumbled Online Chatter." *Journal of Information Technology and Politics* 6 (2): 87–110.

Hindman, Matthew. 2008. *The Myth of Digital Democracy*. Princeton, NJ: Princeton University Press.

Kloppenberg, James T. 2016. *Toward Democracy: The Struggle for Self Rule in European and American Thought*. Oxford: Oxford University Press.

Kuehn, Daniel. 2017. "Diversity, Ability, and Democracy: A Note on Thompson's Challenge to Hong and Page." *Critical Review* 29 (1): 72–87.

Kuyper, Jonathan W. 2016. "Systemic Representation: Democracy, Deliberation, and Nonelectoral Representatives." *American Political Science Review* 110 (2): 308–24.

Landemore, Hélène. 2012. "Deliberation, Cognitive Diversity, and Democratic Inclusiveness: An Epistemic Argument for the Random Selection of Representatives." *Synthese* 190 (7): 1209–31.

———. 2013. "On Minimal Deliberation, Partisan Activism, and Teaching People How to Disagree," *Critical Review* 25 (2): 210–25.

———. 2017. "Deliberative Democracy as Open, Not (Just) Representative Democracy." *Daedalus* 146 (3): 51–63.

———. 2018. "What Does It Mean to Take Diversity Seriously?" *Georgetown Journal of Law and Public Policy* 16: 795–805.

———. 2020. *Open Democracy: Reinventing Popular Rule for the 21st Century*. Princeton, NJ: Princeton University Press.

Lanier, Jaron. 2013. *Who Owns the Future?* New York: Simon & Schuster.

Manin, Bernard. 1997. *The Principles of Representative Government*. Cambridge: Cambridge University Press.

Montanaro, Laura. 2012. "The Democratic Legitimacy of Self-Appointed Representatives." *Journal of Politics* 74 (4): 1094–1107.

Mutz, Diana. 2006. *Hearing the Other Side*. Cambridge: Cambridge University Press.

Neblo, Michael A., Kevin M. Esterling, and David M. J. Lazer. 2018. *Politics with the People: Building a Directly Representative Democracy*. Cambridge: Cambridge University Press.

Norbäck, Per. 2012. *The Little Horse from Athens*. N.p.: Amazon Kindle.

O'Reilly, Tim. 2011. "Government as Platform." *Innovations: Technology, Governance, Globalization* 6 (1): 13–40.

Page, Scott. E. 2007. *The Difference*. Princeton, NJ: Princeton University Press.

Peters, Bernhard. 2008. "Law, State and the Political Public Sphere as Forms of Social Self-Organisation." In *Public Deliberation and Public Culture: The Writings of Bernhard Peters, 1993–2005*, edited by H. Wessler, 17–32. Basingstoke, UK: Palgrave Macmillan.

Peters, Bernhard. 1993. *Die Integration moderner Gesellschaften*. Frankfurt: Suhrkamp.

Pitkin, Hannah. 1967. *The Concept of Representation*. Berkeley: University of California Press.

Plotke, David. 1997. "Representation Is Democracy." *Constellations* 4 (1): 19–34

Popper, Karl. (1945) 2013. *The Open Society and Its Enemies.* Princeton, NJ: Princeton University Press.

Posner, Eric, and Glen Weyl. 2018. *Radical Markets: Uprooting Capitalism and Democracy for a Just Society.* Princeton, NJ: Princeton University Press.

Quirk, Paul. 2015. "Making It Up on Volume: Are Larger Groups Really Smarter?" *Critical Review* 26 (1–2): 129–50.

Rehfeld, Andrew. 2006. "Towards a General Theory of Political Representation." *Journal of Politics* 68: 1–21.

———. 2009. "Representation Rethought: On Trustees, Delegates, and Gyroscopes in the Study of Political Representation and Democracy." *American Political Science Review* 103 (2): 214–30.

Rose, Julie. 2016. *Free Time.* Princeton, NJ: Princeton University Press.

Singer, Daniel. 2018. "Diversity, Not Randomness, Trumps Ability." *Philosophy of Science* 86 (1): 178–91.

Sintomer, Yves. 2018. *From Radical to Deliberative Democracy: Random Selection in Politics from Athens to the Present.* Cambridge: Cambridge University Press.

Spada, Paolo, Mark Klein, Raffaele Calabretta, Luca Iandoli, and Ivana Quinto. 2016. "A First Step toward Scaling-up Deliberation: Optimizing Large Group E-Deliberation Using Argument Maps." https://www.researchgate.net/publication/278028114_A_First_Step_toward_Scaling-up_Deliberation_Optimizing_Large_Group_E-Deliberation_using_Argument_Maps.

Swierczek, Björn. 2014. "Five Years of Liquid Democracy in Germany." *Liquid Democracy Journal* 1. https://liquid-democracy-journal.org/issue/1/The_Liquid_Democracy_Journal-Issue001-02-Five_years_of_Liquid_Democracy_in_Germany.html.

Thompson, Abigail. 2016. "Does Diversity Trump Ability?" *Notices of the American Mathematical Society* 61 (9): 1024–30.

Urbinati, Nadia. 2008. *Representative Democracy: Principles and Genealogy.* Chicago: University of Chicago Press.

Urbinati, Nadia, and Mark Warren. 2008. "The Concept of Representation in Contemporary Democratic Theory." *Annual Review of Political Science* 11: 387–412.

Velikanov, Cyril. 2012. "Mass Online Deliberation: Requirements, Metrics, Tools and Procedures." Working paper, https://www.academia.edu/12031548/Mass_Online_Deliberation.

Warren, Mark E. 2008. "Citizen Representatives." In *Designing Deliberative Democracy: The British Columbia Citizens Assembly*, edited by Mark E. Warren and Hilary Pearse, 50–69. Cambridge: Cambridge University Press.

3

Purpose-Built Digital Associations

Lucy Bernholz

Digitized networked data is a fundamentally different economic resource than either time or money. As a nonrival, generative good it presents new challenges to existing institutional forms that were primarily established to manage financial resources. This chapter examines institutional adaptations driven by the aspiration to manage digitized data for public benefit. While innovation is happening in democracies everywhere, this chapter focuses specifically on the nonprofit sector in the United States. Within this institutionally dominant slice of civil society we find a common reliance on licensing schemes and the use of contract law to define how digitized data will be applied to a specific public purpose. This represents an important shift for civil society—away from an era in which corporate, tax, and charity law have served as the primary defining bounds of the sector to one in which intellectual property law and data protections are increasingly important. The creation of purpose-built organizations dedicated to managing digital resources for public benefit will require the development of regulatory and oversight mechanisms such as those necessitated by the introduction of nonprofit corporations and may prove as significant in shaping civil society in democracies.

We are surrounded by the institutional implications of digitized data in the commercial sector. Business models built on data aggregation and monetization have produced globally dominant companies such as Amazon, Facebook, and Google. Each of these companies not only holds majority market share in its original product spaces (online retail, social media, and search, respectively) but also has expanded in influence across other domains (e.g., cloud storage, news delivery, mapping, translation). In the age of analog data these products and services were distinct. Today, the connection between the original product (e.g., book sales) and the secondary domain (e.g., enterprise software) is the company's ability to build efficient systems for capturing and using unprecedented quantities of digitized data. This shift in how market power is attained has catalyzed new theories about markets, corporate structure, and antitrust law.[1]

The increasing dependence of civil society actors on digitized data and networks should catalyze the same kind of attention from scholars. I argue that the institutional evolution currently under way, driven by the use of networked digital devices and digitized data, should inform a new research agenda for civil society.

The animating question of this chapter is how are we changing our associational structures as we organize around digitized data and software code?[2] It speaks directly to some of the sectoral possibilities raised by Julia Cagé in chapter 9 and to the individual organizational characteristics explored in chapter 8 by Lee, Levy, and Seely Brown. I begin with a brief description of why digitized data matters and how it differs from material resources. I then consider three different examples in which we find participants seeking to manage digital data for public benefit. The first example, administrative data trusts, emerge to manage administrative data—that which is generated in the course of everyday government business. This seemingly mundane resource, the click trails and data generated from issuance of driver's licenses, tracking of prisoners, or keeping of school records, is of significant interest to commercial and public-interest actors, and usage rights are hotly contested. The second example is firmly anchored in the nonprofit sector, where we find foundations and nonprofits experimenting with ways to make cultural knowledge and scholarship more widely available at lower cost, with market interests never far away. The third example, software code repositories, are a central feature of the open source software community. These very different models, and the balancing of public and private interests that each exemplifies, provide a wide-angle lens on the challenges to watch for as we experience a proliferation of purpose-built digital associational forms. The chapter concludes with implications for civil society, democracy, and scholarship.

Digitized Data as Objects of Exchange

There is a rich literature arguing against the deterministic nature of tech-centric analysis.[3] Gangadharan, Ananny, and Caplan (in their chapters in this volume) present analyses that recognize the recursive relationships between environment, people, laws, institutions, and technology. I agree with these analyses in terms of their contributions to understanding the nature of digitized data and the interactions between existing systems and new digital capacities. I am interested here in how people are building new associational forms as they seek to generate public benefit from networked digitized data. For simplicity's sake I refer to the complex weave of digital systems (e.g., networks, devices) and data (everything transported over those networks) as *networked digital data* or, occasionally, simply, *data*.

The nonrival characteristics of networked digitized data distinguishes it from material resources such as money.[4] This characteristic makes networked digitized data more similar to classic public goods than private goods. However, the pervasive reach of networked computers, mobile phones, and internet use has been driven, in the United States and beyond, by commercial innovation, and thus networked digital data are produced primarily within corporate settings. Legal scholars including Yochai Benkler, Larry Lessig, Pamela Samuelson, and Jason Schultz have documented the economic potential of digitized resources as well as the challenges they present to existing commercial and legal standards with particular regard to ownership.[5] Elinor Ostrom, Brett Frischmann, and others have examined the possibilities that digitized data bring for reinvigorating a commons approach to resource management.[6] Since the mid-1980s, when networked computers began to be used beyond the bounds of governments and academia, most of the digitized data that we have produced and needed to manage have come into being on hardware, software, and networked systems built, owned, and operated by corporations. Efforts to control digitized data abound—from digital rights management software to contractual claims via terms of service. Intellectual property law is one of the key areas of legal expertise regarding the use of digitized data. There has been, effectively, forty years of legal and policy action to make digitized data fit into the extant legal and commercial practices built for analog resources.[7]

This has not happened without significant pushback and the development of numerous alternatives. Software developers do their work in and with digitized data and produce code that is both of and for the management of digitized data. Software developers have debated resource governance options from the beginning of their profession. Arguments over the nature of software code as a resource that should be "free like speech, not like beer," open for all to use, or patentable and proprietary take place continuously in classrooms and coding communities and, more episodically, in courts of law. Disagreements over how software code should or should not be used, whether as a shareable resource available for everyone to modify or as closed systems owned and sold as property, are at the root of what was first called free software and is now commonly known as open source.

As Mark Andreesen's 2011 quip about "software [eating] the world" has come to pass, the implications of these arguments have also expanded.[8] Mobile phones, networked computers, text messaging, email, social media, as well as the networks of data and algorithmic analysis that power in-home personal assistants, office-building security systems, public-policy making, public transit, and private cars mean that people are generating and using, or being guided by, digitized data and digital networks in every aspect of their lives.[9] The legal, economic, and institutional arguments that accompany them are now important issues for all of us. By looking at different examples of institutional arrangements to manage these resources, I intend to reveal how these matters manifest for people who are trying to use digitized resources for various civic and democratic purposes.

Digital Institutions

This section contains examples of institutional forms designed to dedicate digital data to public purposes. The examples are deliberately diverse in the type of data they manage. They are also cross-sectoral, representing a variety of relationships among individuals, associations, governments, and companies. All of the examples reveal the importance of licensing regimes and contract law as part of the emerging apparatus for governing digitized data. From the mid-1980s to the present we see licenses and contracts become the primary means by which we manage digital data. In some cases these mechanisms tie disparate organizations together. In others, they stand alone, forming "contract stacks" that are the singular manifestation of data governance. In the case of open-source software, the license structure is definitional to the community of participants, more so than any one organizational choice. As the importance of licenses and contracts emerges, we need to consider how these regulating mechanisms interact with different corporate forms (particularly nonprofit versus commercial), as there are many ways they can be mixed and matched.

ADMINISTRATIVE DATA RESEARCH TRUSTS

The value of information as a resource for democratic governance has deep roots, and discussions about the right information conditions for democracy can be found in the introduction to this volume and in chapters by Caplan,

Schwartzberg and Farrell, and Ananny. It makes sense to begin this survey of digital purpose associations by looking at how digitized government data can be made available for public benefit. The roots of this story are as old as the Republic, and the interweaving of public interest with commercial opportunity winds throughout.

Data have always mattered to governing the United States. The first call for governmental data collection comes as early as the sixth paragraph of the US Constitution, in article 1, section 2:

> Representatives and direct Taxes shall be apportioned among the several States which may be included within this Union, according to their respective Numbers, which shall be determined by adding to the whole Number of free Persons, including those bound to Service for a Term of Years, and excluding Indians not taxed, three fifths of all other Persons.
>
> The actual Enumeration shall be made within three Years after the first Meeting of the Congress of the United States, and within every subsequent Term of ten Years, in such Manner as they shall by Law direct.

As they were written, these two sentences are revelatory. Data collection is part of the government's job. The data are intended to serve two distinct purposes—taxation and representation. And race matters.[10]

Collecting information and sharing it are two very different tasks. The government's role as data collector has expanded significantly since the first census in 1790. The census also has grown in scope. The role of racial categorization and other demographic designations has been—and continues to be—hotly debated. The act of collecting data for purposes of governance has been continuous and contentious since the nation's founding.

But making government information available to the people has a more uneven history. The colonies tightly controlled access to and licenses for printing presses. Before statehood, the legislative record, including individual vote counts in the Commonwealth of Massachusetts, was confidential.[11] The first law requiring government to print and make information readily available came in 1813. Until 1860 this work was done via private contractors, despite the common understanding that such contracts were secured via political bribes. In 1860 the Government Printing Office was established by law, although it took another thirty-five years before funds were appropriated to run the office. The *Federal Register* was not created until 1934.[12]

From the census of 1880 through the turn of the twentieth century there was a coterminous development of statistical methodology, computational equipment, and government interest in collecting data on citizens and their activities. The 1880 census collected information on 215 different subject areas (compared to five in 1870).[13] By 1935, to implement the Social Security Act, the federal government would need the computational storage and payment processing ability to maintain employment information on twenty-six million people.[14] Today, the federal government collects data and produces official statistics on topics ranging from agriculture to veterans. In April 2018, there were more than three hundred thousand data sets available for free to the public on the federal government's Data.gov website.

For most of the nineteenth and twentieth centuries, the government collected data for its own purposes. Businesses were active partners in collecting, printing, and distributing it, but not necessarily in using it. But in the mid-twentieth century this began to change, as businesses and citizens began to take an interest not just in the collection of the data but also in the actual content of it. The historian Michael Schudson locates the federal government's shift from data collector to data source in the mid- to late twentieth century. The "rise of the right to know," has roots, Schudson argues, in postwar global demands for government accountability, battles over the balance of power between Congress and the executive branch, and a more educated US population that was better equipped for critical engagement with government. The thirty years from war to Watergate produced the Freedom of Information Act, fair labeling practices, and environmental impact statements. In Schudson's telling, these policy actions set the stage for direct citizen access to government information, although their origins were rooted in shifting information roles and responsibilities across government branches.

Notably, Schudson argues that technology is not a driver of openness. The cultural and political expectations about information access, government oversight, the role of journalism, and public participation throughout the democratic process, and not just at election time, predate and reach more broadly to a more diverse cross section of the population than will the open government or open data movement of the 2000s. This is important for several reasons, two of which I prioritize here. First, political and cultural discussions about openness highlight the converse importance of secrecy in democracies. The postwar period includes a landmark Supreme Court case, the 1958 *NAACP v. Alabama* decision, which established the right to keep associational membership lists private. Second, by rooting openness in this broader, nontechnological history, we bring a more diverse set of perspectives into the debate from the beginning. The voices of women and racial and sexual minorities make clear that not all people—individuals or communities—experience information transparency the same way.

Although technology did not create openness, digital networked technologies do influence how government information is managed, shared, stored, and made open or kept closed. There are two aspects of this, the deliberate efforts to open and share government data and the threats, omnipresent in a digital networked age, of indiscriminate, unintentional, and/or malicious access. Administrative data trusts, like other efforts to share government data, are intentional structures. But like all well-considered institutions in a digital networked age, one of their design constraints is the pervasive, persistent threats of unintentional access.

Administrative data trusts have three sets of roots—government printing office and repository library practices, government accounting practices and policies for funded research, and a growing interest in and demand for "evidence-based" policy making. Historians of numeracy and statistics point to the role that the census plays in congressional apportionment as a decennially repeating spark for public interest in this foundational government data set.[15] The federal government's role as data collector has expanded steadily over time, accelerating throughout the twentieth century as government departments and programs grew. Roosevelt's New Deal created several new funding mechanisms, many of which also introduced detailed reporting requirements, and World War II fueled investment in vast new computational power and strategies.[16]

Between 1980 and 2018 the world of information processing shifted dramatically. What started as mainframe computer businesses processing raw government data for sale to broadcasters, law firms, and investment banks shifted first to a world of desktop publishing, then to the age of connected laptops, and finally to our current moment of ever-present internet-connected devices. Each of these technological shifts changes the market for and economics of government information.

As cloud computing became common and *big data* the buzzword of business and governments, the interest in using administrative data—both from public agencies and from companies—boomed. Administrative data—information collected in the process of enrolling someone in a service or conducting basic transactions—abounds inside of government departments and corporations. It is not collected for research purposes, but medical researchers and social scientists have long seen its potential for establishing baseline parameters or as complementary data to that collected for specific purposes. Some areas of research and advocacy, such as youth development, started pulling together disparate government data as early as 2006.[17] The idea was to build data sets for specific research projects that could also be added to and reinterrogated as new questions or partners emerged.[18]

In 2011 three funding agencies in the United Kingdom, one private and two public, set up the Administrative Data Task Force, partially in response to fear that the kingdom was falling behind its European Union neighbors.[19] Nordic countries had been using administrative data to complement their census surveys, and in the course of doing so were leading the way in realizing the value of these data.[20] In 2012 the task force released a report calling for the creation of the UK Administrative Research Network, which established Administrative Data Research Centers in each of four UK countries and a network to coordinate them. By 2013 the Ministry of Justice, in partnership with New Philanthropy Capital, was experimenting with the Justice Data Lab. The lab helps nonprofit organizations working with criminal offenders assess their work against national data sources on reoffenders and recidivism. Through a set of contractual relationships, NPC helps nonprofits share their data on clients with the Justice Data Lab, which then analyzes the nonprofit's list against its national data. It then informs the nonprofit of which percentage of their clients are to be found in the national data set of reoffenders. The public data never leaves the public servers, nor does any identifying information. All of the analysis is done in-house by civil servants. The service was made a permanent part of the Ministry of Justice in 2015.[21]

In the United States, a different approach to collecting and using administrative data was emerging. With an opening investment of more than $83 million, the Bill and Melinda Gates Foundation announced the launch of InBloom. This nonprofit data platform promised to collect, clean, curate, and make available student data on K–12 students from public school districts across the country. One of the distinguishing features of InBloom was its commitment to make the data available for research, policy, and commercial innovation. InBloom collapsed less than three years later, imploding the $100 million of philanthropic investment and generating a student-data privacy backlash that led to dozens of states passing new legislation about how data about students could be used.[22]

The InBloom collapse unintentionally awakened broader concern about data privacy than had previously been expressed in the United States. Public unease with government and corporate data collection and collaboration also grew as a result of Edward Snowden's revelations in 2013. Both of these events led to legislative and regulatory changes regarding data collection, but only insofar as certain types of data were concerned (e.g., student information, email metadata). Some scholars, such as those at Carnegie Mellon and Stanford who banded together to create LearnSphere, developed alternatives to InBloom that avoided administrative data altogether. Meanwhile, scholars and policy makers interested in using administrative data from governments and the vast stores of search, social media, and call-record data generated by internet users only continued to grow.

The US Congress threw its support behind the idea of using administrative data in a 2016 report on evidence-based policy making issued with much fanfare by Representative Paul Ryan.[23] The National Academies of Science took up the topic in a series of workshops from 2015 to 2016.[24] New York University's Governance Lab, known as GovLab, published numerous papers on data collaboratives, its term for cross-sector research and policy-focused initiatives built around shared data.[25] GovLab also created a list of more than one hundred data collaboratives, including several that involved partnerships of industry, academia, and government.[26] In 2017, the county offices of education in Santa Clara and San Mateo, California, along with the University of California, launched the Silicon Valley Data Trust—a shared repository for public school data and research, with an eye toward policy makers. The Alfred P. Sloan Foundation has invested more than $6.5 million in thirteen projects intended to study or experiment with the scholarly use of administrative and corporate data.[27]

TRUSTED NONPROFIT DATA INTERMEDIARIES

Government is not the only source of publicly beneficial data. Most libraries and museums exist to present and protect information and cultural artifacts for public benefit. Libraries were early movers when it came to digital networks. They exist to manage information, and text was the first type of "content" to be widely digitized.[28] Academic libraries, and publishers, also faced a number of market challenges that made digital resources attractive. Academic journals, the backbone of scholarship and an arena of publishing critical to the modern academic career, are bespoke publications with small audiences, and a mostly closed market of contributors, subscribers, buyers, and readers. The Andrew Mellon Foundation, a longtime supporter of scholarship in the arts, humanities, and sciences, began looking hard at the challenges facing academic publishing in 1989.[29] In 1992, the foundation began funding "experiments" in using electronic technologies (digitization and networks) in various stages of the production cycle of academic research—from the sharing of initial findings, to producing digital versions instead of printed journals, to new approaches to peer review, to storage and distribution.[30] Most of the experiments built on practices that scholars in the natural sciences had been trying for some time but that hadn't yet made it into the humanities.[31]

Two of the Mellon Foundation's experiments, JSTOR and Artstor, have become global intermediaries for digital data.[32] JSTOR is an online digital database of academic journals that provides access via library subscriptions.

Artstor followed on this model and intermediates the use of digital copies of fine art from museum, gallery, and library collections around the world.

When the Mellon Foundation first launched JSTOR, it did so as a series of grants to the University of Michigan, which was already at work on relevant technology. From the beginning the foundation recognized that the technology would be one challenge, but that the program would rise or fall based on how well it navigated numerous intellectual property challenges. These included negotiating the licensing rights between publishers and the project, as well as the contractual terms for library subscriptions. There were legal rights to the software itself and to the output of the software, and licensing choices to be made for each. The foundation had anticipated the importance of licensing and intellectual property and modified its standard grant agreement with the University of Michigan several times. The final agreement allocated the intellectual property rights to the database software to the University of Michigan but provided the foundation with a nonexclusive license to use it for noncommercial purposes. The grant agreement also gave the foundation rights to all equipment funded by the grant. These decisions enabled the foundation to keep its options open regarding a longer-term home for the project.[33]

After a year of building a wonky but working prototype of the database, the foundation decided that the organizational challenges of running the project required a dedicated, independent organization. Mellon staff spun out the journal database project into an independent nonprofit, named JSTOR (Journal Storage) in 1995.[34] Working out the details of licensing, publishers' arrangements, library subscriptions, and a revenue model demanded as much, if not more, attention than building functioning software. Two years after starting, only seventeen journals had been digitized. Four years and twelve thousand hours of work later (equally divided between humans and optical-character-reading software) the JSTOR team had digitized 333 years of publishing history from the collection of the Royal Society of London.[35] The librarians at JSTOR had made progress in organizing a general science collection and the art and science collection, and they were working on general ecology. JSTOR reached the millennium with more than 190 library subscribers, functioning collections, and momentum behind it.

This accomplishment encouraged the Mellon Foundation to turn its attention from academic journals to fine art. Through the course of the work to understand the implications of digitization on academic publishing, the foundation's team had learned of the challenges facing other educational institutions, including museums, libraries, and galleries, regarding their image collections. Once again, technological opportunity and legal rights and risks rose to the top. Mellon responded by dedicating two staff people with programmatic and legal expertise to launch a pilot project focused on a small image collection. From this, plans were drawn up for a more complete set of services—legal and technological—that would allow institutions to manage their collections internally, with each other, and with a general public. In 2001 Neil Rudenstine, outgoing president of Harvard, joined James Shulman (from Mellon) to take the image project, now called Artstor, out on its own as a nonprofit, independent from the Mellon Foundation.[36]

Over the next several years, both Artstor and JSTOR grew their catalogs of materials, numbers of subscribers, and general users. The two organizations manage collections of information (images and text-based journals) that are different enough in the physical world that they each exist within a distinct ecosystem of suppliers, users, and revenue models. Journals, for example, involve individual scholars, professional associations, peer-review processes, scholarly credentialing and tenure practices, publishing houses, libraries, and subscribers. Images are created and valued by an ecosystem that involves artists, curators, cultural heritage bodies, galleries, museums, libraries, collectors, auction houses, and publishers. Once digitized, however, journals and images become more like any other digital data and less like material objects that need these distinct ecosystems. Digitizing the object challenges these surrounding systems. Artstor and JSTOR's successes come as much from ongoing attention to these systems, their needs and incentives, as they do from providing functioning technology and appropriate legal support, especially in the arena of intellectual property.

These categories of work—legal (intellectual property, licensing, and contracts), technological (network infrastructure, metadata standards, interoperability, and operating system adaptation), and ecosystem facilitation (engaging partners across the supply chain, negotiating compromises across institutions, developing standards and incentives that engage different stakeholders)—account for the core of Artstor and JSTOR's work. The practices that have been developed in each of these categories constitute part of the apparatus for using digital objects in learning environments at scale and across borders.

JSTOR and Artstor's leaders also needed to create a revenue model to sustain these shared resources. Both Artstor and JSTOR were built with significant philanthropic investment. Between September 1999 and June 2017, the Mellon Foundation alone invested more than $54 million in Artstor and its partners.[37] The Mellon Foundation's initial grant to start Artstor (after the pilot project) had been $5 million.[38] Over time both have come to depend on institutional subscriptions as part of their sustaining revenue. In 2016, for example, Artstor reported more than $10 million in subscriber revenue and received about $2 million in philanthropic grants. In 2015 Artstor formed a strategic alliance with Ithaka S+R, a nonprofit that had been instrumental in creating, and is now also the home of, JSTOR.

JSTOR and Artstor demonstrate how the digitization of objects and efforts to take advantage of information networks change much more than the form of objects. New legal contracts, new types of partnerships, and new revenue models must be created. Once materials are digitized and can be easily shared, the institutional lines between libraries and museums begin to blur. International networks of galleries, libraries, and museums have developed. The interchangeability of digitized text and images enables us to imagine institutions without borders. At the same time, the need to generate revenue and respect intellectual property rights catalyzes licensing and legal innovation. JSTOR and Artstor manage these multiple-stakeholder relationships—serving as both a broker of licenses, tracking data, and revenue and a repository that helps people find the digital file they need, wherever the analog journal or image might actually live.

Where previous institutions collected, protected, preserved, and presented their physical objects through a combination of walls, vaults, access policies, and admission fees, new versions of these functions were necessary for the digital journals or images. JSTOR and Artstor emerged to manage the tensions between access and revenue, property rights and knowledge advancement. They do this largely by brokering licensing rights, managing subscription contracts, and building and maintaining the technological infrastructure for people to find what they want and access it according to several tiers of opportunity. Key to their success is the building and managing of trust between diverse parties and the provision of systems that make it easier for everyone involved to do their work.

The intermediary role is complicated. Because they sit in between creators, distributors, owners, and users of information, JSTOR and Artstor have access to information on who uses what, where, and for how long. Their relationship with academic libraries means that public access is limited to those with an appropriate institutional affiliation. Artstor can log which images are most popular with which kinds of users and use that to encourage new collections or charge users for it when they are making programming decisions. Artstor has had to find ways to use this data in ways that preserve the trust of the original contributors and users.[39]

These aren't tasks or roles to be taken lightly. JSTOR discovered just how sharp both edges of this data sword are when its user logs revealed the presence of a single account using MIT's institutional subscription to download several million articles from JSTOR's servers. Who or why was not immediately clear, but soon enough a well-known computer prodigy named Aaron Swartz was identified. Charges were filed by the Federal Bureau of Investigation, the Massachusetts Institute of Technology, and JSTOR, although eventually JSTOR dropped its claims as the legal battle under the Computer Fraud and Abuse Act grew out of proportion to the act itself. In 2013, defeated by the scale of the legal challenge, Swartz committed suicide. This sad tale shows how the digital opportunity to reimagine learning and scholarship requires expertise in legal domains, namely intellectual property and telecommunications law, that were not previously core to running nonprofits.

I conclude with two institutions designed to facilitate and organize the highly distributed, punctuated, and fractious (as a positive) ways of working that networked, digitized data enable. Unlike activities built around physical resources, working with digitized data can be done at a distance, on one's own schedule, and in ways that expand far beyond hierarchical organizational structures. The first example, open-source code repositories, have been a mainstay of software code development for decades. The second, open collectives, are a more recent innovation, launched in 2018 by a group of democracy activists.

OPEN-SOURCE SOFTWARE REPOSITORIES

Software code is the paradigmatic digital resource. It is created in digital form and is used, in turn, to collect, store, organize, and analyze the other key digital resource, data.[40] If you are going to share software, you need to have a place to store it so others can find it. Software code is modular. If one person writes a few lines of code in a particular software language, such as the instructions to turn a button in an app blue when it is selected by the user, then there's no reason someone else needs to write those lines again. They can use the first person's code, insert it into their program, and go from there. Of course, if they have a more elegant or simpler version of the same instruction, they can share that as well. But to do this, you need to be able to find, use, and "put back" the software modules that you need. Simply put, software communities need a place to store all the "modules." These places are called repositories.

In the early days of writing software, sharing the code was the norm. Coders shared their creations with others as mathematicians would share proofs—each contributing a bit to a greater system of knowledge. Proprietary and closed systems came second. The 1980s saw the split between those who thought software should be open for anyone to use and those who sought fortunes by locking down their code. Decades of debate and activism have ensued among characters such as Richard Stallman, founder of the Free Software Foundation (FSF), later evangelists who replaced the word *free* with *open source*, and those who seek to lock down their software code as fully proprietary.[41]

In creating a 501(c)(3) charitable nonprofit, Stallman veered away from the more decentralized, hosted project model that other internet-related groups of the mid-1980s favored. Siobhan O'Mahoney, who has studied the organizational forms associated with community-managed software, notes that the FSF opted for the nonprofit structure precisely because it was less democratic and participatory: "We don't invite all the people who work on GNU to vote on what our goals should be, because a lot of people contribute to GNU programs without sharing our ultimate ideals about why we are working on this."[42] As O'Mahoney explains, FSF chose to prioritize its commitment to its political goals by sacrificing democratic goals.[43] In the decade following the creation of FSF, at least fifteen other software projects launched nonprofit foundations.[44]

By the early 2000s several of these nonprofits were providing "incubation" services to other open-source projects. As late as 2008 lawyers specializing in open-source software were still providing guidance on how to establish a nonprofit to host open-source projects.[45] However, five years later the tide had changed. In 2014 the Internal Revenue Service denied tax-exempt status under two different subsections of the exemption code to two different open-source projects.[46] At least part of the reason was the profit potential inherent in open-source code.[47]

The initial proposal for the language of open source was intended to end the confusion about whether *free* meant "no money." In 1998 Netscape, which was in the midst of a browser war with Microsoft, opened up the source code for its popular browser. That it could do so and still seek to function as a company necessitated an alternative to the language of *free*. Software coders needed to be able to explain to investors and the general public the commercial potential of publicly modifiable software code. The Open Source Initiative was born of these efforts, promoting language generally credited to Eric Raymond.[48] The Open Source Initiative focused its efforts on developing and promoting a broader set of software license choices than either FSF or the marketplace were using. There are now more than seventy commonly used open-source approved licenses.[49]

In 1998 a company called VA Linux launched a public code repository called SourceForge. Since then, several others have come and gone. Today, the most widely used repository is GitHub, which launched as Logical Awesome LLC in 2008. It offers tools to manage version control, making it easier for groups to work together and to share within and across communities. Many groups now use the platform not just for their software code but also to share organizational policies, to document how they worked together, and to host the contractual (legal code) that shapes their work. GitHub has become a *de facto* repository for and a shaper of community norms, common language, and shared process for groups working together to build software as well as people managing other kinds of projects.

Both SourceForge and GitHub are commercial platforms built on proprietary software. This might reflect the "symbiotic balancing act between property and a commons" that Larry Lessig argues is key to both open code and open societies.[50] O'Mahoney notes that external pressures about liability, negotiation, and protecting group norms in the face of rapid growth spurred many open-source project managers to create institutions. At the same time, she concludes, their institutional choices represent their "desire to maintain pluralism," in terms of both who participates and what gets created. Open software standards and licenses serve to generate interoperable, portable applications, and the nonprofit foundation–commercial repository hybrids prevent any one party from becoming too influential.[51]

Perhaps no other event more clearly demonstrates the distinction between the norms of open-source communities and the organizational choices that support them than the 2018 purchase of GitHub by Microsoft. The digital hub for open-source code and communities, defined by their licenses and norm of sharing, became the property of a company that defined more than a decade of closed, proprietary business practices. But this irony also makes clear the degree to which licensing agreements and community norms are sufficient to the task of governing digitized data. Simply put, the world's largest repository of publicly accessible, shared digital data was born on closed systems, as a profitable commercial enterprise, and is now owned by one of the world's largest commercial corporations.

OPEN COLLECTIVES

Since the 1990s open-source software has become critical to widely used services such as email and web security protocols. Although some software is supported by institutionalized projects such as those discussed earlier, much is dependent on the continued volunteer efforts of distributed communities. Sustaining these communities and the software that they support can be difficult. Money is only one of a diverse set of incentives that drive open-source software coders.[52] Finding ways to sustain the products of distributed volunteers who produce goods that serve many purposes is challenging.[53] In 2014 the Internal Revenue Service, which previously had approved applications for tax exemption by organizations supporting open source projects, rejected two such applications.[54] It is unclear whether or how such applications will be considered in the future.

The allure of writing new code and building new tools often outweighs the attraction of doing the core maintenance work on critical software. Finding ways to finance these "infrastructural" software projects has bedeviled the open-source community for years. In 2017 Pia Mancini, an Argentinian democracy reformer, launched an online system for shared governance and payment processing that would make it much easier for existing organizations to host the costs of these small, but critical, open-source development teams. Her solution, Open Collective, allows small projects to easily open bank accounts, manage their finances, and pay their taxes while letting funders see where every penny goes. Within six months of launching, Open Collective was hosting more than five hundred projects and had to put a temporary freeze on hosting any more. While it was quickly adopted within the open-source community, the model of OpenCollective.com—a mix of fiscal sponsorship, crowdfunding, transparent budgeting, and online payment systems—is meant to provide an associational form to any kind of collective action.[55] For example, it is currently being used by open-source coding projects, independent podcasts, climate-change activists, independent political groups, local chapters of Women Who Code, and a community of expatriates living in Belgium.

Networked digitized data is the critical enabler of Open Collective, as well as part of the problem it seeks to solve. Collective governance is as old as time, and groups of people coming together to solve shared problems is not new. Open Collective enables these groups to quickly and efficiently manage their financial contributions in a transparent way that seeks to promote trust and accountability. It is designed as a way to let groups raise, track, and share money for their collective action without having to establish a new, hierarchical nonprofit organization for each such project.

Other experiments abound. In 2018, in response to congressional investigations into the use of the Facebook platform during the 2016 presidential election, a complicated deal was announced to make data from the company available to researchers. The model for managing the data involved creating a new effort, called Social Science One, is a multimillion-dollar, experimental partnership between the company and the Social Science Research Council, funded by foundations. In December 2019 the funders pulled out of the arrangement, citing Facebook's failure to provide the necessary data. Elsewhere, corporate behemoth Alphabet is promoting data trusts as a mechanism for managing data collected by a subsidiary company's efforts to install proprietary systems throughout the Quayside neighborhood in Toronto. The public outcry to such corporate control over public data and governance killed the project in 2020.

Conclusion

How to collect and manage data in democracies is a question for the ages. It's been asked and addressed in a variety of ways in the United States since the nation's founding. The advent of the internet has changed the nature and source of data that matters to democracies. As a result, the nature of associational forms to manage and govern digital data are changing, with old forms taking on new responsibilities and new forms coming into being. The nonprofit corporation has served as a standard-bearer of civil society in the United States throughout the twentieth century, but as digital data becomes central to associational life new institutional arrangements are emerging to manage it. Licenses and contracts tied directly to data and software code are growing in importance. These mechanisms can be used within or by any corporate form—be it a public agency, a commercial enterprise, or a nonprofit.

This use of licenses complicates what was once a rather straightforward assignment of purpose with corporate form; there are now numerous possible permutations of corporate form and license type. These licenses and contracts require existing organizations to take on new responsibilities, allow for organizations to work across sectors, and are beginning to give rise to new kinds of organizations. These modified forms of associational structure and resource governance are likely to become more prominent in the decades ahead.

Drawing from open-source software repositories, Open Collectives, and trusted data intermediaries, we can identify an emerging apparatus, made up of licensing agreements and systems, organizational governance documents and practices, and community norms, upon which civil society in the digital age is building itself. All of the examples rely on licensing agreements (sometimes attached directly to software code, other times written alongside it) and contracts to govern the use of the digital resources and the relationships between different parties. These can be used independently or together. What is notable is how the communities of participants and the license agreements are "mixed and matched" across traditional organizational forms. Licenses that permit the open sharing of software by distributed communities of volunteers can be created by professionally staffed nonprofit corporations, as we see with the Free Software Foundation or Creative Commons.

Both GitHub and SourceForge are examples of commercial enterprises providing storage space, tools, and processes to steward the community vernacular and norms of open-source coders. As the project management practices designed on these platforms become familiar, they start to be applied beyond software code. GitHub, for example, is where you might find the code of conduct for staff and volunteers of the nonprofit Code for America (CodeforAmerica.org). In keeping with the behavioral practice of building on others' work, the code notes that it is an adaption of antiharassment policies written by other nonprofits.

Data contracts and software licenses have become a key part of organizational governance and interorganizational relations since 1995. These tools are used within and across sectors. In some cases, as in parts of the open-source software communities, they are both legal agreements and key representations of the group's identity. These tools challenge existing assumptions about nonprofits, enable new partnerships, and spark new possibilities such as civic data trusts and Open Collective. Because these are governance tools tied directly to a resource, not to an organizational form, they challenge us to look beyond familiar institutions of civil society to emerging new forms of association.

This shift, from corporate structure to "contract stacks," is reminiscent of the nineteenth-century creation of the nonprofit corporation. Jonathan Levy, an economic historian, has done extensive work on the evolution of the nonprofit corporation. His work, along with that of John Havens, Naomi Lemoreaux, and Olivier Zunz, notes the important role that trusts, estates, and probate laws played in shaping modern philanthropy.[56] As the economy industrialized, states became increasingly involved in setting the bounds for incorporation. Commercial enterprises advocated for a corporate form that could focus on private (investor) benefits. This required sheaving off a private corporate form from its previous entanglement with public purpose. This began in earnest in the 1870s, driven largely by interest in developing transcontinental railroads.[57] Levy argues that the development of a modern nonprofit corporation was not a deliberate act; rather, it developed as an artifact of creating a corporate form dedicated to profit. It would take until after World War II for the nonprofit corporation to rise to dominance, and until the 1970s before state governments truly "opened the form to all," largely in response to the demands of the civil rights movement.[58]

When the opportunity to build great private fortunes drove the legal separation of profit from public purpose, a new commercial corporation was born and a nonprofit counterpart created alongside it. Today, with the many different approaches to safeguarding data and/or managing the mediating contracts and licenses, we are seeing a similar process of enterprise reinvention.

An important difference may be the source of experimentation, which is coming from communities and organizations using legal tools (licenses and contracts) rather than seeking to rewrite state corporate law (at least yet). However, standardization of these "stacks" is the likely next stage of development. In 2018, after a year of wrangling, community residents in Toronto successfully convinced Sidewalk Labs (an Alphabet-owned company) and the City of Toronto to discuss a data trust for all of the data that would result from the Quayside Project, which has been touted as the "smartest" of smart cities. Residents were unimpressed with the grandiose plans and angry at the thought that a commercial enterprise (based in the United States) with no limits would sweep up the digital data from anyone and everyone in the area. However, when Sidewalk Labs put forward the first draft of the trust documents, it was clear that the residents' interests in governance, privacy, and redress were not taken seriously.[59] It took two more years of community activism before pressure (and pandemic-related business considerations) led Sidewalk Labs to scrap its plans. The power of existing law rests with data companies, not with communities.

The nineteenth-century creation of the nonprofit corporation shaped the direction of the next 125 years of US associational life. Their growth was intertwined with funding relationships between the federal government and local communities that in many ways defined twentieth-century American politics. It also gave rise to the modern philanthropic foundation and a uniquely American story of private wealth being used for public purposes. The growth of nonprofits also depended on new regulatory structures to monitor and enforce the laws that enabled them. As we did a century ago, we once again find ourselves needing new institutional forms and regulatory regime. In the late nineteenth century the challenge was creating organizations and rules that could delineate the role of the state, the marketplace, and a third sector. Today the challenge is to design governance rules for digital data that preserve civil society's purposes and democratic principles.

Notes

1. Tim Wu, *The Curse of Bigness: Antitrust in the New Gilded Age* (New York: Columbia Global Reports, 2018); Lina M. Khan, "Amazon's Antitrust Paradox," *Yale Law Review* 126, no. 3 (2017): 564–907. Mathew Powers, "In Forms That Are Familiar and Yet-to-Be Invented: American Journalism and the Discourse of Technologically Specific Work," *Journal of Communication Inquiry* 36, no. 1 (2012): 24–43; and Meredith Broussard, *Artificial Unintelligence: How Computers Misunderstand the World* (Cambridge, MA: MIT Press, 2018).

2. There is a rich and important literature on the meaning of the word *information*, the rise of the "information age," and the differences among data, information, and wisdom. See, e.g.,

Ronald E. Day, *The Modern Invention of Information: Discourse, History, and Power* (Carbondale: Southern Illinois University Press, 2001). There is a similarly important literature on the cultural meaning of the internet. In this chapter I use the term *digitized data* to refer to anything captured in binary code (ones and zeros) and stored or shared on networked computing devices. The term *data* in this chapter is meant to any matter that has been digitized, including photos, videos, text, numbers, analysis and analytic methods, algorithmic instructions, and software code.

3. See, for example, the social determinism of scholars such as Langdon Winner, particularly "Do Artifacts Have Politics?," *Daedalus* 109, no. 1 (Winter 1980): 121–36.

4. Charles I. Jones and Christopher Tonetti, "Nonrivalry and the Economics of Data" (Working Paper No. 3716), July 31, 2018, Graduate School of Business, Stanford University, https://www.gsb.stanford.edu/faculty-research/working-papers/nonrivalry-economics-data.

5. Yochai Benkler, *The Wealth of Networks: How Social Product Transforms Markets and Freedom* (New Haven, CT: Yale University Press, 2006); Larry Lessig, *Code: and Other Laws of Cyberspace, Version 2.0* (New York: Basic Books, 2006); Aaron Perzanowski and Jason Schultz, *The End of Ownership: Personal Property in the Digital Age* (Cambridge, MA: MIT Press, 2016). Pamela Samuelson is prolific on these themes and her writings can be found at https://www.law.berkeley.edu/library/ir/faculty/?id=5476; see, in particular, Pamela Samuelson, Randall Davis, Mitchell D. Kapor, and J. H. Reichman, "A Manifesto Concerning the Legal Protection of Computer Programs," Berkeley Law Scholarship Repository, 1994, https://scholarship.law.berkeley.edu/cgi/viewcontent.cgi?article=2717&context=facpubs.

6. Charlotte Hess and Elinor Ostrom, eds., *Understanding Knowledge as a Commons: From Theory to Practice* (Cambridge, MA: MIT Press, 2006); Brett M. Frischmann, *Infrastructure: The Social Value of Shared Resources* (London: Oxford University Press, 2016).

7. Lessig, "Open Code and Open Societies" (keynote address, Free Software—A Model for Society?), June 1, 2000, Tutzing, Germany, https://cyber.harvard.edu/ilaw/Contract/Lessig%20on%20Open%20Code.pdf.

8. Mark Andreessen, "Why Software Is Eating the World," *Wall Street Journal*, August 20, 2011, https://www.wsj.com/articles/SB10001424053111903480904576512250915629460.

9. Many parts of the world, and many communities, do not have ready access to functioning, always on networks. The digital divide is real and important. For the purposes of this chapter, however, I am focusing on areas and communities where the systems are robust, although the policies and pricing of service can still create vast discrepancies in access.

10. For a sweeping review of how governments seek—and often fail—to deliver on goals this lofty, see James C. Scott, *Seeing Like a State: How Certain Schemes to Improve the Human Condition Have Failed* (New Haven, CT: Yale University Press, 1998).

11. Michael Schudson, *The Rise of the Right to Know: Politics and the Culture of Transparency, 1945–1974* (Cambridge, MA: Harvard University Press, 2015), 2.

12. Sandra Braman, *Change of State: Information, Policy and Power* (Cambridge, MA: MIT Press, 2006), 82.

13. James R. Beniger, *The Control Revolution: Technological and Economic Origins of the Information Society* (Cambridge, MA: Harvard University Press, 1986), 408–9.

14. Beniger.

15. Margo A. Conk, "The 1980 Census in Perspective," in *The Politics of Numbers*, ed. William Alonso and Paul Starr (New York: The Russell Sage Foundation, 1987), 155–86, 160; James Davis, "The Beginnings of American Social Research," in *Nineteenth Century American Science: A Reappraisal*, ed. George Daniels (Evanston, IL: Northwestern University Press, 1972), 152–78.

16. Mark Perlman, "Political Purpose and the National Accounts," in *The Politics of Numbers*, ed. William Alonso and Paul Starr (New York: The Russell Sage Foundation, 1987), 137–51.

17. The John Gardner Center for Youth Development launched the Youth Data Archive project in 2006.

18. Stefaan G. Verhulst, Data Collaboratives: Exchanging Data to Improve People's Lives, posted April 22, 2015, https://medium.com/@sverhulst/data-collaboratives-exchanging-data-to-improve-people-s-lives-d0fcfc1bdd9a.

19. Administrative Data Task Force, *The UK Administrative Data Research Network: Improving Access for Research and Policy*, December 2012, 2, https://www.adrn.ac.uk/media/1376/improving-access-for-research-and-policy.pdf.

20. *Register-Based Statistics in the Nordic Countries: Review of Best Practices with Focus on Population and Social Statistics* (New York: United Nations, 2007), https://www.unece.org/fileadmin/DAM/stats/publications/Register_based_statistics_in_Nordic_countries.pdf.

21. "Accessing the Justice Data Lab Service," May 25, 2018, https://www.gov.uk/government/publications/justice-data-lab.

22. Monica Bulger, Patrick McCormick, and Mikaela Pitcan, "The Legacy of InBloom" (working paper, February 2, 2017, Data and Society, New York), https://datasociety.net/pubs/ecl/InBloom_feb_2017.pdf.

23. *The Promise of Evidence-Based Policy Making: Report of the Commission on Evidence-Based Policymaking*, September, 2017, https://www.cep.gov/content/dam/cep/report/cep-final-report.pdf.

24. The National Academies of Sciences, Engineering and Medicine, Committee on National Statistics, list of Multiple Data Sources Workshops and Presentations, https://sites.nationalacademies.org/DBASSE/CNSTAT/DBASSE_170269#Feb2016workshop.

25. See GovLab's Data Lab case studies at GovLab, *DataLab: Case Studies*, December 12, 2017, https://medium.com/data-labs/case-studies/home. See also GovLab, *Data Collaboratives*, http://datacollaboratives.org/.

26. GovLab, *Data Collaboratives*, http://datacollaboratives.org/explorer.html. See also Jasmine McNealy, "An Ecological Approach to Data Governance," 2020, New York: Data & Society, Data Bite 127. Online at https://youtu.be/jB5_NrdWH7k.

27. Alfred E. Sloan Foundation grants database, https://sloan.org/search?q=administrative+data.

28. Lucy Bernholz, "Creating Digital Civil Society: The Digital Public Library of America," in *Philanthropy in Democratic Societies*, ed. Rob Reich, Chiara Cordelli, and Lucy Bernholz (Chicago: University of Chicago Press, 2016), 178–206.

29. Richard H. Ekman, "Technology and the University Press: A New Reality for Scholarly Communications," *Change* 28, no. 5 (September–October 1996): 34–39.

30. Anthony M. Cummings, Marcia L. Witte, William G. Bowen, Laura O. Lazarus, and Richard H. Ekman, *University Libraries and Scholarly Communication: A Study Prepared for The Andrew W. Mellon Foundation*, November 1992, https://babel.hathitrust.org/cgi/pt?id=mdp.39015026956246;view=1up;seq=13.

31. Ekman, "Technology and the University Press," 36.

32. Roger C. Schonfeld, *JSTOR: A History* (Princeton, NJ: Princeton University Press, 2003).

33. Schonfeld, 66.

34. Schonfeld, 95–118.

35. John Taylor, "JSTOR: An Electronic Archive from 1665," *Notes and Records of the Royal*

Society London 55, no. 1 (2001): 179–81, http://www.jstor.org.stanford.idm.oclc.org/stable/pdf/532157.pdf?refreqid=search%3A5adc28579c350bcf00a01a10751a941c.

36. Ithaka, *Artsor*, (2020) http://www.artstor.org/mission-history.

37. Ithaka.

38. See the Mellon Foundation's grants database, at https://mellon.org/grants/grants-database/grants/artstor-inc/40200639/.

39. James Shulman, "Spanning Today's Chasms: Seven Steps toward Building Trusted Data Intermediaries," *Shared Experiences* (blog), January 2017, https://mellon.org/resources/shared-experiences-blog/spanning-todays-chasms-seven-steps-building-trusted-data-intermediaries/.

40. Lessig, "Open Code and Open Societies."

41. Steven Weber, *The Success of Open Source* (Cambridge, MA: Harvard University Press, 2004), 139.

42. Richard Stallman, interview with Siobhan O'Mahoney, April 26, 2001, quoted in "Non-profit Foundations and Their Role in Community-Firm Software Collaboration," in *Perspectives on Free and Open Source Software*, ed. Joseph Feller, Brian Fitzgerald, Scott A. Hassam and Karim R. Lakhani (Cambridge, MA: MIT Press, 2005) 400.

43. O'Mahoney.

44. See O'Mahoney, 400, noting Apache, Debian, Gnu, Gnome, FreeBSD, Jabber, Perl, Python, KDE, BIND, Samba, the Linux Kernel, Linux Standards Base, Mozilla, and Chandler.

45. See the FOSS Legal Primer, published by the Software Freedom Foundation, and available at https://www.softwarefreedom.org/resources/2008/foss-primer.pdf.

46. Ernie Smith, "Open Source Projects Failing to Pass IRS Nonprofit Muster," *Associations Now*, July 22, 2014, https://associationsnow.com/2014/07/open-source-projects-failing-pass-irs-nonprofit-muster/ and Ryan Paul, "IRS policy that targeted political groups also aimed at open source projects, *Arstechnica*, July 2, 2014, https://arstechnica.com/tech-policy/2014/07/irs-policy-that-targeted-tea-party-groups-also-aimed-at-open-source-projects/

47. Nadia Eghbal, "Governance without Foundations," October 31, 2018, https://nadiaeghbal.com/foundations.

48. Christopher Kelty, *Two Bits: The Cultural Significance of Free Software* (Durham, NC: Duke University Press, 2008).

49. Open Source Initiative, "Approved Licenses by Name," https://opensource.org/licenses/alphabetical.

50. Lessig, "Open Code and Open Societies."

51. O'Mahoney, "Nonprofit Foundations and Their Role in Community-Firm Software Collaboration," 410.

52. O'Mahoney.

53. Nadia Eghbal, *Roads and Bridges: The Unseen Labor behind Our Digital Infrastructure* (New York: Ford Foundation, July 14, 2016), https://www.fordfoundation.org/about/library/reports-and-studies/roads-and-bridges-the-unseen-labor-behind-our-digital-infrastructure/.

54. Yorba and OpenStack rejected these in 2014. Ryan Paul, "IRS policy that targeted political groups also aimed at open source projects, *Arstechnica*, July 2, 2014, https://arstechnica.com/tech-policy/2014/07/irs-policy-that-targeted-tea-party-groups-also-aimed-at-open-source-projects/ and Kendra Albert, "Open Source Madness," *EFF Deeplinks*, July 16, 2014, https://www.eff.org/deeplinks/2014/07/open-source-madness.

55. Interview by the author with Pia Mancini, June 2018.

56. Jonathan Levy, "Altruism and the Origins of Nonprofit Philanthropy," in *Philanthropy in Democratic Societies: History, Institutions, Values*, ed. Rob Reich, Chiara Cordelli, and Lucy Bernholz (Chicago: University of Chicago Press, 2016), 19–43; Naomi R. Lemoreaux and William J. Novak, eds., *Corporations and American Democracy* (Cambridge, MA: Harvard University Press, 2017); Olivier Zunz, *Philanthropy in America: A History* (Princeton, NJ: Princeton University Press, 2012).

57. Levy, "Altruism," 213.

58. Levy, 215.

59. The proposal from Sidewalk Labs is online at Alyssa Harvey Dawson, "An Update on Data Governance for Sidewalk Toronto," *Sidewalk Labs*, October 15, 2018, https://www.sidewalklabs.com/blog/an-update-on-data-governance-for-sidewalk-toronto/ Response from some members of the community is online at Chris Rattan, "Torontonians should take control of their data," *NowToronto*, May 23, 2018, https://nowtoronto.com/news/owns-data-toronto-smart-city/ and Bianca Wylie, "Sidewalk Toronto: Gaslighting Toronto Residents Backfired—Capacity's Built and Power's Shifted," October 16, 2018, *Medium*, https://medium.com/@biancawylie/sidewalk-toronto-gaslighting-toronto-residents-backfired-capacitys-built-and-power-s-shifted-77c455b150a3.

4

Digital Exclusion: A Politics of Refusal

Seeta Peña Gangadharan

Over the past few decades, the term *digital exclusion* has been linked to debates about access to internet infrastructures, adoption of internet-enabled technologies, and conditions of social and economic marginalization and historical forms of oppression. Adding to these concerns are considerations of privacy and surveillance and their consequences for members of marginalized communities, as well as political and economic factors that shape technology companies' relationship to processes of marginalization. Taken together, these exclusionary problems amount to what the feminist political philosopher Iris Marion Young (2001, 16) would refer to as a "plausible structural story," or history of patterned inequality and structural injustice.

But do these accounts of harmful exclusion obscure the agentic possibilities of willful self-exclusion in technologically mediated society? Like Ananny's examination of silence and absences in digital platforms as productive political communication (chapter 5), I take the opportunity in this chapter to rethink exclusion as an important mechanism for transforming political discourse. Specifically, I reevaluate digital exclusion as a form of active refusal of technologies' seemingly inevitable uses and ends. Refusal does not mean dropping out or rejecting wholesale involvement with digital devices, internet infrastructures, or internet-based technologies. Instead, grounding the discussion with attention to marginality, "informed refusal" (Benjamin 2016, 970), and abolition of tech or of digital technologies that punish and police marginalized people (Benjamin 2019; Roberts 2019), I argue that when marginalized people refuse technologies, they imagine new ways of being and relating to one another in a technologically mediated society. Refusal serves as a means for individuals and groups, especially members of historically marginalized groups, to assert themselves and collectively determine a technologically mediated world in which they wish to belong.

To argue this, the chapter locates itself in a normative framework that updates Young's theory of communicative justice for an era of data-driven technologies. The framework marries Young's discussions of marginality, discourse, and the problem of "internal exclusion" with critical debates about technology, its politics, and the importance of refusal in the lives of members of marginalized communities. I then use this framework to evaluate and extend conventional debates about digital exclusion. Specifically, I present six kinds of digital exclusion. While the first five characterize exclusion as negative or harmful to marginalized groups, the sixth posits exclusion in affirmative terms, in which members of marginalized groups assert their agency in the face of structural injustice and refuse technology in different ways in order to develop and determine their own technologically mediated lives. By the end of the chapter, I will show the generative aspects of digital exclusion as refusal and its importance to communicative justice in the twenty-first century.

Rethinking Exclusion in Communicative Justice: Refusal of Technology

YOUNG'S THEORY OF DISCURSIVE INEQUALITIES, INCLUSION, AND EXCLUSION

With its conspicuous attention to equity and justice, feminist political theory provides a useful place to begin thinking about digital exclusion. Iris Marion Young's work is especially critical, because it examines the centrality of communication to a democratic society through the discursive dimensions of being included and excluded. Young's work—broadly definable as a theory of communicative justice—challenges a long-standing liberal belief in distributive justice, namely the idea that a procedurally fair distribution of resources or material goods in society will suffice to address all objectionable forms of inequality. Much political theory of the past several generations has focused on distributive justice. Or, as John Rawls (1971, 9) put it in *A Theory of Justice*, "a conception of social justice . . . is to be regarded as providing in the first instance a standard whereby the distributive aspects of the basic structure of society are to be assessed." Young (1990, 39), by contrast, argues that "justice should refer not only to distribution, but also to the institutional conditions necessary for the development and exercise of individual capacities and collective communication and cooperation."

For Young (1990, 2000), justice requires attention to institutionalized communicative practices and their transformation. It also requires attention to the way in which individuals are engaged in a process of recognizing one another's humanity. In a world where certain groups are marginalized, suffer from exploitation, are made to feel powerless or culturally inferior, or face violence in institutionalized, systematic ways, more inclusive forms of communication can provide the essential basis for alleviating entrenched systems of oppression.[1] Rather than focus on individualized processes of reflection or information gathering, as is the case with liberal democratic norms, Young argues that society needs norms of speaking that ensure recognition of all individuals in society, including and especially oppressed groups. With norms of communicative rather than only distributive justice, those who have suffered due to processes of marginalization, exploitation, powerlessness, cultural imperialism, and violence will become visible, audible, and knowable. Their needs can more readily become part of what are considered legitimate political claims.

Young's attention to different forms of discursive inequalities provides a way to open up the idea of digital exclusion. When describing the importance of new conversational rules, she identifies problems of both external exclusion—failing to be included in a group—and internal exclusion in deliberative settings. Even when part of a deliberative group, it is possible for dynamics of exclusion to be at work. This delineation helps illuminate the conspicuous and inconspicuous ways in which exclusion takes place, even when one is formally included in a space of deliberation. Applied to online worlds or what is described elsewhere in this volume as the digital public sphere, Young's work allows us to scrutinize the tricky, multifarious ways in which marginalized groups experience exclusion in digital or technologically mediated terms.

To grasp Young's relevance to debates about the meaning and value of digital exclusion in society, a more detailed explanation of external exclusion and internal exclusion is in order. Young names the practice of external exclusion as a means by which power holders or elites who lead decision-making processes keep members of marginalized groups out of deliberative fora in which critical decisions are being made about problems that affect them. The marginalized are simply not to be found in such spaces of political consequence. Such geographical (e.g., physical, spatial obstacles), informational (e.g., lack of means to learn about when a forum is taking place), and material (e.g., impediments due to cost) barriers prevent certain groups from contributing to key discussions, which has an impact on their well-being and also violates commonly held standards of fairness. In the case of external (discursive) exclusion, individuals' and groups' inability to participate in a key forum connects to other patterns or forms of social exclusion. Ghettos are disconnected from city centers by a lack of roads. Meeting announcements are communicated via channels that only a select kind of consumers have access to. Meetings take place in venues that require a costly transit ticket, incur extra costs of childcare, or involve some form of payment affordable only to some, not all.

In describing the concept of internal exclusion, Young (2000) argues that it is not enough to rectify the problem of external exclusion, important as that may be. Spaces for discussion and collective problem solving can become accessible, such as by building new roads, being announced in ways that reach all affected stakeholders, or lowering or eliminating the cost of participation. However, even with these changes, these fora are often structured in ways that are unwelcoming and impenetrable to participants who come from historically marginalized communities. These fora follow a particular set of rules or conventions that make sense and are familiar to some but not to all. As a result, some, more marginal participants are included but unwelcome and effectively silenced in the process.

Young's vision of inclusive communication matters because it reminds us that process needs substance as much as substance needs process for a communicative democracy to thrive and for democracy's promise of equality to be fulfilled. Access to spaces of political deliberation (which can be understood in procedural terms) and design of spaces of political deliberation (which Young insists is historically and contextually bound and thus concerns the substantive) both matter. Without paying attention to and inviting other forms of discursive practice, access alone will be an incomplete remedy to the problem of exclusion. Conversely, without connecting members of marginalized groups, members of privileged groups, and everyone in between to one another, dialogue and political decision making will remain in an enclave and the domain of powerful elites.

SOME CLARIFICATIONS: WHY COMMUNICATIVE JUSTICE NEEDS AN UPDATE

Before we apply the theory of communicative justice to digital and data-driven contexts, we need to acknowledge the limitations of Young's perspective and update it. In fact, while Young's communicative justice brings attention to those individuals made or kept invisible by status quo communicative practices, the theory is limited by two key factors. First, and completely independent of technology, Young did not give much consideration to the concept of self-exclusion and its democratic value to members of marginalized communities. Second, Young's work could not anticipate complex forms of technologically mediated communication in democratic societies. To make use of her theory thus requires that we modify her understanding of exclusion to adequately take into account self-exclusion, diverse technologies that affect how we communicate and relate to one another, and refusal of technologies that mediate our modern communication today.

As regards the first limitation, Young's theory of communicative justice may be faulted for a type of conservatism, as opposed to transformative vision of politics (see also Fraser 1997). At its core, Young's analysis is concerned chiefly with sites of formal political decision making, such as town halls or legislative chambers, and the exclusion of the marginalized from such spaces as well as a lack of inclusive communication within them. If we extend Young's concern and envisage the agents of change who might bring about internal inclusion, she appears to place the onus of responsibility on people with the power to set the terms of debate and discussion. In other words, discursive changes appear to depend on power holders or elites who design, manage, or maintain norms of communication within spaces of political deliberation. Powerful people, not necessarily the marginalized, must be motivated and persuaded to enact new communicative norms and bring about a more just and democratic society. The privileged shoulder responsibility for rectifying exclusion.

Yet individuals and groups who suffer from exclusionary practices must also play a part in transforming social values, social structures, and democratic processes. As Fraser (1990) suggests, a collective awareness of a history of oppression and domination may motivate members of marginalized groups to self-exclude in order to formulate a political vision and make political demands. Marginalized groups might prefer to enclave themselves for the purposes of self-creation and self-determination. They retreat in order to develop a sense of themselves, to recognize the oppression they have endured, to understand its structural features, to communicate their grievances to each other, and to develop political will to transform social conditions. As Fraser's work establishes, refusal does not necessarily imply a desire to withdraw or drop out of political deliberation, but it can be an essential component of a transformative politics that allows for a diversity of political communication to bubble up from within marginalized communities.

Admittedly, refusal does not follow a linear path, from self-creation to self-determination. As discussed by Cannon (1995), people who confront hardship on a daily basis may not be able, even if they wish, to fully retreat or separate themselves from a dominant culture. As Cannon explains, Black women who contested the brutality of slavery had to compromise or be accommodating (to slave owners) in order to survive, in part because they did not have the luxury of doing otherwise. Thus, acts of refusal may necessarily involve compromise as opposed to some kind of principled disobedience or full-fledged opposition to dominant culture (Howe 2003). Yet as Fraser (1990) demonstrates, refusal can also evolve into collective communicative action. Members of marginalized groups can exclude themselves together in counterpublic spheres. In these spaces, individuals engage in processes of collective self-creation, begin to represent themselves to external audiences, and learn thereby how to contest dominant political culture. Refusal, in Fraser's (1990, 67) eyes, means broadening how we contest and transform political discourse.

To be clear, the role of refusal in generating transformative politics does not negate the role that privileged power holders need to play or the importance of institutional transformations that Young suggests.[2] The perpetrators of injustice do bear responsibility for repair and rectification. The path to justice need not wait for the organized response of the oppressed. Nevertheless, we cannot just rely on the enlightened goodwill of privileged elites to recognize and rectify communicative injustices. Thus, following Fraser, a critical first amendment to Young's theory of communicative justice requires that we include refusal and value self-exclusion as one important means by which members of marginalized communities assert their agency and strive to achieve recognition as equals. Refusal by the marginalized, for example, to join conventional discourse, matters just as much as recognition of the marginalized by privileged people or institutions.

As regards the second limitation, Young's theory of communicative justice begs a more complex reading of communication that includes technologically mediated communication. Young focuses on face-to-face communication with respect to democratic deliberation and says very little about mediated communication, digital or otherwise. This omission is notable, given that Young's notion of communicative democracy tries to update Habermas's deliberative democratic theory, which includes a prominent critique of mass media and, to a lesser extent, digital technology (Feenberg 1996; Habermas 1970, 1989, 2006). Whereas Habermas acknowledges the importance of institutions of public communication (Habermas 2006; Wessler 2018), Young avoids any sustained examination of complex communication systems or the extent to which they exclude or include members of historically marginalized groups. Arguably, those who are responsible for welcoming different kinds of speakers involve not just designers or hosts of town-hall debates but also managers of media institutions and engineers of digitally mediated communication in big tech companies, as well as regulators of each.

My aim here, therefore, is to supplement Young's theory of communicative justice with insights from critical theories of technology. Such theories help clarify the formative role that technologies play in determining the conditions for communicative justice. As Feenberg (1996, 2002) explains, technologies serve as building blocks of our social world. They not only affect how we live our lives but also change what we value in life. In this sense, technologies help to construct the very nature and structures of the social world, or what Feenberg refers to as social ontologies. This approach allows us to think beyond "good" or "bad" uses of technologies, on the one hand, or to assume technologies' neutral position in society, on the other. Instead, Feenberg states, when we understand technologies as social practices, we acknowledge that they create particular possibilities of being with or relating to others while simultaneously diminishing or eliminating other possibilities.

Given the ubiquity of digital technologies and their use in communication, we should orient our attention there as well as to the kinds of face-to-face communication theorized by Young. The design of digital technologies and, in particular, the norms that inform their design will be as important as the (analog) discursive norms found in Young's original conceptualization. Thus, a second, critical amendment to the theory of communicative justice requires we acknowledge that communication is technologically mediated, not only face-to-face.

TECHNOLOGIES OF COMMUNICATIVE JUSTICE, REFUSAL, AND REINVENTION

If we combine these two criticisms, we arrive at the crux of this chapter: any serious rethinking of digital exclusion must consider the place of refusal—of refusing technologies and the ways they shape our being and relating to others. Doing this goes against the grain of many critical appraisals of technology and democracy. Historically, refusal does not represent the first object of analysis in the consideration of the impact of technology on democracy. Rather, these theories of technology tend to expose differences between technologies, democratic institutions, and processes of accountability and draw our attention to design and engineering requirements for making technologies democratic. We can see this most obviously in Feenberg's work. Feenberg (2002) focuses on design, not processes of dissent. He explains that once we recognize technologies as consequential in the creation of social possibilities, we need to focus on their construction or development. Hildebrandt (2015) also draws attention to engineering or design stages. She argues that digital technologies and, in particular, automated technologies, unlike the law, lack a court of public opinion for evaluating the legitimacy of decisions they make. To ensure that these technologies do not undermine due process or supplant the rule of law, they must be engineered in ways that reflect democratic values (see also Yeung 2015).

While Feenberg, Hildebrandt, and others celebrate various forms of democratic design and engineering, this literature tends to neglect the role of refusal in transforming the "terms and conditions" of technologically mediated communication in democratic societies. But just as marginalized groups can play essential roles in mobilizing the terms of their inclusion in face-to-face communication, members of marginalized groups who face exclusion from elite processes of innovating, launching, and maintaining technologies and technology infrastructures have a role to play, too, in technological transformation.

However, while the engineering and design of technologies can be made more democratic by installing democratic values in design stages, we should not depend solely on the goodwill of privileged engineers or designers, or the elite institutions that support them. Rather, communicative justice in the twenty-first century requires that we acknowledge the importance of refusal by members of marginalized groups of the ways in which technologies—or the engineers, designers, and institutions that create them—want us to communicate, want us to be and relate to others. The upshot is clear: even when technologies have already been developed, marginalized groups can still assert themselves and play a role in transforming the terms and conditions of their—and our—technologically mediated lives.

Technological refusal is not about dropping out or wholesale rejection of all things digital. Members of marginalized communities may not have the luxury of refusing a technology or dropping out of technological infrastructures (Vaidhyanathan 2019; Lanier 2018; Vertesi 2014). Instead, they might accommodate or be a part of sociotechnical systems, because they lack alternatives. But they can still assert themselves and communicate their views amid the oppression of the systems to which they must belong. In cases where individuals self-exclude together, their refusal can result in challenges to the dominant articulation of political possibilities. Technological refusal can thus be understood as about self-creation and self-determination.

Benjamin's (2016, 2019) work further clarifies the idea of technological refusal by identifying instances of where individuals act alone and together. In an examination of individual and collective responses to biomedical research,

Benjamin (2016, 970) discovered that her study participants—African Americans, members of the San tribe in South Africa, and refugees in the United Kingdom—practice what she calls "informed refusal." In their interactions with doctors, they opted out of medical research with acute awareness of their surroundings and, in many cases, presented alternative bases for understanding health and well-being to determine what was medically right for them. She notes the collective efforts of the San people, who disrupted meetings convened by medical authorities and challenged research representatives in their native language (without a translator).

The collective dimension of refusal is also evident in the concept of tech abolitionism. In more recent writing about digital "resisters," Benjamin (2019) has brought attention to marginalized communities working in the tradition of prison abolitionists. Although she emphasizes the why over the how of local efforts, she captures organizing efforts to block governments from adopting technologies that exacerbate racist and discriminatory treatment of Black and Brown communities. By bringing up community-led movements against surveillance technologies, her work shows how members of marginalized communities collectively develop and advocate alternative technological futures that seek to hold governments and technology companies to account. In a similar vein, Roberts (2019) outlines a tech-abolitionist agenda that highlights ways in which racial justice movements use technology to facilitate social change, document rising inequalities, and propose policies to end state violence.

Applied to an expanded vision of communicative justice, the ideas of informed refusal and abolition are essential. They acknowledge both how people develop their own identities (self-creation) in spite of how they might be controlled or categorized by technology and how members of marginalized groups can work together and identify on their own terms (self-determination) what they want from technological systems. In other words, engineers and computer scientists may have already designed and diffused a digital device or data-driven technology, with (or without) democratic values in the design and deployment. Important as designing with democratic values may be, those who are most vulnerable to technologies' exclusionary power also have a critical role to play. They (too) must decide and articulate what technologies make and should make possible or impossible in their lives. People refuse technologies in order to put into motion the kinds of relationships that otherwise-excluded individuals and groups need to realize their full humanity and claim recognition as an equal in a technologically mediated world.

My goal here is to build on and adapt Young's theory for a digital age. In essence, I have begun to conceptualize what amounts to a theory of technologies of communicative justice. I have done so in ways that put the agency of individuals and groups at the center and position technological refusal (by the marginalized) alongside design or engineering (by the elite or privileged). By proposing that refusal can involve both self-creation and self-determination, I am urging us to rethink digital exclusion in new and expanded ways. Doing so, I believe, is critical to confronting the challenges and opportunities of digital and data-driven infrastructures in modern democracies.

In the remainder of the chapter, I expand the argument that refusal can be understood as an affirmative form of digital exclusion by introducing several examples of refusal in action. To clarify the agentic qualities of digital exclusion, I walk through other, more conventional meanings of the term. That is, I first highlight ways in which this term evokes Young's ideas of external and internal exclusion, drawing attention to newer and underacknowledged dimensions of such exclusion that result from inequalities in and of the information economy, surveillance capitalism, or data economy.[3] I then turn toward the idea of refusal and the ways in which this idea of refusal allows us to reinterpret exclusion in a way that affirms the agency and political potency of the excluded. In the third section, I reflect on differences between conceptualizations of digital exclusion that prioritize or neglect the agency of the excluded, and I affirm the importance of technological refusal to a twenty-first-century vision of communicative justice.

Five Faces of Digital Exclusion

FAMILIAR ARGUMENTS

Digital Exclusion as External Exclusion: The "Digital Divide"

In conventional policy debates, efforts to close the so-called digital divide echo a logic found in Young's concepts of external exclusion. If we understand digital technology primarily in terms of internet infrastructure, then we understand that when people lack the opportunity or means to go online, they fail to reach decision-making spaces that have an impact on their everyday lives. This barrier to entry typically derives from some form of market failure, whereby broadband service providers (e.g., cable and telecommunications companies) lack the incentives to provide free or low-cost access to all people. This market failure, so the conventional argument goes, prevents individuals from joining the information economy and online spaces in which individuals can engage in democratic activities.

As aggressive efforts to construct a national information infrastructure moved forward at the turn of the twenty-first century, some began to acknowledge the unique ways in which members of historically marginalized populations lagged behind other groups in getting online (Irving 1999). For example, Baynes's (2004) legal analysis demonstrates the barriers in telecommunications markets faced by Native Americans, Latinos, and African Americans in the 1990s and early 2000s, including legacy regulations that hindered these populations from first accessing basic telephony services. These (legacy) problems amount to a form of geographical de facto segregation based on race, leading to what Baynes refers to as electronic redlining: the experience of economic, racial, and social marginalization due to lack of telephone access repeats itself in the digital era, as telecommunication providers avoid these so-called underperforming markets. Furthermore, as members of marginalized groups lack a basic connection to the internet, including fast and continuous broadband service, their experience of social, racial, and economic marginalization is further compounded by exclusion from the information economy, civic activities, or cultural life, all of which are increasingly technologically mediated (Powell, Bryne, and Dailey 2010).

Digital Exclusion as Internal Exclusion: Meaningful Broadband Use, Digital Illiteracy, and Elite Internet Culture

While arguments about lack of connectivity evoke Young's concept of external exclusion, arguments about lack of digital literacy evoke her writings on internal exclusion. Even when connected, members of marginalized communities experience digital exclusion as internal exclusion. That is, marginalized people might be able to access the internet or successfully adopt broadband, but a combination of socioeconomic factors interferes with their ability to use broadband and participate meaningfully in internet culture.

Such questions are first and foremost an empirical undertaking that understands digital exclusion by asking "the pressing question . . . [of] 'what are people doing, and what are they able to do, when they go online'" (DiMaggio and Hargittai 2001, 28) and examining what factors or resources "allow people to use technology well" and "engage in meaningful social practices" (Warschauer 2002, n.p.). Since the publication of formative studies in the early 2000s, user studies on digital inequalities have mushroomed, with an intense focus on differences in digital skills, attitudes towards the internet, engagement on the internet, and well-being factors against other forms of social and economic advantage or disadvantage (Helsper 2012).

A corollary argument to this kind of digital exclusion focuses on elitism in internet culture. After more than a decade of optimistic characterizations of the internet and online civic action (Froomkin 2004; Norris 2008), Hindman (2009) examined "A-list" netizens, their institutional affiliations or connections, and popular blogs, commonly portrayed at the time as internet democracy in action. While Hindman does not connect elitism with marginalization or other forms of oppression in online spaces, his work suggests that deliberative spaces encode the values of those who host or inhabit them (here, Harvard and other Ivy League graduates) and that the so-called democratic internet allows those with know-how and existing privilege to "speak" and dominate (in this case, the "blogosphere"). Other studies highlighting internet history point to the exclusive nature of those "architecting" the internet, the exclusive coterie of engineers making choices about internet design, security, or applications (Abbate 1999; Turner 2006; Zittrain 2009).

NEW DIGITAL EXCLUSIONS IN A DATA-DRIVEN ERA

Additional Internal Exclusion: Predation and Privacy, or a Lack Thereof

In the context of surveillance capitalism or the data economy, additional hurdles for meaningful participation online exist. They stem from lack of or low digital privacy and amount to predation of marginalized people.

For members of marginalized communities, data-driven technologies pose threats to privacy, exposing them to new vulnerabilities previously unaddressed or ill addressed in conventional policy debates about the so-called digital divide or digital illiteracy (Gangadharan 2013, 2015). It is worth noting that by the 2000s, as efforts to tackle lack of connectivity and improve marginalized populations' digital skills increased, internet tracking and targeting and the internet advertising industry were expanding. With little regulatory oversight, companies advanced techniques of consumer data profiling and explored their predatory potential. Internet tracking and targeting were integral to subprime lending throughout the 2000s, revealing how the lack of consumer internet privacy connects to marginalization (Gangadharan 2012; Fisher 2009). Lenders, online marketing companies, and data brokers all contributed to the compilation of data profiles, which were then used to effectively target buyers of "ghetto loans" (quoted in Fisher 2009). A significant majority of these risky consumers came from Black or Latino communities—that is, neighborhoods with low socioeconomic indicators and urgent need for quick cash.

This case points to an emergent internal contradiction regarding efforts to bridge the so-called digital divide—and a striking one, given that the financial crisis led the US Congress to invest an historic $4.7 billion for broadband adoption and infrastructure development (through the American Recovery and Reinvestment Act of 2009). Connecting to the internet is as much about being accessed as it is gaining access to services, resources, and experiences to meet basic human needs or participate meaningfully in society. The terms of accessing the internet are profoundly lacking in reciprocity between user and corporate actor. As several scholars have written (Acquisti, Taylor, and Wagman 2016; Brown 2015; Pasquale 2015; Zuboff 2019), technology companies continually collect and process people's personal information, and they thrive on information asymmetries, whereby value is extracted from tracked-and-targeted consumers who know least rather than best.[4]

The threat of tracking and targeting adds nuance to the discussion of digital exclusion. From the perspective of internal exclusion and digital illiteracy, the complexity of tracking and targeting adds to a hierarchy of privacy-skilled versus privacy-unskilled users who experience very different kinds of content and services on the basis of data profiles (Gandy 2009). This threat of privacy deficits puts a premium on being privacy literate, not only digitally literate, and adds to a growing list of skills or knowledge that users need when online. The privacy deficit threatens to create new problems of access as well. As members of marginalized communities begin to realize that using the internet requires some measure of data protection, obfuscation, or encryption, they could very well be priced out of broadband markets. The cost of additional services to protect against the data mining and profiling might deter individuals from going online altogether (Wiley et al. 2018).

Be that as it may, the attendant solutions to privacy problems nevertheless suggest a type of (self-preserving) conservatism toward technologically mediated life. A lack of or low privacy leads to measures to increase privacy literacy for marginalized populations or the introduction of affordable privacy-by-design solutions. These measures imply that as long as these protections are in place, the promise of meaningful participation online can be fulfilled. But what if certain groups are forced to adopt technologies whether they like it or not?

Digital Exclusion and Coercion: Forced Adoption of Surveillance Technologies

It is worth reminding the reader that the previous sections began by equating digital technologies with the internet. However, clearly not all digital technologies are consumer-facing or consumer-oriented internet applications. Some technologies might be networked and use internet infrastructure, but they can operate as business-to-business products, invisible to the consumer and, most importantly, function for a prescribed purpose that is not evident or open to challenge from users. For simplicity's sake, we might refer to these as data-driven technologies. Similarly, not all surveillance depends on internet use, and in this sense, we can distinguish between digital surveillance and internet surveillance. Internet-enabled tracking and targeting, as described earlier, depends on people voluntarily using broadband or having to use it for a purpose. However, with digital surveillance, adoption of digital technology is not a choice. It is coerced.

Under the condition of forced adoption, individuals are, technically, digitally included, but the terms of their inclusion are not "consentful" (Lee 2017). Rather, in the case of involuntary adoption, the use of data-driven technologies reflects sociotechnical systems of oppression that target groups marked by social difference. Monahan (2008, 220) calls this "marginalizing surveillance" and explains that it contributes to the "creation or enforcement of conditions of marginality through the application of different surveillance systems for different populations." So while all individuals in contemporary society might suffer from pervasive forms of information collection, classifying, and sorting, the kinds of surveillance practiced on the marginalized involve distinctive techniques that intensify their marginality.

Such forms of digital surveillance have profound impacts on marginalized people's lives. From workplace performance technologies to computerized welfare to law enforcement monitoring systems, digital technologies permeate the lives of marginalized populations in ways that amplify their experience of everyday oppression. As Gilliom (2001) and Eubanks (2011, 2018) show, digital technologies follow welfare recipients' behavior closely, requiring individuals to report intimate details of their lives in often stigmatizing ways. The goal of computerized welfare assistance ties to a broader institutional goal: help welfare agencies reduce, if not eliminate, poor people's dependence on welfare altogether. In the process, the constant monitoring of intimate lives leaves poor people bereft of privacy and emotionally exhausted, if not materially deprived, when classified as ineligible for public assistance. Similarly, Bridges (2017) offers insights into the lack of privacy for low-income mothers as they navigate public assistance, are "databased," and are deemed responsible for their own indigence.

In each of these cases, digital technologies exact an emotional and material toll on marginalized people, in part because they cannot choose not to subject themselves to these tools or systems. They must use—or more appropriately, be used by—these technologies or risk further consequences, such as loss of health care or reduction in unemployment insurance. Each of these projected outcomes lies as a threat that individuals must evaluate against other potential outcomes, including the experience of intimidation, shame, and distress or even bodily violence. Far from creating so-called digital opportunities, such technologies present marginalized people with choices of lesser evils: privacy or health care, or dignity or unemployment insurance.

Marginalizing surveillance is not accidental or arbitrary. As legal history in Eubanks's (2018) and Roberts's (2019) scholarship shows, welfare reform over the past several decades has turned public benefits from an entitlements system to a punitive mechanism that attempts to control or modify welfare recipients' behavior. Privacy laws also disfavor poor, marginalized people, permitting exceptions that allow, for example, police or welfare workers to intrude into the lives in ways that would not be permissible for wealthier, more advantaged populations (Bridges 2017; Gilman 2012; Slogobin 2003). This legal history echoes the negative sentiment held by members of marginalized communities that data-driven systems are extractive, taking more than they give (Petty et al. 2018; Eubanks 2018; Bridges 2017). Given that under conditions of digital surveillance, marginalized people are digitally included, but not on consensual terms, the opportunity to refuse to participate or be included seems an understandably desirable preference or strategy.

Powerlessness: Economic and Social Divides and Tech Elites

Before returning to refusal as an agentic form of digital exclusion, I turn to one last consideration of exclusion: technology companies (or the elites that manage them) and social and economic divides that they exacerbate. I introduce this discussion as an important addendum to the previous reflections on digital exclusion. Some might call this unnecessary, because the form of exclusion I present here does not refer to an existing discourse on digital exclusion, nor is it a problem particular to the technology industry.[5] Yet as the example of marginalizing surveillance technologies suggests, there is more to understanding exclusion than exclusion from or within so-called online worlds. By talking about technology companies and marginality, I similarly bracket the notion of online and offline and the idea that digital and data technologies comprise a space where one must "go" or "visit."

However, unlike the case of marginalizing surveillance, a critique of tech elites centers on the role of digital and data-driven technology companies in processes of economic and social marginalization. This critique implies that digital divides are not between haves and have-nots but between marginalized members of society and privileged tech elites or power holders whose elite status happens to derive from economically powerful technology companies. In this rehashing of digital divide terminology, tech elites are viewed as directly hindering the ability of affected groups to determine their lives.

For example, in the United States, technology companies routinely move money offshore, write down corporate debt, or lobby for tax breaks. In 2016, Microsoft exemplified this process when it borrowed billions of dollars to fund the acquisition of LinkedIn, despite having $100 billion in cash reserves. As Foroohar (2016) explains, Microsoft avoided having to move money from overseas to the United States and, thus, paying US corporate tax at a rate of roughly 35 percent. Because debt is tax deductible, Microsoft claimed savings of approximately $9 billion, monies that otherwise would have been tax revenue (see also Poon 2017). While tax avoidance affects all individuals, disinvestment in public infrastructure disproportionately affects members of vulnerable groups who live precarious lives and depend on such support infrastructures.

As they contribute to the contraction of the welfare state at the national level, technology companies also have an impact on the welfare of marginalized populations at the local level. Silicon Valley serves as the site of some of the country's highest homelessness rates (Campbell and Flores 2014; Levin 2016), and rents are skyrocketing, forcing lower-income residents out of the region (Urban Displacement Project 2015). In the wake of these crises, technology companies have moved into real estate and housing finance. In Palo Alto, Facebook is building mixed-income housing units, a small percentage of which is allotted for qualifying low- and middle-income residents (Nunez 2016). Meanwhile, in Seattle, one of the country's other major tech hubs, Microsoft pledged a half billion dollars to support low-income housing development, a move that marks a new source of profitability for the company.[6] Such a move (into the private sector) forecloses the policy work of antipoverty and housing advocates and the populations they represent and bodes of increased struggles by residents and advocates to navigate an increasingly complex financialized housing market (Fields 2017; Kim 2019).

Under these conditions, marginalized groups lack political opportunities to advocate for public policies aimed at eliminating inequality. This lack of political maneuverability is due to shifting governance strategies, whereby managing poverty and inequality moves from public decision making, which affords a measure of transparency, accountability, and democratic legitimacy, to private sectors, which typically lack all three. This form of digital exclusion is thus defined in terms of powerlessness, whereby the "powerless are those who lack authority or power . . . those over whom power is exercised without their exercising it" (Young 1990, 21).

With these last two arguments about exclusions and technology, the salience of refusal in marginalized communities should be evident. Questions of power, inevitability, and alternatives are begging for answers when surveillance systems coerce and the material impacts of network effects exacerbate social and economic inequalities. Are technologies of social control avoidable? Are wealth disparities incurred by rapid growth in data-driven economies desirable? In the following section, I address these questions by rethinking digital exclusion in the affirmative terms of refusal of digital technologies.

Rethinking Digital Exclusion in Affirmative Terms

Digital exclusion, thus far, has referenced five different problems: unavailability of digital technologies, inability to meaningfully use digital technologies due to digital illiteracy, inability to meaningfully use technologies due to low or nonexistent digital privacy, forced or coerced use of surveillant digital technologies, and inequalities between privileged tech elites and members of marginalized groups. The common theme across these arguments is that digital exclusion counts as a democratic negative or problem, impeding the ability of marginalized people to collectively self-determine and lead the lives they have reason to value (Sen 1999; Young 2000).

These distinct forms of exclusion interact in important ways. For example, as already alluded to with the problem of predation and being "online," solving the unavailability or access problem can be a boon to digital and data-driven technology companies, by allowing them to monetize customer data.[7] The power of tech elites in local housing markets might also be a boon for digital surveillance of marginalized groups—both those who are welcome and those whom tech elites wish to keep out. The point here is not to draw a causal link between different kinds of digital exclusion but rather consider the ways they can complement and reinforce one another, align interests of the powerful, and accrue consequences for members of marginalized communities.

But as mentioned earlier, the negative impacts of such exclusions beg the question of alternatives versus inevitabilities. If we return to critiques of communicative justice and, in particular, Fraser's argument about the potential benefits of self-exclusion, we can begin to consider how digital exclusion might represent a collective, agentic form of action. This type of refusal does not entail a simple rejection or translate to a person being antitechnology. Digital exclusion, when done as a form of informed refusal, implies agency, activeness, and willingness to confront and transform the terms and conditions of technology adoption and technological control.

In this final section, I explore digital exclusion in an affirmative, democratically valuable sixth sense by presenting a case from the city of Detroit. This case, I suggest, underscores the ways in which refusal appears in many different forms. Refusal is willfully chosen by both self-preserving individuals as well as collectivities—in the case of Detroit, efforts mobilized primarily by community-based organizations. Behind these diverse examples of refusal lies a common theme: refusal carries with it the potential for self-creation and self-determination. Refusal represents a critical deviation from the negative appraisals of digital exclusion and, instead, involves a reconfiguration of the conventional terms and conditions that private or government actors present to members of marginalized communities. Refusal reveals not a democratic problem but a democratically valuable mechanism to overcome oppression, resist structural injustice, and pursue self-determination.

A SIXTH SENSE: REFUSAL ON THE PATH TO "WHAT WE WANT"

My case puts Detroit residents and community-based organizations at the center of attention.[8] Detroit provides an opportunity to reflect on the normative basis upon which to broaden the idea of digital exclusion and to further conceptualize technologies of communicative justice. The case is both complex and different from the examples I have offered so far. It is complex because work in the community has a prehistory and continues to evolve. The case also differs in that it requires a deeper level of specificity to help explain the nuances of a politics of technological refusal.

Three points about Detroit's history provide helpful context. First, Detroit—the fourth-largest city in the United States—has been a crucible of transformative politics and participatory democracy (J. Boggs and Boggs 2008), where residents routinely invoke the saying that African American folk have been making a way out of no way (Swann-Wright 2002). This "no way" has resulted from what community leaders identify as fissures in a capitalist political economy that declined throughout the twentieth century (J. Boggs and Ward 2011), the Great Recession in 2008, and the city's bankruptcy filing in 2013. Detroit's strong culture of community organizing is reflected in its urban gardening movement, citizenship schools, Detroit Summer, and prodigious arts and music community (G. L. Boggs and Kurashige 2012; Kurashige 2017).

Second, and related, Detroiters continually fight a narrative that excludes the city's residents, obscures their presence, or distorts their voice. For example, in 1967, following what is known as the Detroit Rebellion, the city transitioned from majority white to majority black and soon after elected the city's first Black mayor (Kurashige 2017). Amid these historic changes, Detroit residents have been the subject of media stereotypes and a dominant narrative that portrays them as "dumb, lazy, happy, and rich" (Clifford 2017). In the wake of the Great Recession of 2008 and the city's bankruptcy, when the city convulsed from rapid foreclosures, evictions, and deterioration of public services and social support systems,[9] community leaders noted that this disparaging narrative reached near global acceptance. Meanwhile, systematic disinvestment and discrimination against residents occurred with little notice from the rest of the world (Howell 2014).

Third, a more recent narrative around rebuilding Detroit touts the city's "comeback" or "rebirth." This narrative neglects the city's most marginalized populations while celebrating corporate-led initiatives, especially those led by billionaire Dan Gilbert. A longtime Detroiter, Gilbert moved Quicken Loans, the country's largest online mortgage lender, a data-driven technology company, to downtown Detroit. News media and policy makers credit Gilbert with attracting businesses and preserving architectural landmarks in the downtown area, as well as generating much-needed tax revenue for the city (Felton 2014). But, as community leaders argue, this narrative omits important details about the dependency of redevelopment initiatives on the extraction of public resources, Quicken Loans' fraudulent actions, and the return of redlining practices (Smith, LaFond, and Moehlman 2018; Creswell 2017; Felton 2014). Gilbert has also spearheaded the deployment of a privately governed surveillance system called Project Green Light in the city's downtown (Hunter 2015; Kaffer 2015; see also Detroit Community Technology Project 2019a).

Against this backdrop, several community-based projects or organizations have emerged as agents of technological refusal. Launched in 2009, the Detroit Digital Justice Coalition serves as a space for residents and organizational representatives to develop and enact strategies supporting communication as a human right. One of its signature initiatives involved a community visioning and education project focused on the future of Detroit, including Detroit Future Media, a digital literacy program (Allied Media Projects 2015). In time, Detroit Future Media evolved into the Detroit Community Technology Project, which focuses on using and developing "technology rooted in community needs that strengthens human connections to each other and the planet" (Detroit Community Technology Project 2019c). As the quote suggests, Detroit Community Technology Project embraces an affirmative view of technology, especially the ways in which technology can help individuals and communities be and relate to one another. It is an affirmation that comes with refusal of the terms and conditions of technological access, technology adoption (both forced and voluntary), and of technology companies' power.

Four kinds of refusal are evident in the history of these projects and groups. The first refusal refers to contesting the dominant political discourse and redefining narratives about the city. Shortly after the financial crisis, the city's bankruptcy, and the hemorrhaging of the local real estate market due to foreclosures and evictions, Detroit Future Media launched an initiative that combined access and literacy arguments with a distinctive counternarrative to the city's redevelopment frame. With the Detroit Digital Justice Coalition, Detroit Future Media engaged in a widespread social media campaign urging residents to articulate their visions of Detroit's future online. Concurrently, members of the coalition developed and promulgated a set of digital justice principles about the value of broadband infrastructure in their communities. Coming from education, environmental justice, and digital storytelling backgrounds, members injected intergenerational perspectives into the principles and infused them with an awareness of distorting narratives, economic deprivation, and residents' historical struggles in dealing with government and private industry. Together, they helped to generate neighborhood interest in community-built wireless networks, which would alleviate dependence on conventional internet service providers, and encouraged Detroiters to digitally narrate the kind of city they envisioned for the future.

A second kind of refusal focuses on community ownership of digital infrastructure. This refusal builds from efforts to challenge mainstream narratives and offer alternative visions of Detroit's future. In the years following the creation of digital justice principles, Detroit Community Technology Project advocated for and began to develop its own wireless infrastructure, whose main aim is community self-sufficiency, and to bypass broadband service providers. Over time, the project has evidenced a commitment to building autochthonous infrastructure on the community's own terms while refusing the imposed infrastructure of others. With philanthropic funding, the organization established the Equitable Internet Initiative, purchasing high-speed fiber with the aim of redistributing connectivity to digitally redlined neighborhoods. Expanding digital literacy efforts described earlier, Detroit Community Technology Project also developed a community stewardship model that trains residents in these same neighborhoods to build and maintain networks and teach and share digital literacy skills. While the Equitable Internet Initiative continues to evolve its infrastructure efforts, it remains committed to confronting the challenges of broadband unavailability and use while adhering to principles of community health and sustainability as a path to community self-reliance.

The third refusal refers to rejecting and resisting one's data profile. As part of Our Data Bodies, a collaborative research project (of which I am part), Detroit Community Technology Project interviewed people to understand how data and data-driven systems have an impact on their daily lives. Several Detroiters talked about data errors and the way they refused to let those errors define them. Maria, a research participant, was traumatized and threatened with financial ruin after tax assessors misclassified the property she lives on. Eventually, she succeeded in penetrating city bureaucracy and found someone who would correct the error, but not without considerable stress and anxiety. For Detroiters like Maria, the asymmetrical power of data-driven systems is inescapable. These systems are physically occupying or tied up in the process of planning to occupy swatches of land where Detroiters live. These systems are interfering with material needs, including access to shelter and access to water.

A fourth refusal refers to blocking technology deployment. As mentioned earlier, Dan Gilbert spearheaded Project Green Light, an extensive surveillance network throughout the downtown area. Recently, touting the success of this project, the city announced that it would install five hundred surveillance cameras to monitor traffic. Along with allies and collaborators, and influenced by the insights of Our Data Bodies, Detroit Community Technology Project alongside the Detroit Digital Justice Coalition has mobilized to protest the deployment and resist the inevitability of a surveilled cityscape. While, at the time of writing, the case is still unfolding, it nevertheless demonstrates the early, if not ongoing, steps of refusing to adopt coercive surveillance technologies.

In the examples presented here, the mode of refusal is by no means an extreme, an act of pure disengagement or full-scale rejection of digital technology. Rather, the refusal that organizers and residents engage in rests on the idea of autonomy and agency. The projects of digital literacy, of infrastructure building, of contesting data errors, and of blocking technology deployment challenge an array of problems that tie to the logics of the broadband marketplace, the data economy, surveillance, and marginalization more generally. Detroit residents are refusing the terms and conditions by which technologies are introduced and then operated in relation to processes of social and economic marginalization.

At the same time, the examples suggest different styles of self-exclusion and self-determination. Refusal to accept dominant narratives evokes Fraser's idea of counterpublics, where members of marginalized groups create an enclave in order to develop and actualize a different vision of the world. Refusal as resisting and righting data profiles occurs at an individual level and represents a form of individual self-creation. Meanwhile, refusing the mainstream (broadband) networks and rejecting (surveillance) technology attempts to interrupt dominant technological infrastructures rather than narratives or histories. Overall, the differences seem to underscore a diversity and inventiveness in acts of refusal.

Consistent with other writings on refusal, we can also view these refusals as necessarily incomplete and sometimes accommodating as opposed to all-encompassing. One could easily argue that building out one's own network or developing an awareness of tracking, targeting, and surveillance fails to tackle problems of misinformation and social media or does not dent the social, economic, or political power of companies like Facebook. One could also argue that blocking surveillance camera deployment will do little to eliminate surveillance technology used by other government and commercial-run services that marginalized people. But Detroit's story provides an opening. It inspires reflection about the political possibilities emergent from communities of users rather than the inevitability of technology as dictated by elites. And it challenges arguments that claim that civil disobedience is being engineered out of technological designs (Hildebrandt 2015). Detroit's story teaches us that there is more to discover in how we contest and challenge technologies and the institutions and people behind them.

In total, the sentiments and efforts of community-based groups and of individuals and communities in their orbit demonstrate the power of refusal. Their many acts of refusal present an opportunity to reflect on what kinds of technologies and supportive infrastructures ought to be created or what kinds of strategies individuals and groups can use to counter subordinating and controlling impacts of technologies in our lives. For members of marginalized communities who have experienced a variety of digital exclusions over the past few decades, refusal implies creativeness or generativity and speaks to both immediate and more long-term visions about how the world can be and what needs to change in order to transform these visions into reality.

Conclusion

Digital exclusion remains a multifaceted problem in the twenty-first century. This is due in no small part to the fact that digital technologies have transformed, become more sophisticated, and intersect in complicated ways with how we communicate and relate to one another in society. Within this environment, the prospects for communicative justice have become more complex as well.

But although it is tempting to focus on the negative, by arguing that members of marginalized communities face greater hardship when they become embedded in the vast array of sociotechnical systems that crisscross our daily lives, or advocating that members of marginalized communities should drop out or stay away from digital technologies, I have proposed an alternative pathway to communicative justice in the twenty-first century. Specifically, I have shown digital exclusion in a different light, where people willfully and knowingly exclude themselves and refuse technologies in the terms and conditions presented to them by governments or private actors. Such acts function as a generative force of self-creation and self-determination. By making refusal become a part of the lexicon of digital exclusion, we can begin to understand the ways in which exclusion can help us imagine other ways of being and relating to one another in a technologically mediated society.

Democratizing digital technologies is not only about policy makers figuring out how to broaden access to technology or increase people's skills to meaningfully use technology. Neither is it only about benevolent designers, engineers, or data scientists capable of installing norms and values into technological systems. Those most affected by technologies' potentially negative and marginalizing impacts have a role to play, too, in intervening in the social, economic, and political contexts in which digital technologies are embedded. We should recognize, anticipate, and learn from messy and diverse acts of technological refusal. To ignore this is to deny marginalized people's agency and their capacity to reinvent the world around them.

Notes

1. Young (1990, 39) calls exploitation, marginalization, powerlessness, cultural imperialism, and violence "five faces of oppression."

2. See Freire (1996) for a larger discussion of how the privileged and the marginalized both play a part in rectifying historical inequities.

3. As Morozov (2019) points out, these terms draw from very different intellectual traditions. I present them together here only to acknowledge them over time.

4. Although all consumers face the perils of being tracked and targeted, for marginal internet users, the risk of injury or harm due to such tracking and targeting runs high. For example, as Gangadharan (2015) writes, marginal internet users (e.g., first-time users coming from marginalized communities) have difficulty discerning the credibility of different services, following one click to the next, not realizing they do not or should not divulge personal information. A form purporting to be from a potential employer, online educational company, or resume-building service can serve as a pathway to identify theft (Petty et al. 2018). Given that the institutions that help to welcome members of marginalized communities to the internet and computers lack the means to address the downsides of being connected and online, the possibilities for rebounding or recovering are daunting (Gangadharan 2015).

5. We have to look only to previous eras and remember the ability of railroad, oil, or automobile companies to affect the distribution of wealth and privilege in society. Technology compa-

nies (internet or otherwise) bear resemblance to these older titans in that they routinely marshal resources and circulate industry narratives in ways that deeply affect conditions of marginality.

6. Although approximately 5 percent of the $500 million qualifies as charitable grant giving, Microsoft stands to make a profit on the remaining $475 million, earmarked for both market-rate and below-market-rate loans for housing construction. Five percent equals $25 million, which is approximately one-quarter of 1 percent of the $9 billion in corporate tax that Microsoft could have paid during its acquisition of LinkedIn.

7. For example, investigative journalists found that AT&T, T-Mobile, and Sprint, companies that support bridging the so-called digital divide, engaged in practices that hit its disadvantaged customers. Specifically, these companies sold cell-phone tower data and Global Positioning System (GPS) data to an intermediary company, which sold data to a reseller, which sold data to bounty hunters and bail agents (Koebler and Cox 2019). Additionally, and at the time of writing, federal regulations permit broadband providers to sell web browsing history to third parties such as data aggregators and data brokers. Privacy advocates have claimed that such lax rules make it easier for companies to profit off poor and marginalized customers, whose data profiles are valuable to predatory businesses (Moy 2017).

8. Following Stake's (2008) framework for qualitative case study research, I focus on Detroit to deepen understanding of technological refusal. I draw from a variety of sources, including in-depth interviews, news reports, and documents produced by community organizations (see, e.g., Detroit Community Technology Project 2019a, 2019b; Allied Media Projects 2013, 2015, n.d.; Detroit Digital Justice Coalition 2010; We the People of Detroit 2016; Petty et al. 2018).

9. The city had already been in decline due to the slow collapse of the automobile industry. When the city filed for bankruptcy, it faced additional surmounting challenges. Public infrastructure fell into emergency management, forcing utility companies, for example, to pursue aggressive cost-saving measures, the burden of which fell on low-income residents (Beydoun 2014; Goodman 2014; Gottesdiener 2014; Hackman 2014).

References

Abbate, Janet. 1999. *Inventing the Internet.* Cambridge, MA: MIT Press.

Acquisti, Alessandro, Curtis R. Taylor, and Liad Wagman. 2016. "The Economics of Privacy." *Journal of Economic Literature* 54 (2): 442–92. https://doi.org/10.1257/jel.54.2.442.

Allied Media Projects. N.d. "Detroit Digital Justice Coalition: A Vision for Digital Justice." https://www.alliedmedia.org/ddjc/story.

———. 2013. *DFM Story.* YouTube video. Posted by alliedmedia. https://www.youtube.com/watch?feature=player_embedded&v=E-POp-f2XHU.

———. 2015. "Detroit Future." Allied Media Projects. January 16, 2015. https://www.alliedmedia.org/detroit-future.

American Recovery and Reinvestment Act. 2009. Pub. L. No. 111-5. https://www.congress.gov/bill/111th-congress/house-bill/1/text.

Baynes, Leonard M. 2004. "Deregulatory Injustice and Electronic Redlining: The Color of Access to Telecommunications." *Administrative Law Review* 56 (2): 263–352.

Benjamin, Ruha. 2016. "Informed Refusal: Toward a Justice-Based Bioethics." *Science, Technology, & Human Values* 41 (6): 967–90. https://doi.org/10.1177/0162243916656059.

———. 2019. *Race after Technology: Abolitionist Tools for the New Jim Code.* Cambridge, MA: Polity Books.

Beydoun, Khaled A. 2014. "Detroit's Water: Behind the Crisis." *Al Jazeera*, July 27, 2014.

https://www.aljazeera.com/indepth/opinion/2014/07/detroit-water-behind-crisis-2014727123440725511.html.

Boggs, Grace Lee, and Scott Kurashige. 2012. *The Next American Revolution: Sustainable Activism for the Twenty-First Century.* Berkeley: University of California Press.

Boggs, James, and Grace Lee Boggs. 2008. *Revolution and Evolution in the Twentieth Century.* New York: Monthly Review Press.

Boggs, James, and Stephen M. Ward. 2011. *Pages from a Black Radical's Notebook: A James Boggs Reader.* Detroit: Wayne State University Press.

Bridges, Khiara M. 2017. *The Poverty of Privacy Rights.* Stanford, CA: Stanford Law Books.

Brown, Ian. 2015. "The Economics of Privacy, Data Protection and Surveillance." In *Handbook on the Economics of the Internet*, edited by Johannes M. Bauer, 247–61. Cheltenham, UK: Edward Elgar. http://papers.ssrn.com/sol3/papers.cfm?abstract_id=2358392.

Campbell, Alexia Fernández, and Reena Flores. 2014. "How Silicon Valley Created America's Largest Homeless Camp." *The Atlantic*, November 25. https://www.theatlantic.com/politics/archive/2014/11/how-silicon-valley-created-americas-largest-homeless-camp/431739/.

Cannon, Katie G. 1995. *Katie's Canon: Womanism and the Soul of the Black Community.* New York: Continuum.

Clifford, Carolyn. 2017. "White Flight and What It Meant to Detroit in the Wake of the 1967 Riots." WXYZ (Detroit), July 11. https://www.wxyz.com/news/detroit1967/white-flight-and-what-it-meant-to-detroit-in-the-wake-of-the-1967-riots.

Creswell, Julie. 2017. "Quicken Loans, the New Mortgage Machine." *New York Times*, January 21. https://www.nytimes.com/2017/01/21/business/dealbook/quicken-loans-dan-gilbert-mortgage-lender.html.

Detroit Community Technology Project. 2019a. *A Critical Summary of Detroit's Project Green Light and Its Greater Context.* Detroit: Detroit Community Technology Project.

———. 2019b. "Equitable Internet Initiative." https://www.detroitcommunitytech.org/eii.

———. 2019c. "Technology Rooted in Community Needs." https://www.detroitcommunitytech.org.

Detroit Digital Justice Coalition. 2010. "Principles." November 24. http://detroitdjc.org/principles/.

DiMaggio, Paul, and Eszter Hargittai. 2001. "From the 'Digital Divide' to 'Digital Inequality': Studying Internet Use as Penetration Increases." Working Paper No. 15, Center for Arts and Cultural Policy Studies, Woodrow Wilson School of Public Administration, Princeton University, Princeton, NJ. https://culturalpolicy.princeton.edu/sites/culturalpolicy/files/wp15_dimaggio_hargittai.pdf.

Eubanks, Virginia. 2011. *Digital Dead End: Fighting for Social Justice in the Information Age.* Cambridge, MA: MIT Press.

———. 2018. *Automating Inequality: How High-Tech Tools Profile, Police, and Punish the Poor.* New York: St. Martin's Press.

Feenberg, Andrew. 1996. "Marcuse or Habermas: Two Critiques of Technology." *Inquiry* 39: 45–70.

———. 2002. *Transforming Technology: A Critical Theory Revisited.* New York: Oxford University Press.

Felton, Ryan. 2014. "What Kind of Track Record Does Quicken Loans Have in Detroit? Does Anyone Really Care?" *Detroit Metro Times*, November 12. https://www.metrotimes.com/detroit/what-kind-of-track-record-does-quicken-loans-have-in-detroit-does-anyone-really-care/Content?oid=2266383.

Fields, Desiree. 2017. "Urban Struggles with Financialization." *Geography Compass* 11 (11): e12334. https://doi.org/10.1111/gec3.12334.

Fisher, Linda E. 2009. "Target Marketing of Subprime Loans: Racialized Consumer Fraud and Reverse Redlining." *Brooklyn Journal of Law and Policy* 18 (1): 101–35.

Foroohar, Rana. 2016. "Microsoft's Massive LinkedIn Deal Is a Sign of Something Dangerous." *Time*, June 14. http://time.com/4368047/microsoft-linkedin-deal-merger-debt/.

Fraser, Nancy. 1990. "Rethinking the Public Sphere: A Contribution to the Critique of Actually Existing Democracy." *Social Text* 25–26: 56–80.

———. 1997. "A Rejoinder to Iris Young." *New Left Review* 1 (223): 126–29.

Freire, Paulo. 1996. "Chapter 1." In *Pedagogy of the Oppressed*, 25–42. London: Penguin.

Froomkin, A. Michael. 2004. "Technologies for Democracy." In *Democracy Online: The Prospects for Democratic Renewal Through the Internet*, 3–20. New York: Routledge.

Gandy, Oscar H. 2009. *Coming to Terms with Chance: Engaging Rational Discrimination and Cumulative Disadvantage*. Burlington, VT: Ashgate.

Gangadharan, Seeta Peña. 2012. "Digital Inclusion and Data Profiling." *First Monday* 17:5–7. https://doi.org/10.5210/fm.v17i5.3821.

———. 2013. *Joining the Surveillance Society?* Washington, DC: New America Foundation. http://web.archive.org/web/20140312042501/http://newamerica.net/sites/newamerica.net/files/policydocs/JoiningtheSurveillanceSociety_1.pdf.

———. 2015. "The Downside of Digital Inclusion: Expectations and Experiences of Privacy and Surveillance among Marginal Internet Users." *New Media & Society* 19 (4): 1–19. https://doi.org/10.1177/1461444815614053.

Gilliom, John. 2001. *Overseers of the Poor: Surveillance, Resistance, and the Limits of Privacy*. Chicago: University of Chicago Press.

Gilman, Michele E. 2012. "The Class Differential in Privacy Law." *Brooklyn Law Review* 77 (4): 1389–1445.

Goodman, Amy. 2014. "Detroit Faces 'Humanitarian Crisis' as City Shuts off Water Access for Thousands of Residents." *Democracy Now!* http://www.democracynow.org/2014/10/10/detroit_faces_humanitarian_crisis_as_city.

Gottesdiener, Laura. 2014. "UN Officials 'Shocked' by Detroit's Mass Water Shutoffs." *Al-Jazeera America*, October 20. http://america.aljazeera.com/articles/2014/10/20/detroit-water-un.html.

Habermas, Jürgen. 1970. *Toward a Rational Society: Student Protest, Science, and Politics*. Boston: Beacon Press.

———. 1989. *The Structural Transformation of the Public Sphere: An Inquiry into a Category of Bourgeois Society*. Translated by Thomas Burger and Frederick Lawrence. Boston: MIT Press.

———. 2006. "Political Communication in Media Society: Does Democracy Still Enjoy an Epistemic Dimension? The Impact of Normative Theory on Empirical Research." *Communication Theory* 16 (4): 411–26.

Hackman, Rose. 2014. "What Happens When Detroit Shuts Off the Water of 100,000 People." *The Atlantic*, July 17.

Helsper, Ellen Johanna. 2012. "A Corresponding Fields Model for the Links between Social and Digital Exclusion." *Communication Theory* 22 (4): 403–26.

Hildebrandt, Mireille. 2015. *Smart Technologies and the End(s) of Law: Novel Entanglements of Law and Technology*. Cheltenham, UK: Edward Elgar.

Hindman, Matthew. 2009. *The Myth of Digital Democracy*. Princeton, NJ: Princeton University Press.

Howe, Louis E. 2003. "Ontology and Refusal in Subaltern Ethics." *Administrative Theory & Praxis* 25 (2): 277–98.

Howell, Shea. 2014. "Distorted Reality in Detroit." *Common Dreams* (blog). July 21. https://www.commondreams.org/views/2014/07/21/distorted-reality-detroit.

Hunter, George. 2015. "Detroit Cops to Get Body, Dash Cams." *Detroit News*, August 18. http://www.detroitnews.com/story/news/local/detroit-city/2015/08/18/detroit-police-body-cameras/31900641/.

Irving, Larry. 1999. *Falling through the Net: Defining the Digital Divide*. Washington, DC: United States National Telecommunications and Information Administration. http://www.ntia.doc.gov/ntiahome/fttn99/.

Kaffer, Nancy. 2015. "Who's Watching the Detroit Watchmen?" *Detroit Free Press*, March 21. http://www.freep.com/story/opinion/columnists/nancy-kaffer/2015/03/21/private-downtown-security-quicken/25117481/.

Kim, E. Tammy. 2019. "Microsoft Cannot Fix Seattle's Housing Crisis." *New York Times*, January 18. https://www.nytimes.com/2019/01/18/opinion/microsoft-seattle-housing.html.

Koebler, Jason, and Joseph Cox. 2019. "Hundreds of Bounty Hunters Had Access to AT&T, T-Mobile, and Sprint Customer Location Data for Years." *Motherboard* (blog), February 6. https://motherboard.vice.com/en_us/article/43z3dn/hundreds-bounty-hunters-att-tmobile-sprint-customer-location-data-years.

Kurashige, Scott. 2017. *The Fifty-Year Rebellion: How the US Political Crisis Began in Detroit*. Berkeley: University of California Press.

Lanier, Jaron. 2018. *Ten Arguments for Deleting Your Social Media Accounts Now*. New York: Henry Holt and Co.

Lee, Una. 2017. *Building a Consentful Tech* (zine). https://www.andalsotoo.net/wp-content/uploads/2018/10/Building-Consentful-Tech-Zine-SPREADS.pdf.

Levin, Sam. 2016. "'Largest-Ever' Silicon Valley Eviction to Displace Hundreds of Tenants." *The Guardian*, July 7. http://www.theguardian.com/technology/2016/jul/07/silicon-valley-largest-eviction-rent-controlled-tenants-income-inequality.

Monahan, Torin. 2008. "Editorial: Surveillance and Inequality." *Surveillance & Society* 5 (3): 11. https://doi.org/10.24908/ss.v5i3.3421.

Morozov, Evgeny. 2019. "Capitalism's New Clothes." *The Baffler*, February 4. https://thebaffler.com/latest/capitalisms-new-clothes-morozov.

Moy, Laura. 2017. *Algorithms: How Companies' Decisions about Data and Content Impact Consumers*. Washington, DC: US House of Representatives. http://docs.house.gov/meetings/IF/IF17/20171129/106659/HHRG-115-IF17-Wstate-MoyL-20171129.pdf.

Norris, Pippa. 2008. *Digital Divide: Civic Engagement, Information Poverty, and the Internet Worldwide*. Cambridge: Cambridge University Press.

Nunez, Michael. 2016. "Facebook Is Building Apartments Anyone Can Rent—But There's a Huge Catch." *Gizmodo*, July 26. https://gizmodo.com/facebook-is-building-apartments-anyone-can-rent-but-the-1784309208.

Pasquale, Frank. 2015. *The Black Box Society: The Secret Algorithms That Control Money and Information*. Cambridge, MA: Harvard University Press.

Petty, Tawana, Mariella Saba, Tamika Lewis, Seeta Peña Gangadharan, and Virginia Eubanks. 2018. *Our Data Bodies: Reclaiming Our Data*. Detroit: Our Data Bodies Project.

Poon, Martha. 2017. "Microsoft's Pivot." Presentation at the Computers, Privacy, and Data Protection conference "Rethinking Big Data," Brussels. Available at https://www.youtube.com/watch?v=bg39dliGlNE.

Powell, Alison, Amelia Bryne, and Dharma Dailey. 2010. "The Essential Internet: Digital Exclusion in Low-Income American Communities." *Policy & Internet* 2 (7). https://doi.org/10.2202/1944-2866.1058.

Roberts, Dorothy. 2019. "Digitizing the Carceral State." *Harvard Law Review* 132 (6): 1695–1732. https://harvardlawreview.org/2019/04/digitizing-the-carceral-state/.

Sen, Amartya. 1999. "Freedom and Foundations of Justice." In *Development as Freedom*, 54–86. New York: Knopf.

Slogobin, Christopher. 2003. "The Poverty Exception to the Fourth Amendment." *Florida Law Review* 55 (1): 391–412.

Smith, Lindsey, Kaye LaFond, and Lara Moehlman. 2018. "Data Analysis: 'Modern-Day Redlining' Happening in Detroit and Lansing." *Michigan Radio*. https://www.michiganradio.org/post/data-analysis-modern-day-redlining-happening-detroit-and-lansing.

Stake, Robert E. 2008. "Qualitative Case Studies." In *Strategies of Qualitative Inquiry*, edited by Norman K. Denzin and Yvonne S. Lincoln, 119–49. Los Angeles: Sage Publications.

Swann-Wright, Dianne. 2002. *A Way out of No Way: Claiming Family and Freedom in the New South*. Charlottesville: University of Virginia Press.

Turner, Fred. 2006. *From Counterculture to Cyberculture: Stewart Brand, the Whole Earth Network, and the Rise of Digital Utopianism*. Chicago: University of Chicago Press.

Urban Displacement Project. 2015. "Urban Displacement San Francisco Map." http://www.urbandisplacement.org/map/sf.

Vaidhyanathan, Siva. 2019. *Antisocial Media: How Facebook Disconnects Us and Undermines Democracy*. New York: Oxford University Press.

Vertesi, Janet. 2014. "Internet Privacy and What Happens When You Try to Opt Out." *Time*, May 1. http://time.com/83200/privacy-internet-big-data-opt-out/.

Warschauer, Marc. 2002. "Reconceptualizing the Digital Divide." *First Monday* 7 (December). https://doi.org/10.5210/fm.v7i7.967.

Wessler, Hartmut. 2018. *Habermas and the Media*. Medford, MA: Polity Press. http://public.eblib.com/choice/PublicFullRecord.aspx?p=5630504.

We the People of Detroit. 2016. *Mapping the Water Crisis*. Detroit: We the People of Detroit. https://wethepeopleofdetroit.com/communityresearch/water/.

Wiley, Maya, Greta Byrum, Michelle Ponce, and Oscar Romero. 2018. *Take It or Leave It: How NYC Residents Are Forced to Sacrifice Online Service for Internet Service*. New York: Digital Equity Lab at the New School. https://www.digitalequitylab.org/take-it-or-leave-it/.

Yeung, Karen. 2015. "Design for the Value of Regulation." In *Handbook of Ethics, Values, and Technological Design: Sources, Theory, Values and Application Domains*, edited by Jeroen van den Hoven, Pieter E. Vermaas, and Ibo van de Poel, 447–72. Dordrecht, Netherlands: Springer.

Young, Iris Marion. 1990. *Justice and the Politics of Difference*. Princeton, NJ: Princeton University Press.

———. 2000. *Inclusion and Democracy*. New York: Oxford University Press.

———. 2001. "Equality of Whom? Social Groups and Judgments of Injustice." *Journal of Political Philosophy* 9 (1): 1–18. https://doi.org/10.1111/1467-9760.00115.

Zittrain, Jonathan. 2009. *The Future of the Internet and How to Stop It*. London: Penguin.

Zuboff, Shoshana. 2019. *The Age of Surveillance Capitalism: The Fight for a Human Future at the New Frontier of Power*. New York: PublicAffairs.

5

Presence of Absence: Exploring the Democratic Significance of Silence

Mike Ananny

A theme seems to run through democratic self-governance that goes like this: people discover and manage shared social conditions by deliberating among themselves, devising solutions, and choosing representatives to speak on their behalf. A key feature of this idea is that people have a duty to express themselves in one or more ways: they speak face-to-face, create media (letters, television shows, social media posts), circulate their ideas, and build institutions that encourage and protect expression. The answer to bad speech is more speech; if you aren't speaking you aren't participating; the onus is on individuals to speak up.

My focus in this chapter is what this dominant view neglects and sometimes dismisses: that the silences and absences that accompany listening, observation, and reflection are also key to communicative self-governance. In particular, I want to critique the assumption that participation means speaking and sketch the idea—using the theories of material publics increasingly common among communication and science and technology scholars—that legitimate democratic participation includes making publics in ways that do not always leave visible, material traces. Speakers need listeners, but because listeners are often invisible, they can go unseen by political theorists and technologists alike and fail to appear in the communication systems and institutions that communicative self-governance requires.

This means seeing silence as political communication. Sometimes that silence is coerced and oppressive, with speakers and their voices actively censored, but sometimes that silence is voluntary and part of listening. Both types of silence—coerced and voluntary—are critical to understanding the democratic power of absence, but I try here to be somewhat agnostic and disengaged from the question of when silence should be forced or chosen.

Instead, I try to explicate absence as a little-seen type of political participation that underpins the communication infrastructures shaping much of contemporary online life. I argue that absence is a type of public participation that—when seen and valued—offers new ways to think about what's being communicated, by whom, to whom, in what contexts, and with what effects. Appreciating the complexity and force of silence can help us see digital publics differently. Since its early incarnations, scholars and advocates alike have heralded the internet as a new type of public space in which participation through digital information represents new types of constituencies and forms of representation.

The internet has often and optimistically been seen as a way to speak—to share opinions, form associations, experiment with new ideas, uncover forms of power, and advocate for change. Despite the early warnings of often marginalized critics about the internet's structural inequalities and the challenges of too neatly mapping digital technology, network associations, and distributed mediation onto political agency (Agre, 2002; Baker, 2002; Barber, 2004; Brook & Boal, 1995; Dahlberg, 2001; Gandy, 1993; Hindman, 2008; Silver & Massanari, 2006), the early, dominant image of the internet was as a flat, open, networked, marketplace-like space (Benkler, 2003, 2006; Shirky, 2008) in which low costs of information production and circulation were presumed to translate into robust political life (Sunstein, 2001). A perfect storm of undervaluing listening arose when these celebrations of digital speech were joined by a burst of scholarly interest in participatory cultures and online communities (Gillmor, 2004; Jenkins, 2006; Rheingold, 1993); the idea that online listening was unethical "lurking," free-riding, and a participation problem to be solved (Bishop, 2007; Preece, Nonnecke, & Andrews, 2004; Sun, Rau, & Ma, 2014); and commercial imperatives of advertising-fueled technology companies eager to mine user-generated content for patterns that they could use to commodify users. Speaking, expression, and actions that left visible traces were privileged over listening, observation, and participation that couldn't be easily commodified. Online participation was framed as the visible traces—texts, sounds, images—of the content creator. Silences and absences were politically suspicious, evidence of Luddites, went against the grain of participatory culture, and were commercially unprofitable. The dominant image of the digital public sphere was as a place for speakers, not listeners.

What would it mean to take seriously silence and absence as critical dynamics of political participation—and try to find evidence for them in today's communication infrastructures? Aligned with other authors in this volume—such as Ford's use of self-organizing technological systems to reimagine core principles of democratic governance (chapter 10), and Lee, Levi, and Brown's vision of a community-centered and computationally supported collaborative ecosystem (chapter 8)—I want to explore the silences and absences of sociotechnical systems as ways to think differently about democratic, communicative self-governance.

I try here to trace the significance of silence by showing how it appears in both voluntary and involuntary dimensions of communication systems. In the first part of the chapter, through a series of examples, I try to show that—intentional or otherwise—silence exists in the actions of individuals, media-making cultures, and communication infrastructures. The second part of the chapter frames political participation as sociotechnical communication—as intertwined moments of individual expression and technological engagement. Seen in this way, we might reconsider contemporary, sociotechnical silences as legitimate forms of political participation that exist alongside more visible, marketplace models of communicative self-governance that tend to focus on the production and circulation of speech.

The Presence of Absence

Before examining absence as a sociotechnical phenomenon, it is worth reviewing how absence has appeared historically in different forms of political participation and association.

For example, conscientious objectors purposefully violate the law motivated by the belief that they are "morally prohibited to follow the law because the law is either bad or wrong." (Brownlee, 2013, n.p.) When in the context of refusing to obey a law, examples include draft dodging or refusing to pay taxes; or absence may surface in objections against court directives, such as Jehovah's Witnesses opting out of saluting a flag. The key dimension of conscientiously refusing a law or legal directive—versus conscientiously evading the same—is that there is a visible, explicit declaration of the refusal's meaning. The absence of compliance appears in a statement or behavior, either as an individual action or as a coordinated, collective refusal (Brownlee, 2013; Raz, 1979). Such absences and refusals are concrete and visible counterforces designed to visibly resist dominant forms of power and rally others into conscientious objection.

Another type of political absence exists as associational invisibility or self-isolation. Catherine Squires (2002) describes the creation and sustenance of black "enclave publics" by abolitionists and slaves in the US South who needed privacy and seclusion to develop secure systems of communication. She also finds "counterpublics" in the US civil rights movement who wanted to tightly control their visibility and legibility in order to foster strategic communities of resistance that could ensure solidarity and minimize debate. And she further discovered "satellite publics" created by, for example, the Nation of Islam, to purposefully avoid inclusion in wider debates that they saw as incompatible with the threats of violence and civil disruption that they advocated. In each case, these publics used strategic invisibility and isolation to nurture memberships and associations, and achieve specific forms of social change. What seemed like silence, invisibility, or isolation was actually strategically and selectively (in)visible participatory politics.

Absence can also appear in voting behaviors. Adopting an economic model of participation and information, Downs (1957) suggests that people may rationally choose to be ignorant because the "saving a consumer could make by becoming informed about how government policy affects any one product he purchases simply does not recompense him for the cost of informing himself—particularly since his personal influence on government policy would probably be slight" (p. 149).[1] Schumpeter similarly anticipated the high cost of navigating surfeits of political information, suggesting that nonvoting was an understandably apathetic response in an environment that was simply too complex to navigate and that individuals stood a small chance of influencing. It made sense not to participate. A lack of voting and broader political participation can also be read as trust in governing elites, involuntary structural exclusion (e.g., based on race, gender, or class), oppressive voter registration requirements, the absence of a mobilizing community, a perceived lack of relevance of political life and parties, and skepticism about the efficacy of participation (DeLuca, 1995; Held, 2006). Even the visibility of voting and campaigning has changed. Voting first appeared as vocal, public pronouncements, then moved to secret ballots (Schudson, 1998), and now is signaled through "I Voted" stickers and Facebook badges as well as rules barring campaign activities around polling places and photographing ballots. As DeLuca (1995) and Eliasoph (1998) show, entire models of power and politics can be developed by tracing the absence of voting, antipathy toward political action, or retreat from civic communication. Indeed, as Wells et al. (2017) found, there came a point during the recall of Wisconsin governor Scott Walker when "many citizens found it impossible to continue political discussion . . . call[ing] into question the ability of talk to bridge political and social differences in periods of polarization and fragmentation." (p. 131)

Conscientious objection, enclave publics, and voting abstentions are largely voluntary forms of absence—people choose to opt out of a political regime, association, representation, or electoral process. But there are also involuntary political absences. Most obviously, people can be disenfranchised, as was the case in many Western democracies that long disallowed voting by anyone but white, landowning males. At various points in the history of national censuses (M. J. Anderson, 2015), survey techniques (Igo, 2007), and information archiving (Carter, 2006), the creation, deletion, and wording of bureaucratic categories effectively erased many people's identities, histories, and demographics from official public records. They were involuntarily made invisible to information systems and not counted as political constituencies (B. Anderson, 1983). In these cases, absences and silences were not evidence of rational ignorance, conscientious protest, or strategic self-isolation. They were officially sanctioned erasures that showed who had the power to define, categorize, and include—a kind of "associational suppression" that created political constituencies and set the terms of political participation.[2]

This kind of structural, associational suppression appears in Gangadharan's account of digital exclusion (chapter 4 in this volume). With contemporary Detroit's socioeconomic transformations as her backdrop, she argues that instead of seeing digital exclusion as simply a lack of broadband connectivity—a problem to be overcome with more technology—the online invisibility of marginalized groups is actually evidence of more complex exclusionary dynamics. Through "internal exclusion" processes, even though "marginalized people might be able to access the internet or successfully adopt broadband . . . a combination of socioeconomic factors interferes with their ability to use broadband and participate meaningfully in internet culture," Gangadharan shows. Their ways of communicating are not supported, their concerns are minimized, their associations are not represented. They may have broadband connections, devices, and accounts, but their presence is always potential—or a small fraction of what it could be—because their inclusion is narrowly defined as technology access, not communicative power. Their invisibility and suppression can never fully be redressed in purely technological images of connectivity.

Although such absences have different motivations and forms, they share orientations to agency and structure, best explainable as structuration: agency is not an inalienable feature of an individual but rather a set of dynamic forces that create an individual "capacity for agency" out of intertwined social structures, cognitive schemas, and organizational resources (Giddens, 1984; Sewell, 1992). We might see absence similarly. You have both an individual capacity to make absences and an inescapable experience of having absences imposed upon you. Some absences emerge from individual choices not to participate, such as calculating that participation is simply irrelevant or not worth the effort. Others come from institutions and broader structural forces: voting policies that exclude, political parties that devalue constituents, ballots that must be secret, and campaigning that is barred at certain times and places.

In an age when much political participation happens through digital communication systems, how should we think about the voluntary and involuntary dynamics of absence?

Before moving on to a broader discussion of publics and public participation as intertwined social and technological absences, it is worth looking closely at how digital communication systems create silences. As much as digital materials, objects, and infrastructures can be ways to trace political power (e.g., how data create identities, how algorithms structure choices, how platforms moderate content), they can also be read for absences: moments when people, places, and perspectives are not rendered in information infrastructures, do not take visible or recognizable forms, or are hidden from or silenced within public discussions. I elsewhere call this a "whitespace press" (Ananny, 2017) in which a mix of journalistic practices, traditions, institutions, and digital infrastructures—explicitly and implicitly, intentionally and unintentionally—create absences. Such absences—everything from pauses in interviews and uncovered topics to outright censorship and algorithmically blocked content—are not simply failures to speak but are, most broadly, opportunities to rethink political participation in digital communication contexts.

Such absence is participation and political communication that is best understood by tracing how human and computational power work together. These sociotechnical absences can appear as individual actions (behaviors, devices, and choices of media audiences and technology users), decisions by communication creators (journalists who create news and social media platform designers who create the conditions under which news circulates and is interpreted), and infrastructural dynamics (largely invisible intersections of semi-autonomous technologies, design assumptions, and relational databases that regulate how information is produced and made visible). At each site, we might consider the work behind these sociotechnical absences and ask what they mean for communicative self-governance and political participation.

In the following three sections I examine these sociotechnical absences and, using examples, consider the role that each plays in structuring contemporary, networked, digital political participation.

INDIVIDUAL BEHAVIORS

Some sociotechnical silences are created because people disconnect from technology, isolate themselves from communication systems, and seek out media-free spaces. Such absences take different forms. Approximately 70 percent of Americans report some kind of "news fatigue" (Gottfried & Barthel, 2018), and many report complex feelings of exhilaration, guilt, and relapse when they abandon and return to social media technologies (Baumer, Guha, Quan, Mimno, & Gay, 2015). There is a growing community of reformed technologists calling on technology companies to fix the addictive qualities of their products (Lewis, 2017); device makers are beginning to help people limit technology through designs that create pauses and "media prophylaxis" that anticipates and ameliorates the harms of addiction (Mulvin, 2018); and scholars are increasingly following the suggestions of early internet researchers (Oudshoorn & Pinch, 2003; Wyatt, Thomas, & Terranova, 2002) to map the complex history and phenomenology of individuals' technology avoidance, reuptake, and addiction (Baumer, Ames, Burrell, Brubaker, & Dourish, 2015; Plaut, 2015).

The communication spaces that ostensibly create rich democratic self-governance are targets of nonuse. Some young people avoid in-depth engagements with news, instead seeking out "functional information alternatives" to headlines they see and topics they see journalists covering. For example, instead of reading news stories about international affairs that headlines tell them are relevant and timely, they might instead watch a television show or movie on a similar topic or read a Wikipedia page about the relevant area (Edgerly, 2016). Often called "media sabbaticals," these can be triggered by "message fatigue" and a desire to avoid persuasive environments (So, Kim, & Cohen, 2016) and news contexts altogether (Ksiazek, Malthouse, & Webster, 2010). And such news avoidance is gendered, with working- and middle-class women often reporting that "news is for men," that divisions of labor within their households leave them with less time to consume news, and that the care work they are disproportionately responsible for harms their ability to keep up with news that is often seen as negative or depressing (Toff & Palmer, 2018). Women may be less likely to appear in surveys of news audiences because news as a genre is incompatible with highly gendered work and home environments.

Recent work also finds that a fear of online social isolation—of saying the wrong thing and finding oneself banished from Facebook spaces (Chan, 2017), or talking about topics thought to be too politically sensitive (Hampton et al., 2014)—drives people to self-censor, retreat from online forums, seek out positive expressions, and create entirely new "spiral of silence" dynamics that make them less likely to speak if they think their opinions are in the minority or are soon to be (Askay, 2015; Soffer & Gordoni, 2017). In the aftermath of Trump's election, Philippa Lockwood told the *Guardian*:

> Immediately after the election, I had to take a break from social media. These days, I'm back on, but making an effort to focus my time on positive people and reputable, newsworthy posts. As counter-productive as this may seem, I'm also surrounding myself with people who share my liberal political beliefs, who see the value in immigrants, who believe in women's rights, etc. I know we're supposed to be 'bursting our political bubbles', but I find it more rewarding to share with likeminded souls, and use our time and energy to mobilize those like us to shake things up in the next election cycle. (Eberspacher & *Guardian* Readers, 2017)

Other more seemingly trivial examples illustrate the extent to which individuals are creating or choosing low- or no-media environments. People are asking for and patronizing facilities that ban cable news from its televisions; Life Time Fitness removed CNN, Fox News, MSNBC and CNBC from all of its 128 gyms in the United States and Canada (Schmidt, 2018). People engage in a wide variety of selective avoidance techniques online, unfollowing, unfriending, and blocking people they disagree with, especially during times of acute political conflict (Bode, 2016; John & Dvir-Gvirsman, 2015; Zhu, Skoric, & Shen, 2017)—although a federal judge recently ruled that such blocking by Trump and other public officials unconstitutionally sequesters them from the publics they are meant to hear (Herrman & Savage, 2018). Finally, *Disconnect* magazine (Ingram, 2018) is an online publication that is free only to readers who disconnect their devices from the internet. It simply fails to load until people disable the WiFi on their laptops, phones, or tablets, giving the magazine their entire attention, effectively banishing multitasking and other forms of mediated interactivity while reading.

Idiosyncratic and fun, deeply structural and politically grave, these are new sociotechnical forces of absence. From news avoidance and technology sabbaticals to self-censorship and selective exposure, new forms of demediation give individuals a great deal of power to tailor their media environments. Such absences are related to, but distinct from, personalized echo chambers and algorithmic filter bubbles—they are, instead, spaces of silence through which people can isolate themselves.

MEDIA-MAKING CULTURES

A second type of sociotechnical absence centers on cultures of professionalized media making. These are populated and defined by a broad set of actors I call "communication creators": professional producers who create online content (e.g., journalists who write stories for mainstream news outlets, advertisers who create messages) and designers who create conditions under which online content circulates and is interpreted (e.g., user experience and news-feed algorithm designers at Facebook or Google). In the traditions, practices, and cultures of such creators and designers, we can see the formalization and normalization of decisions about what not to show, what not to algorithmically surface.

In her study of journalists' engagements with online troll communities, hate groups, and disinformation networks, Phillips (2018) found that the "choices reporters and editors make about what to cover and how to cover it play a key part in regulating the amount of oxygen supplied to the falsehoods, antagonisms, and manipulations that threaten to overrun the contemporary media ecosystem" (p. 2). This question of whether to oxygenate or suffocate such harmful content played out as professional dilemmas in which journalists expressed deep ambivalence. Covering such communities may increase their popularity and harmfulness of them and their tactics, but ignoring them makes it possible for worse actors to take their place, gives the impression that such speech is acceptable, and misses an opportunity to educate audiences and mobilize counterpublics. More recently, journalists have debated whether to report on Trump's tweets at all, fearing that they are fueling the visibility of material that actually works against a free press (Pope, 2017)—a dilemma about whether journalists should always cover a president's communication or whether they should refuse to cover material that they think harms the public.[3]

Some news organizations also look critically at what kind of sources are systematically absent from their coverage. For example, a strong gender bias that runs throughout journalistic practices that routinely silence women: on Twitter, journalists primarily follow and retweet men, soliciting them as sources for stories and story ideas; and on Wikipedia (a source that many journalists routinely draw upon for background research and authoritative sources), the vast majority of contributors and editors are men. Men are the primary online speakers that journalists are listening to and quoting; because women are not speaking or being listened to on these platforms, they often fail to appear in online journalists' stories (Matias, Szalavitz, & Zuckerman,

2017; Usher, Holcomb, & Littman, 2018). Spurred on by this research, popular Twitter hashtags like #WomenAlsoKnow have arisen. And after observing that, on average, only 21 percent of the people it was quoting were women, the *Financial Times* recently tried to curb its gender biases by creating a computer program to monitor the gender patterns in its sources' names and alert senior editors if they were systematically underquoting women.

Sometimes, though, news organizations explicitly try to silence sources and limit audience participation. The *Los Angeles Times* does not allow letters to the editor that deny the human role in climate change (Thornton, 2013); *Popular Science* discontinued comments on many of its stories after experimenters found that people who read stories with comments actually learned less about science and technology than they did in stories without comments (LaBarre, 2013) because their viewpoints became so polarized (A. A. Anderson, Brossard, Scheufele, Xenos, & Ladwig, 2013); and the *Atlantic* discontinued online comments on most of its stories, instead reverting to the older, curated letters-to-the-editor format (Goldberg, 2018). The overarching logic is that sometimes, according a journalists' own conceptions of what publics need, some voices need to be silenced and some topics need to be limited.

Facebook faces similar dilemmas in designing its content moderation strategies and battling misinformation at the size, scope, and speed it demands. Should it expose its users to potentially false and harmful information, letting them see a full range of claims and counterclaims, rarely censoring content or silencing speakers? The social psychology—much less its operationalization in interface design—is not yet settled on whether people benefit from seeing counterclaims, contextualizing misinformation, or being insulated entirely from content judged erroneous (Bode & Vraga, 2015; M. Karlsson, Clerwall, & Nord, 2014; M. Karlsson, Clerwall, & Nord, 2016; Lewandowsky, Ecker, Seifert, Schwarz, & Cook, 2012; Lyons, 2017). Or, should it take a more aggressive moderation stance on banning content? Such moves can silence actors the company defines as bad—for example, it explicitly banned Russia's Internet Research Agency after public pressure (Stamos, 2018)—but they can also reveal the unworkable scale of its moderation regime, reliance on imperfect algorithms, and the unaccountable nature of its moderation policies (Gillespie, 2018). The opacity and ambiguities of Facebook's moderation systems mean that it routinely bans actors it later decides to be legitimate: journalists (Liptak, 2017), public officials (Scott & Isaac, 2016), activists (Levin, 2016), anthropologists (Alexander, 2016), and entire racial groups (Angwin & Grassegger, 2017). Facebook silences come in the form of one-off bans, policy implementations, and false-positive accidents—none of which is governed by formal public accountability mechanisms, but all of which have an impact on which content is seen or unseen on the world's largest social media platform.

Other choices by communication creators may seem tangential or unrelated to the idea of political participation but they suggest other, related domains of absence in authorship and design. For example, advertisers can effectively influence which online content is financially viable when they set the terms under which they allow their advertisements to be seen alongside news content: some will avoid stories with what they see as controversial keywords, others refuse to have their ads appear near user comments on controversial stories, and still others refuse to appear on certain news sites altogether (Dwoskin & Timberg, 2017; Vranica, 2017).

And technology can be a playful arbiter of scarcity. Impossible Labs recently experimented with producing a "limited edition digital book": only one hundred people can own the original, but before any owner can transfer ownership to other person, she must "remove two words and add one to every page," meaning that "frankly, after 20 owners it will be unreadable" (Stinson, 2017).

These sites and practices all point to an emerging domain of content production and platform design characterized by a concern with absence: when to create it, what professional and public rationales underpin its existence, who will be excluded, and how to mix computational and curatorial systems in its production.

INFRASTRUCTURAL DYNAMICS

Absences are also created through a mix of infrastructural forces that control the visibility and unseen nature of media. By "infrastructure," I mean Susan Leigh Star's and Karen Ruhleder's model of artifacts, practices, and values that are embedded in sociotechnical structures, standardized, largely invisible, intertwined with expertise, inseparable from individuals' perspectives, learned through communities of practice, and extended in space and time beyond single occurrences (Star & Ruhleder, 1996). The presence and power of such infrastructures are often seen only when they break down or when analysts engage in what Bowker and Star (1999) call "infrastructural inversion": surfacing the unseen, invisible work that always entails a mix of hidden artifacts and de-emphasized and often gendered labor (Shapin, 1989; Star & Strauss, 1999).

In contemporary media spaces, such infrastructural dynamics play out in a number of ways. One of these centers on ways that "big data" systems are designed to make some people more visible than others—to implicitly enroll certain people in data-based publics while effectively ignoring the existence of others. For example, in his critique of the idea that public information systems predicated on "big data" offer more egalitarian access to public goods and services, Lerman (2013) shows how the information of many people "is not regularly collected or analyzed, because they do not routinely engage in activities that big data is designed to capture. Consequently, their preferences and needs risk being routinely ignored when governments and private industry use big data and advanced analytics to shape public policy and the marketplace" (p. 55). He goes on to construct a hypothetical but plausible persona of a person who will never appear in the data sets used to model and optimize public transit: "He lives two hours southwest of Manhattan, in Camden, New Jersey, America's poorest city. He is underemployed, working part-time at a restaurant, paid under the table in cash. He has no cell phone, no computer, no cable. He rarely travels and has no passport, car, or GPS. He uses the Internet, but only at the local library on public terminals. When he rides the bus, he pays the fare in cash." But it is not the right answer to simply equip low-income people with such technologies and make sure that they are included in public data systems. This point is taken up by Gangadharan (2012), who notes that when "chronically underserved communities"—poor people, communities of color, indigenous, and migrants—are included in online data sets, through broadband inclusion initiatives, they become more vulnerable to surveillance and privacy infringements. That is, making chronically underserved communities more visible online actually disproportionally exposes them to profiling and persuasion techniques that further drive their disempowerment. It may make more sense for such groups to avoid broadband inclusion initiatives and the "opportunities" they afford, if the privacy costs are too high. Paradoxically and tragically, it may actually be in the self-interests of disempowered groups to continue being invisible to such infrastructures—to avoid being enrolled in data publics.

Such concerns about privacy and visibility also play out more broadly for all platform users. For example, through its Pixel technology, Facebook tracks web users and includes their behavioral analytics in its advertising models, regardless of whether they have a Facebook profile (Brandom, 2018). There is, effectively, no opting out of participating in Facebook's data profiling.

Infrastructural invisibility can also appear in the design and deployment of communication infrastructures. For example, placing a seemingly simple piece of code on a web server—a file named "robots.txt"—renders that server invisible to search engine crawlers like Google. A mix of social and technological protocols (Elmer, 2008)—by convention, crawlers are understood to ignore servers with this file and thus render them invisible in indexes—effectively creates a shadow layer of online content that is inaccessible to search-engine users.

Other types of invisibility are embedded within an online community's technology. For example, the software development community GitHub recently introduced a "cooling off" period as a way of helping people take a break when online interactions become heated: "sometimes a solution can't be reached until everyone has had time to cool down. Now, with interaction limits, maintainers can temporarily limit who can comment, create pull requests, and open issues among existing users, collaborators, and prior contributors. After 24 hours, limits expire and allow users to go back to participating in the conversation" (Bieda, 2017, n.p.). And, seemingly inspired by a Habermasian ideal of bracketing reputations in communication contexts and a desire to experience analytics-free social media, artist Benjamin Grosser created Demetricator: a browser plug-in for Facebook and Twitter that "removes all visible metrics from the platform" (Zweig, 2018, n.p.). When installed, the plug-in lets users see Facebook and Twitter profiles as they normally would but removes all numeric indicators of popularity and scale—tallies of likes, retweets, friends, followers are all erased. This is a kind of reputational whitespace, an attempt to focus conditions of participation on the content of messages rather than the quantitatively perceived authority of the messengers.

A final dimension of infrastructural absence relates to the often-invisible labor required to maintain and repair digital communication platforms that are often perceived as algorithmic, automated, and largely the product of computational processes that require little human involvement. In fact, much of the seeming consistency of online life is thanks to a large set of often unseen laborers: for example, the "microworkers" of Amazon's Mechanical Turk underpinning many artificial intelligence training systems and much academic research (Irani, 2015a), the human attendants who turned pages for Google's supposedly automatic book scanners (Irani, 2015b), and the vast armies of commercial content moderators who must quickly—they often only have a few seconds—review psychologically and emotionally taxing content, and apply technology companies' often-vague censorship policies (Roberts, 2016). The machinery of online participation may seem like it offers endless opportunities for everyone to express themselves and build community through communication, but such associational life rests upon a vast and often invisible infrastructure of human labor.

Participation as Sociotechnical Action

Although much scholarship has traditionally seen political participation in voting behaviors, civic associations, political campaigns, media forums, fan cultures, and many other settings, it can also be seen as a sociotechnical system that regulates the (in)visibility of expression. Today, those sociotechnical forces and communicative phenomena live in social media platforms, information algorithms, relational databases, and artificial intelligence systems. What might absences in these systems and structures mean, and how we might see a seeming lack of media representations not as a failure to participate but as evidence of political participation based on listening, observation, purposeful retreat, and interpretations of relevance?

Participation is a kind of sociotechnical action that plays out in associations and relationships among humans and nonhumans, people and artifacts. Just as the materiality of digital spaces is different from that of traditionally, face-to-face, physical spaces, so, too, are the material dimensions of digitally mediated political participation. Part of that materiality entails asking what a lack of materiality means: When do people and ideas not appear in online groups? How might their absences be explained not by apathy or exclusion, but by an assemblage of human and nonhuman actors? As people's contemporary communication is increasingly intertwined with computational devices and platforms, how might their absences be understood as functions of how they feel about those objects and spaces? When someone leaves Facebook—or fails to show up through a mix of personal choice and algorithmic exclusion—do they also leave a political constituency, a set of people who are similarly choose to be or are made invisible and silent?

Although the term *participation* is central to much of democratic theory, it is still highly contingent and debatable in both form and aims. It means different things in different contexts, often presumes different democratic ideals, and carries with it different assumptions about what role individual and collective actions play in democratic institutions.

The literature on democratic participation—its meanings, debates, deployments—is too vast to be summarized here, but it is worth briefly sketching a few of its dimensions. As Kelty (2017) outlines in his "Grammar of Participation," the term arguably makes one of its first and most significant contemporary appearances in the 1962 *Port Huron Statement*, spurring a burst of scholarship. The idea has variously been seen as the central concept in democratic theory (Held, 2006; Pateman, 1970); a precondition and product of deliberative democracy (Fishkin, 1991); a key dynamic in creating democratic polities through associational life and civic volunteerism (Fung, 2003;

Putnam, 1993; Verba, Schlozman, & Brady, 1995); evidence of the validity of elections and legitimacy of different forms of representative government (Held, 2006), public opinion measurements (Page & Shapiro, 1992; Salmon & Glasser, 1995), and journalistic practices (Glasser, 1999). Finally, cultural studies scholars have critiqued assumptions about participation to draw attention to forms of popular culture, individual expression, and socially organized participation that have political power and significance but have been historically absent from political theory's dominant models of political participation (Hall, 1992; Jenkins, 1992; Williams, 1983). Such participation is variously defined as evidence of an engaged and legitimate populace, as a literacy to be taught, and as a source of discontent that elites need to manage.

Absences also take material form in other less explicitly political domains that nonetheless are crucial to association making. Rooted in cultural studies' call for broader and subaltern models of political participation (Jenkins, 2006), urban planning's development of participatory professional practices (Forester, 1993; Schön, 1992), and science and technology scholars' arguments that materials play powerful roles in social settings (Law & Mol, 1995; Williams, 2006), a newer approach to political participation has more recently appeared that sees objects, infrastructures, and physical settings not just as places where social participation happens or representations of social processes but as actors in their own right that, inseparably intertwined with humans, have political force and a power to shape participation. Most forcefully advanced by Marres (2012), this type of participation "involves the supplanting of the familiar character of the 'informational citizen'—the one in need of information in order to adequately perform his role of opinionated, decision-making subject—with another figure, which we could call the material public" (pp. 4–5). Marres and her model of material publics is deeply rooted in American pragmatism and, in particular, John Dewey's image of a public as the "conversion of . . . associated behavior into a community of action, saturated and regulated by mutual interest in shared meanings, consequences that are translated into ideas and desired objects by means of symbols" (Dewey, 2007, p. 13).

Very much aligned with visions of social life as human-nonhuman intertwining, Marres and Dewey see materiality—objects, symbols, infrastructures—as inseparable from associational life. This inseparability of humans and nonhumans is evidence of precisely the kind of public they aim to explicate: a public predicated not on informational exchange, private action, or individual choice but on the inextricable consequences of having no choice but to share affiliations and physical settings. For scholars of material publics, it makes less sense to think about participation of humans separate from their nonhuman companions; it makes more sense to try to identify intertwined conditions and consequences that arise from their collective, joint participation.

For example, in his study of the Baku-Tbilisi-Ceyhan pipeline, Barry (2013) finds the "affected public" to be not only a geographic or administrative constituency—people who live in a particular place represented by a certain government—but also a material polity defined by proximity to the pipeline, scientific opinions about the environmental impact of pipeline proximity, economic incentives to produce and transport different volumes of oil, and expert projections about the future significance of the pipeline. For Barry, the pipeline—its route, its physical properties, its history, its projected behavior—participates: it is a constituent in the story of associational life and it plays a role in convening the inextricable, intertwined consequences that define public life.

In the context of digital, networked communication systems, materiality can take on new meanings. While still rooted in the symbol systems and object-relations Dewey saw as core to public action, digital tangibility suggests that something other than simply symbolic language or physical instantiation is at play. Leonardi (2012) offers some helpful distinctions, defining *materiality* as digital forms that "endure across differences in place and time and are important to users," *sociomateriality* as "activities that meld materiality with institutions, norms, discourses," and *material agency* as "the capacity for nonhuman entities to act absent sustained human intervention" (pp. 35–42). Taken together, this typology suggests that digital forms can persist, take on meaning and significance, integrate with human action and socially situated practices, and sometimes achieve a kind of autonomy. In the context of networked, digital communication systems, we might, for example, use this typology to see the material dimensions of data, algorithms, and platforms—and their role in the kind of public-making political participation that Dewey and Marres envision.

For example, consider the role that digital databases play in convening publics. In his study of online identity, Cheney-Lippold (2017) chronicles how myriad dimensions of individuals' identities—race, gender, age, orientation, political affiliations—are constantly derived, updated, categorized through a combination of surveillance, disclosure, and statistical prediction. He argues not just that identities are social and statistical constructs—this is nothing new (Bouk, 2015; Igo, 2007)—but that some image of a person's identity now exists as computationally calculated categories living in largely unsupervised troves of online traces that are periodically polled and rendered to create digitally material, databased representations of identity that endure across times and places, emerge from people's actions and shape the options available to them, and act at semiautonomous scales that resist close scrutiny. Most radically, Cheney-Lippold (2017) argues that we *are* these data; they are not superfluous to our political identities or actions. We are intertwined with the systems that make these digital materials in the same way that Dewey and Marres see publics as people who are inextricably linked via social conditions that are largely beyond their individual control. The material publics that emerge are inseparable from infrastructures of databased identity, making it impossible to ask questions about political standing, association, and agency without considering the construction, assumptions, and power of these database infrastructures.

Algorithms and platforms show evidence a similar materiality. To the extent that algorithms are not just computer code but are institutionally embedded, semiautonomous, computational logics that sustain often inscrutable relationships between people and data (Ananny, 2016; Bucher, 2018), they, too, serve as digital materials of public life. As Gillespie (2014) asserts, they "not only help us find information, they provide a means to know what there is to know and how to know it, to participate in social and political discourse, and to familiarize ourselves with the publics in which we participate" (p. 167). Social media platforms serve similar, and closely aligned public, material functions to data and algorithms. Although platforms do not make content, "they make important choices about that content: what they will distribute and to whom, how they will connect users and broker their interactions, and what they will refuse." (Gillespie, 2017, p. 254) Unlike the common carriage logics of post offices and telecommunication lines, and despite claims to neutrality, platforms have interests (Gillespie, 2018; McKelvey, 2014)—interests that take material form in the moderation policies, algorithmic systems, terms of service, and user experience designs. Indeed, platforms articulate a kind of double public (Langlois, Elmer, McKelvey, & Devereaux, 2009): they are not only digital places where people convene to discover, debate, and coordinate action about their shared social conditions; the governance of platforms themselves is hotly contested and susceptible to extreme power differentials as users often lack knowledge about how exactly platforms shape their communicative practices and lack the agency to leave platforms and recreate their relationships and data elsewhere (Baumer et al., 2013; Portwood-Stacer, 2013).

If the political significance of participation is intertwined with the institutions and material systems that enable individual and collective action, and if these systems in turn are not only linguistic and representational but also material, infrastructural, and digital, then appreciating the different meanings of contemporary political participation means delving more deeply into the details of digital materiality.

Conclusion

Though incomplete and idiosyncratic, this tour of sociotechnical absences and their political significance was meant to offer new ways to think about contemporary, digital, communicative forms of political participation.

Designers and advocates of many communication systems tend to assume that marketplaces of speech are the most effective and equitable ways of achieving communicative self-governance. More speech is always better, more participation will fix political problems, more voices will produce a greater variety of solutions. In many cases this is true, and such marketplaces are democratically essential. But as I have tried to argue, there is another side to communicative self-governance that might take into account the role of silences and absences. Feminist scholars have long argued that seemingly invisible—and often gendered—forms of participation like listening are essential to the health of communicative self-governance. Instead of simply equating public participation with public speaking, we might additionally consider how "listening publics" (Lacey, 2013) might be recognized and nurtured by designing for and valuing silences that are not really absences at all but rather spaces for listening (Bickford, 1996). Such an imperative has political importance because, as Dryzek (2002) writes "the most effective and insidious way to silence others in politics is a refusal to listen" (p. 149). Democracy cannot work if everyone is speaking and no one is listening.

Today and in online contexts, these absences, silences, and invisibilities appear in sociotechnical infrastructures that are often privately controlled, opaque, inscrutable, unaccountable, and driven by economic and advertising logics that value, above all else, more content and networks with more links. Surveillance, profiling, and targeted advertisements do not work very well if people are silent, absent, listening.

And many people, in turn, fear being invisible to such algorithmic systems (Bucher, 2012). If newsfeed algorithms do not see you and your perspectives, if public systems are only designed for those who leave data traces, and if artificial intelligence systems are trained without your data—you run the risk of being invisible without agency. You have not chosen to be invisible or a contemplative listener—you have, instead, been erased as a participant, excluded from a constituency. Such exclusion is very different from taking a "news sabbatical." Indeed, in many ways, being absent is a privilege. If you have the option to rejoin a system at any time and pick up where your silence began, or if you have a trusted and powerful cohort able to represent your interests, the significance of your absences is very different from those who cannot easily slip in and out of polities as their preferences and desires dictate.

The onus is on analysts and creators of contemporary political participation to see absences, to ask who is not present in any given constituency, and to recognize that silences, absences, and exclusions can emerge from a mix of individual behaviors, professional practices, and infrastructural forces. By no means am I arguing that people should be silenced or that silence is a greater form of political participation than speaking. Rather, my hope is that scholars and designers of contemporary, networked political participation see the value of looking for things that don't seem to be there.

Notes

1. For a good overview of agnotology, the emerging field on the sociology of ignorance, see Abbott (2010); Croissant (2014); McGoey (2012); Proctor (2008).

2. Many thanks to Lucy Bernholz for this phrase and concept, which I have tried to do justice to here.

3. I explore more of these journalistic dilemmas of white space in greater detail in Ananny (2017).

References

Abbott, A. (2010). Varieties of ignorance. *American Sociologist, 41*(2), 174–189. https://doi.org/10.1007/s12108-010-9094-x

Agre, P. E. (2002). Real-time politics: The internet and the political process *Information Society, 18*(5), 311–331.

Alexander, L. (2016, March 23). Facebook's censorship of Aboriginal bodies raises troubling ideas of "decency." *The Guardian.* Retrieved from https://www.theguardian.com/technology/2016/mar/23/facebook-censorship-topless-aboriginal-women

Ananny, M. (2016). Toward an ethics of algorithms: Convening, observation, probability, and timeliness. *Science, Technology & Human Values, 41*(1), 93–117. https://doi.org/10.1177/0162243915606523

Ananny, M. (2017). The whitespace press: Designing meaningful absences into networked news. In P. J. Boczkowski & C. W. Anderson (Eds.), *Remaking the news* (pp. 129–146). Cambridge, MA: MIT Press.

Anderson, A. A., Brossard, D., Scheufele, D. A., Xenos, M. A., & Ladwig, P. (2013). Crude comments and concern: Online incivility's effect on risk perceptions of emerging technologies. *Journal of Computer-Mediated Communication.* https://doi.org/10.1111/jcc4.12009

Anderson, B. (1983). *Imagined communities* (Rev. ed.). London, UK: Verso.

Anderson, M. J. (2015). *The American census: A social history* (2nd ed.). New Haven, CT: Yale University Press.

Angwin, J., & Grassegger, H. (2017, June 28). Facebook's secret censorship rules protect white men from hate speech but not black children. *ProPublica.* Retrieved from https://www.propublica.org/article/facebook-hate-speech-censorship-internal-documents-algorithms

Askay, D. A. (2015). Silence in the crowd: The spiral of silence contributing to the positive bias of opinions in an online review system. *New Media & Society, 17*(11), 1811–1829. https://doi.org/10.1177/1461444814535190

Baker, C. E. (2002). *Media, markets, and democracy.* Cambridge, UK: Cambridge University Press.

Barber, B. R. (2004). Which technology and which democracy? In H. Jenkins & D. Thorburn (Eds.), *Democracy and new media* (pp. 33–48). Cambridge, MA: MIT Press.

Barry, A. (2013). *Material politics: Disputes along the pipeline* New York, NY: Wiley-Blackwell.

Baumer, E. P. S., Adams, P., Khovanskaya, V. D., Liao, T. C., Smith, M. E., Sosik, V. S., & Williams, K. (2013). *Limiting, leaving, and (re)lapsing: An exploration of Facebook non-use practices and experiences.* Paper presented at the Proceedings of the SIGCHI Conference on Human Factors in Computing Systems, Paris, France.

Baumer, E. P. S., Ames, M. G., Burrell, J., Brubaker, J. R., & Dourish, P. (2015). Why study technology non-use? *First Monday, 20*(11). Retrieved from http://firstmonday.org/ojs/index.php/fm/article/view/6310/5137

Baumer, E. P. S., Guha, S., Quan, E., Mimno, D., & Gay, G. K. (2015). Missing photos, suffering withdrawal, or finding freedom? How experiences of social media non-use influence the likelihood of reversion. *Social Media + Society, 1*(2). https://doi.org/10.1177/2056305115614851

Benkler, Y. (2003). Freedom in the commons: Towards a political economy of information. *Duke Law Journal, 52,* 1245–1276.

Benkler, Y. (2006). *The wealth of networks: How social production transforms markets and freedom.* New Haven, CT: Yale University Press.

Bickford, S. (1996). *The dissonance of democracy: Listening, conflict, and citizenship.* Ithaca, NY: Cornell University Press.

Bieda, L. (2017, May 30). Introducing temporary interaction limits. *GitHub Blog.* Retrieved from https://blog.github.com/2017-05-30-introducing-temporary-interaction-limits/

Bishop, J. (2007). Increasing participation in online communities: A framework for human-computer interaction. *Computers in Human Behavior, 23*(4), 1881–1893. https://doi.org/10.1016/j.chb.2005.11.004

Bode, L. (2016). Pruning the news feed: Unfriending and unfollowing political content on social media. *Research & Politics, 3*(3), art. 2053168016661873. https://doi.org/10.1177/2053168016661873

Bode, L., & Vraga, E. K. (2015). In related news, that was wrong: The correction of misinformation through related stories functionality in social media. *Journal of Communication, 65*(4), 619–638. https://doi.org/10.1111/jcom.12166

Bouk, D. (2015). *How our days became numbered: Risk and the rise of the statistical individual.* Chicago, IL: University of Chicago Press.

Bowker, G. C., & Star, S. L. (1999). *Sorting things out: Classification and its consequences.* Cambridge, MA: MIT Press.

Brandom, R. (2018, April 11). Shadow profiles are the biggest flaw in Facebook's privacy defense. *The Verge.* Retrieved from https://www.theverge.com/2018/4/11/17225482/facebook-shadow-profiles-zuckerberg-congress-data-privacy

Brook, J., & Boal, I. (Eds.). (1995). *Resisting the virtual life: The culture and politics of information.* San Francisco, CA: City Lights.

Brownlee, K. (2013, December 20). Civil disobedience. *Stanford Encyclopedia of Philosophy*. Retrieved from https://plato.stanford.edu/entries/civil-disobedience/

Bucher, T. (2012). Want to be on top? Algorithmic power and the threat of invisibility on Facebook. *New Media & Society, 14*(7), 1164–1180.

Bucher, T. (2018). *If . . . then: Algorithmic power and politics*. Oxford, UK: Oxford University Press.

Carter, R. G. S. (2006). Of things said and unsaid: Power, archival silences, and power in silence. *Archivaria, 61*, 215–233.

Chan, M. (2017). Reluctance to talk about politics in face-to-face and Facebook settings: Examining the impact of fear of isolation, willingness to self-censor, and peer network characteristics. *Mass Communication and Society*. https://doi.org/10.1080/15205436.2017.1358819

Cheney-Lippold, J. (2017). *We are data: Algorithms and the making of our digital selves*. New York, NY: NYU Press.

Croissant, J. L. (2014). Agnotology: Ignorance and absence or towards a sociology of things that aren't there. *Social Epistemology, 28*(1), 4–25. https://doi.org/10.1080/02691728.2013.862880

Dahlberg, L. (2001). Computer-mediated communication and the public sphere: A critical analysis. *Journal of Computer Mediated Communication, 7*(1).

DeLuca, T. (1995). *The two faces of political apathy*. Philadelphia, PA: Temple University Press.

Dewey, J. (2007). Search for the great community. *Kettering Review, 26*(2), 9–18.

Downs, A. (1957). An economic theory of political action in a democracy. *Journal of Political Economy, 65*(2), 135–150.

Dryzek, J. S. (2002). *Deliberative democracy and beyond: Liberals, critics, contestations*. Oxford, UK: Oxford University Press.

Dwoskin, E., & Timberg, C. (2017, March 24). For advertisers, algorithms can lead to unexpected exposure on sites spewing hate. *Washington Post*. Retrieved from https://www.washingtonpost.com/business/technology/for-advertisers-algorithms-can-lead-to-unexpected-exposure-on-sites-spewing-hate/2017/03/24/046ac164-043d-11e7-b1e9-a05d3c21f7cf_story.html

Eberspacher, S., & Guardian readers. (2017, April 24). Activism, cynicism—and whiskey: How readers are coping with Trump. *The Guardian*. Retrieved from https://www.theguardian.com/us-news/2017/apr/24/trump-presidency-100-days-coping-activism

Edgerly, S. (2016). Seeking out and avoiding the news media: Young adults' proposed strategies for obtaining current events information. *Mass Communication and Society*. https://doi.org/10.1080/15205436.2016.1262424

Eliasoph, N. (1998). *Avoiding politics: How Americans produce apathy in everyday life*. Cambridge, UK: Cambridge University Press.

Elmer, G. (2008). Exclusionary rules? The politics of protocols. In A. Chadwick & P. N. Howard (Eds.), *Routledge handbook of internet politics* (pp. 376–383). London, UK: Routledge.

Fishkin, J. S. (1991). *Democracy and deliberation: New directions for democratic reform*. New Haven, CT: Yale University Press.

Forester, J. (1993). *Critical theory, public policy, and planning practice*. Albany, NY: State University of New York Press.

Fung, A. (2003). Associations and democracy: Between theories, hopes, and realities. *Annual Review of Sociology, 29*, 515–539.

Gandy, O. H. (1993). Toward a political economy of personal information. *Critical Studies in Mass Communication, 10*(1), 70–97. https://doi.org/10.1080/15295039309366849

Gangadharan, S. P. (2012). Digital inclusion and data profiling. *First Monday, 17*(5). https://doi.org/10.5210/fm.v17i5.3821

Giddens, A. (1984). *The constitution of society: Outline of the theory of structuration.* London, UK: Polity Press.

Gillespie, T. (2014). The relevance of algorithms. In T. Gillespie, P. Boczkowski, & K. A. Foot (Eds.), *Media technologies: Essays on communication, materiality, and society* (pp. 167–194). Cambridge, MA: MIT Press.

Gillespie, T. (2017). Governance of and by platforms. In J. Burgess, T. Poell, & A. Marwick (Eds.), *SAGE Handbook of Social Media* (pp. 254–278). London, UK: Sage.

Gillespie, T. (2018). *Custodians of the internet: Platforms, content moderation, and the hidden decisions that shape social media.* New Haven, CT: Yale University Press.

Gillmor, D. (2004). *We the media: Grassroots journalism by the people, for the people.* Sebastopol, CA: O'Reilly Media.

Glasser, T. L. (1999). The idea of public journalism. In T. L. Glasser (Ed.), *The idea of public journalism* (pp. 3–20). New York, NY: Guilford Press.

Goldberg, J. (2018, February 2). We want to hear from you. *The Atlantic.* Retrieved from https://www.theatlantic.com/letters/archive/2018/02/we-want-to-hear-from-you/552170/

Gottfried, J., & Barthel, M. (2018, June 5). *Almost seven-in-ten Americans have news fatigue, more among Republicans.* Pew Research Center. Retrieved from http://www.pewresearch.org/fact-tank/2018/06/05/almost-seven-in-ten-americans-have-news-fatigue-more-among-republicans/

Hall, S. (1992). Cultural studies and its theoretical legacies. In L. Grossberg, C. Nelson, & P. Treichler (Eds.), *Cultural studies* (pp. 277–294). New York, NY: Routledge.

Hampton, K., Rainie, L., Lu, W., Dwyer, M., Inyoung, S., & Purcell, K. (2014, August 26). *Social media and the "spiral of silence."* Pew Research Internet Project. Retrieved from http://www.pewinternet.org/2014/08/26/social-media-and-the-spiral-of-silence/

Held, D. (2006). *Models of democracy* (3rd ed.). Stanford, CA: Stanford University Press.

Herrman, J., & Savage, C. (2018, May 23). Trump's blocking of Twitter users is unconstitutional, judge says. *New York Times.* Retrieved from https://www.nytimes.com/2018/05/23/business/media/trump-twitter-block.html

Hindman, M. (2008). *The myth of digital democracy.* Princeton, NJ: Princeton University Press.

Igo, S. (2007). *The averaged American: Surveys, citizens, and the making of a mass public.* Cambridge, MA: Harvard University Press.

Ingram, M. (2018, February 26). A new digital magazine forces you to unplug from the internet. *Columbia Journalism Review.* Retrieved from https://www.cjr.org/innovations/disconnect-magazine-only-works-offline.php

Irani, L. (2015a). The cultural work of microwork. *New Media & Society, 17*(5), 720–739. https://doi.org/10.1177/1461444813511926

Irani, L. (2015b, January 15, 2015). Justice for "data janitors." *Public Books.* Retrieved from http://www.publicbooks.org/nonfiction/justice-for-data-janitors

Jenkins, H. (1992). *Textual poachers: Television fans and participatory culture.* Oxford, UK: Routledge Press.

Jenkins, H. (2006). *Fans, bloggers, and gamers: Exploring participatory culture.* New York, NY: New York University Press.

John, N. A., & Dvir-Gvirsman, S. (2015). "I don't like you any more": Facebook unfriending by Israelis during the Israel-Gaza conflict of 2014. *Journal of Communication, 65*(6), 953–974. https://doi.org/10.1111/jcom.12188

Karlsson, M., Clerwall, C., & Nord, L. (2014). You ain't seen nothing yet: Transparency's (lack of) effect on source and message credibility. *Journalism Studies.* https://doi.org/10.1080/1461670X.2014.886837

Karlsson, M., Clerwall, C., & Nord, L. (2016). Do not stand corrected: Transparency and users' attitudes to inaccurate news and corrections in online journalism. *Journalism & Mass Communication Quarterly.* https://doi.org/10.1177/1077699016654680

Kelty, C. M. (2017). Too much democracy in all the wrong places: Toward a grammar of participation. *Current Anthropology, 58*(S15), S77–S90. https://doi.org/10.1086/688705

Ksiazek, T. B., Malthouse, E. C., & Webster, J. G. (2010). News-seekers and avoiders: Exploring patterns of total news consumption across media and the relationship to civic participation. *Journal of Broadcasting & Electronic Media, 54*(4), 551–568.

LaBarre, S. (2013, September 24). Why we're shutting off our comments. *Popular Science.* Retrieved from https://www.popsci.com/science/article/2013-09/why-were-shutting-our-comments

Lacey, K. (2013). *Listening publics: The politics and experience of listening in the media age.* Cambridge, UK: Polity.

Langlois, G., Elmer, G., McKelvey, F., & Devereaux, Z. (2009). Networked publics: The double articulation of code and politics on Facebook. *Canadian Journal of Communication, 34,* 415–434.

Law, J., & Mol, A. (1995). Notes on materiality and sociality. *Sociological Review, 43*(2), 274–294.

Leonardi, P. M. (2012). Materiality, sociomateriality, and socio-technical systems: What do these terms mean? How are they related? Do we need them? In P. M. Leonardi, B. A. Nardi, & J. Kallinikos (Eds.), *Materiality and organizing: Social interaction in a technological world* (pp. 25–48). Oxford, UK: Oxford University Press.

Lerman, J. (2013). Big data and its exclusions. *Stanford Law Review, 66,* 55–63.

Levin, S. (2016, September 12). Facebook temporarily blocks Black Lives Matter activist after he posts racist email. *The Guardian.* Retrieved from https://www.theguardian.com/technology/2016/sep/12/facebook-blocks-shaun-king-black-lives-matter

Lewandowsky, S., Ecker, U. K. H., Seifert, C. M., Schwarz, N., & Cook, J. (2012). Misinformation and its correction: Continued influence and successful debiasing. *Psychological Science in the Public Interest, 13*(3), 106–131. https://doi.org/10.1177/1529100612451018

Lewis, P. (2017, October 6). Our minds can be hijacked: The tech insiders who fear a smartphone dystopia. *The Guardian.* Retrieved from https://www.theguardian.com/technology/2017/oct/05/smartphone-addiction-silicon-valley-dystopia

Liptak, A. (2017, May 21). A journalist's account was suspended after he posted allegations of corruption to Facebook. *The Verge.* Retrieved from https://www.theverge.com/2017/5/21/15672666/facebook-matthew-caruana-galizia-allegations-corruption-account-suspended

Lyons, T. (2017, December 20). Replacing disputed flags with related articles. *Facebook Newsroom Blog.* Retrieved from https://newsroom.fb.com/news/2017/12/news-feed-fyi-updates-in-our-fight-against-misinformation/

Marres, N. (2012). *Material participation.* London, UK: Palgrave Macmillan.

Matias, J. N., Szalavitz, S., & Zuckerman, E. (2017). *FollowBias: Supporting behavior change toward gender equality by networked gatekeepers on social media.* Paper presented at the Proceedings of the 2017 ACM Conference on Computer Supported Cooperative Work and Social Computing, Portland, OR.

McGoey, L. (2012). Strategic unknowns: Towards a sociology of ignorance. *Economy and Society, 41*(1), 1–16. https://doi.org/10.1080/03085147.2011.637330

McKelvey, F. (2014). Algorithmic media need democratic methods: Why publics matter to digital media research. *Canadian Journal of Communication, 39*, 597–613.

Mulvin, D. (2018). Media prophylaxis: Night modes and the politics of preventing harm. *Information & Culture, 53*(2), 175–202.

Oudshoorn, N., & Pinch, T. (2003). How users and non-users matter. In N. Oudshoorn & T. Pinch (Eds.), *How users matter: The co-construction of users and technology* (pp. 1–25). Cambridge, MA: MIT Press.

Page, B. I., & Shapiro, R. Y. (1992). *The rational public.* Chicago, IL: University of Chicago Press.

Pateman, C. (1970). *Participation and democratic theory.* Cambridge, UK: Cambridge University Press.

Phillips, W. (2018). The oxygen of amplification: Better practices for reporting on far right extremists, antagonists, and manipulators online. *Data & Society.* https://datasociety.net/wp-content/uploads/2018/05/1_PART_1_Oxygen_of_Amplification_DS.pdf

Plaut, E. R. (2015). Technologies of avoidance: The swear jar and the cell phone. *First Monday, 20*(11). http://dx.doi.org/10.5210/fm.v20i11.6295

Pope, K. (2017, July 5). What we miss when we obsess over Trump's tweets. *Columbia Journalism Review.* Retrieved from https://www.cjr.org/covering_trump/trump_tweets_press_journalists.php

Portwood-Stacer, L. (2013). Media refusal and conspicuous non-consumption: The performative and political dimensions of Facebook abstention. *New Media & Society, 15*(7), 1041–1057. https://doi.org/10.1177/1461444812465139

Preece, J., Nonnecke, B., & Andrews, D. (2004). The top five reasons for lurking: Improving community experiences for everyone. *Computers in Human Behavior, 20*(2), 201–223. https://doi.org/10.1016/j.chb.2003.10.015

Proctor, R. N. (2008). Agnotology: A missing term to describe the cultural production of ignorance (and its study). In R. Proctor & L. Schiebinger (Eds.), *Agnotology: The making and unmaking of ignorance* (pp. 1–33). Stanford, CA: Stanford University Press.

Putnam, R. D. (1993). *Making democracy work: Civic traditions in modern Italy.* Princeton, NJ: Princeton University Press.

Raz, J. (1979). *The authority of law: Essays on law and morality.* Oxford, UK: Clarendon Press.

Rheingold, H. (1993). *The Virtual community: Homesteading on the electronic frontier.* Cambridge, MA: Addison-Wesley.

Roberts, S. T. (2016). Commercial content moderation: Digital laborers' dirty work. In B. M. Tynes & S. U. Noble (Eds.), *The intersectional internet: Race, sex, class, and culture online* (pp. 147–160). New York, NY: Peter Lang.

Salmon, C. T., & Glasser, T. L. (1995). The politics of polling and the limits of consent. In T. L. Glasser & C. T. Salmon (Eds.), *Public opinion and the communication of consent* (pp. 437–458). New York, NY: Guilford Press.

Schmidt, S. (2018, January 9). Fitness chain bans cable news networks as part of "healthy way of life." *Washington Post.* Retrieved from https://www.washingtonpost.com/news/morning-mix/wp/2018/01/09/fitness-chain-has-banned-cable-news-networks-as-part-of-healthy-way-of-life/

Schön, D. (1992). Designing as reflective conversation with the materials of a design situation. *Knowledge-Based Systems, 5*(1), 3–14.

Schudson, M. (1998). *The good citizen: A history of American public life.* New York, NY: Free Press.

Scott, M., & Isaac, M. (2016, September 9). Facebook restores iconic Vietnam War photo it censored for nudity. *New York Times.* Retrieved from https://www.nytimes.com/2016/09/10/technology/facebook-vietnam-war-photo-nudity.html

Sewell, W. (1992). A theory of structure: Duality, agency and transformation. *American Journal of Sociology, 98,* 1–29. Retrieved from http://en.wikipedia.org/wiki/Structuration_theory

Shapin, S. (1989). The invisible technician. *American Scientist, 77*(6), 554–563.

Shirky, C. (2008). *Here comes everybody: The power of organizing without organizations.* New York, NY: Penguin.

Silver, D., & Massanari, A. (Eds.). (2006). *Critical cyberculture studies.* New York, NY: NYU Press.

So, J., Kim, S., & Cohen, H. (2016). Message fatigue: Conceptual definition, operationalization, and correlates. *Communication Monographs,* 1–25. https://doi.org/10.1080/03637751.2016.1250429

Soffer, O., & Gordoni, G. (2017). Opinion expression via user comments on news websites: analysis through the perspective of the spiral of silence. *Information, Communication & Society,* 1–16. https://doi.org/10.1080/1369118X.2017.1281991

Squires, C. R. (2002). Rethinking the black public sphere: An alternative vocabulary for multiple public spheres. *Communication Theory, 12*(4), 446–468. https://doi.org/10.1111/j.1468-2885.2002.tb00278.x

Stamos, A. (2018, April 3). Authenticity matters: The IRA has no place on Facebook. *Facebook Newsroom.* Retrieved from https://newsroom.fb.com/news/2018/04/authenticity-matters/

Star, S. L., & Ruhleder, K. (1996). Steps toward an ecology of infrastructure: Design and access for large information spaces. *Information Systems Research, 7*(1), 111–134.

Star, S. L., & Strauss, A. (1999). Layers of silence, arenas of voice: The ecology of visible and invisible work. *Computer Supported Cooperative Work, 8*(1–2), 9–30. https://doi.org/10.1023/a:1008651105359

Stinson, E. (2017, April 14). The bizarre digital book you must destroy before sharing. *Wired.* Retrieved from https://www.wired.com/2017/04/bizarre-digital-book-must-destroy-sharing/

Sun, N., Rau, P. P.-L., & Ma, L. (2014). Understanding lurkers in online communities: A literature review. *Computers in Human Behavior, 38,* 110–117. https://doi.org/10.1016/j.chb.2014.05.022

Sunstein, C. (2001). *Republic.com.* Princeton, NJ: Princeton University Press.

Thornton, P. (2013, October 8). On letters from climate-change deniers. *Los Angeles Times.* Retrieved from http://www.latimes.com/opinion/opinion-la/la-ol-climate-change-letters-20131008-story.html

Toff, B., & Palmer, R. A. (2018). Explaining the gender gap in news avoidance: "News-is-for-men" perceptions and the burdens of caretaking. *Journalism Studies,* 1–17. https://doi.org/10.1080/1461670X.2018.1528882

Usher, N., Holcomb, J., & Littman, J. (2018). Twitter makes it worse: Political journalists, gendered echo chambers, and the amplification of gender bias. *International Journal of Press/Politics.* https://doi.org/10.1177/1940161218781254

Verba, S., Schlozman, K. L., & Brady, H. E. (1995). *Voice and equality: Civic voluntarism in American politics.* Cambridge, MA: Harvard University Press.

Vranica, S. (2017, June 18). Advertisers try to avoid the web's dark side, from fake news to extremist videos. *Wall Street Journal.* Retrieved from https://www.wsj.com/articles/advertisers-try-to-avoid-the-webs-dark-side-from-fake-news-to-extremist-videos-1497778201

Wells, C., Cramer, K. J., Wagner, M. W., Alvarez, G., Friedland, L. A., Shah, D. V., . . . Franklin, C. (2017). When we stop talking politics: The maintenance and closing of conversation in contentious times. *Journal of Communication*, *67*(1), 131–157. https://doi.org/10.1111/jcom.12280

Williams, R. (1983). *Culture and society: 1780–1950*. New York, NY: Columbia University Press.

Williams, R. (2006). *Culture and materialism*. London, UK: Verso.

Wyatt, S., Thomas, G., & Terranova, T. (2002). They came, they surfed, they went back to the beach: Conceptualising use and non-use of the Internet. In S. Woolgar (Ed.), *Virtual society? Technology, cyberbole, reality* (pp. 23–40). Oxford, UK: Oxford University Press.

Zhu, Q., Skoric, M., & Shen, F. (2017). I shield myself from thee: Selective avoidance on social media during political protests. *Political Communication*, *34*(1), 112–131. https://doi.org/10.1080/10584609.2016.1222471

Zweig, D. (2018, February 27). Escaping Twitter's self-consciousness machine. *New Yorker*. Retrieved from https://www.newyorker.com/tech/elements/escaping-twitters-self-consciousness-machine.

6

The Artisan and the Decision Factory: The Organizational Dynamics of Private Speech Governance

Robyn Caplan

Concerns about the rise of extremist content, hate speech, and false information spreading over social media networks and search engines brought the lawyers of major companies, Google, Twitter, and Facebook, before the US Congress at the end of October 2017.[1] This was followed in the spring of 2018 by testimony given by Mark Zuckerberg, the founder of Facebook.[2] Across the hearings, elected officials from Congress grilled the company representatives as to how they were making decisions about content posted by users across their networks. There was enormous concern across the political spectrum related to potential censorship of political speech by corporate entities, privacy concerns, targeted ads, and the need for platforms to be more accountable for content posted by untraceable entities, driving disinformation campaigns that divide communities here and abroad. The two hearings yielded few, if any, satisfying answers, but they focused public attention on fundamental questions concerning how speech is governed in the digital public sphere and by whom.

Although Facebook has received the brunt of bad press, hate speech, propaganda, and false information is an increasing matter of concern for online platforms, both large and small. Research shows that YouTube is frequently a site for radicalization, as users follow recommendations down rabbit holes that drive them to more extremist positions on views they may already hold.[3] Google's search engine is deployed to spread false information quickly in the wake of breaking news events.[4] Twitter has long been as a space where "computational propaganda" can thrive, with political bots and fake accounts that are tied to state-sponsored efforts to distribute misleading information in countries all over the world.[5] Smaller sites, based in the United States but used globally, like Patreon, Medium, and Discord, also contend with rising hate speech across their networks as well. They have developed new policies in response to events like the Unite the Right rally in Charlottesville, which brought far-right groups, largely organizing online, out into the open for public view.[6] In response, many on the far right now decry the platforms policies as a license to censorship, with some right-wing commentators and media outlets, such as Tucker Carlson and Breitbart, taking up an argument that was once supported by Elizabeth Warren to regulate platforms as "public goods" or utilities.[7]

These platforms are uniformly private companies, usually commercial but sometimes not-for-profit, and First Amendment concerns do not apply directly to private entities. As a result, content moderation falls to internal trust and safety teams across all of these platforms. In this environment, big-tech corporate policy and practice constitute the front lines in policing the false information, hate speech, and broader culture wars taking place across the globe.[8] Concerns about the role social media platforms in particular are playing in information flows typically alternate between worry about potential privatized censorship of citizens and the belief that private companies do too little to limit access to disinformation, hate speech, and extremist content online. Whatever one's position on such questions, the simple fact of the matter is that platform power remains opaque, unaccountable, and uncontestable.

This chapter explores issues related to the private governance of speech by platforms. Although this chapter makes clear that this has been a constant theme of work on governance of and by technologies since at least the late 1990s, the current debate on content moderation has placed the consequences of private control into stark relief. I examine three different strategies of private governance deployed by platforms in order to make the case that governance of content by users is a function of organizational structure and business dynamics. The strategy adopted by platforms varies between a formal and static rule structure and a more flexible, contextual approach and depends further on the platform's scale and relationship with users and moderators. Through a comparative analysis of governance strategies at many of the major content platforms, this chapter asks how we can make all platform governance more participatory and democratic.

Platforms as Privatized Speech Governance

How do companies set content policies and enforce them? Who in the company is responsible for such decisions? There has been research done by scholars and investigative journalists over the past several years,[9] but the world of content moderation and policy remains opaque and hidden from view. For the major companies, like Facebook and Google, scarce details exist about the number of content moderators employed by the companies, and even where those moderators are located.[10] Although transparency reports released by Facebook hint at the extent to which automation is used to detect content that violates its so-called community standards, the mechanisms by which those automated technologies function, the rules they are operating from, and where the content goes if it is deleted, remain a mystery.[11] Sarah T. Roberts describes this as a "logic of opacity" that serves to make platforms appear more objective in making decisions about content, driven more by "machine/machine-like rote behavior that removes any subjectivity and room for nuance" and inoculates companies from "large-scale questioning of the policies and values governing decision-making."[12]

Part of this opacity stems from the fact that what constitutes content moderation, in terms of the practices, rules, and methods of enforcement, is still in flux. Scholars understand the task of content moderation in various ways. Tarleton Gillespie makes the case that moderation is an essential characteristic of what platforms do (in addition to hosting, curating, and circulating user-generated content). James Grimmelman's definition, put forward in 2015, defines content moderation as "the governance mechanisms that structure participation in a community to facilitate cooperation and prevent abuse."[13] Sarah T. Roberts, using the term *commercial content moderation* further specifies the phenomenon by saying it is the "monitoring and vetting" of "user-generated content for social media platforms all types" to ensure "compliance with legal and regulatory exigencies, site/community guidelines, user agreements, and that it falls within norms of taste and acceptability for that site and its cultural context."[14] However, there are myriad ways to understand and conceptualize the work platforms do to structure participation in their networks. Systems of content moderation extend beyond the guidelines and rules for a site, and how they are enforced by human (and increasingly automated) moderators, who review flagged content to determine whether it is permissible on the site or should be removed. Content standards are more than just what is not allowed. Platforms also incentivize the production of certain content, through partnerships with media, or through ad-revenue sharing agreements coupled with "advertiser-friendly guidelines,"[15] and deprioritize content in algorithmically generated feeds to make content less visible.

The rise of interest in content decisions by platforms has generated a growing body of scholarship and stakeholder activity. Scholars like James Grimmelman, Kate Klonick, Tarleton Gillespie, and myself have sought to understand how platforms establish rules used to govern the speech of citizens from around the world. Scholars often differ in the extent to which they view moderation and the role of platforms as necessary for a healthy public sphere, or an overreach of power by private companies. Writing about spaces like Wikimedia and Reddit, Grimmelman views moderation as essential to preserving public spaces online, and preventing them from descending into chaos and anarchy, akin to moderators a town meeting would have in place to maintain order.[16] Klonick argues that platform companies should be considered the "New Governors" of speech.[17] Although she notes the immense power platforms have to shape discourse on their platforms, she makes the case that these unaccountable and "self-regulating entities" are best suited for the job and the most "economically and normatively motivated to reflect the democratic culture and free speech expectations of its users."[18] Gillespie, in his 2018 book *Custodians of the Internet*, acknowledges the difficult position platforms and other online intermediaries (like search engines, which do not fall into his definition of platform) are in, as they balance the need to protect individuals in online spaces from content such as porn and harassment, with a strong defense of freedom of expression. Gillespie calls for platforms to be more honest and explicit about this responsibility, to give up the myth of the "neutral platform," and to change their role as custodian of the internet from cleaning to more active stewardship. For Gillespie, it is up to us as users to demand platforms "take on an expanded sense of responsibility, and that they share the tools to govern collectively."[19] Within all of this work, significant room, particularly by Klonick, has been left for platform companies to self-regulate as private entities, although all acknowledge the importance of external forces, such as media, government, and users, in creating pressures for platforms to change their rules in response to demand or need.

In the United States, the law empowers platforms to self-regulate and make these decisions on their own. Section 230 of the Communications Decency Act (in combination with state-action doctrine) gives platform companies the leeway to govern most content (that is not illegal) as they see fit.[20] The rule provides both a shield from liability for content posted by users and a sword to protect platform rights to take action to voluntarily restrict access to or availability of material determined to be "obscene, lewd, lascivious, filthy, excessively violent, harassing, or otherwise objectionable, whether or not such material is constitutionally protected." The result is significant power in the hands of private platforms. As platforms have grown in scale by lowering barriers to access, production, and circulation of information to many of the world's citizens, many now believe that a small number of platform companies serve as the de facto hosts of a digital public sphere whose health is essential to a flourishing democratic society (see the chapter by Joshua Cohen and Archon Fung). What that means, and the methods by which we steward and preserve these spaces, are now being contested, both in terms of the democratic ideals that should drive rule-making in these networks and in terms of who should be making these decisions—governments, private platforms, civil society organizations, or users themselves. Currently, in most areas outside of content guidelines required by law, platforms are operating as a form of "private government."

Private Platforms, Public Spaces

The idea that the internet could and should be construed as creating a public space is not new. Many of its early advocates pointed to the internet as bringing to fruition the most important concepts of a Habermasian public sphere.[21] Ross Perot's "Electronic Town Hall," and John Perry Barlow's "Declaration of the Independence of Cyberspace" are indicative of this discourse, presenting the internet as a "solution" to the problems created by "top-down, one-way mass mediated systems" through which citizens were communicated to in prior media environments.[22] In the mid-1990s to early 2000s, it appeared as if the internet, with its blogs, chat rooms, and message boards, could be used by citizens like coffee houses and town squares, seemingly fulfilling the Habermasian idea of spaces where citizens could converge and engage in consensus making, where citizens could form ideas, deliberate and debate, free from the control and manipulation by the state.[23] This position was embraced by scholars like Yochai Benkler who argued that a "networked information economy" gave rise to a "networked public sphere," replacing the mass-mediated public sphere of the twentieth century.[24] Benkler argued that this shift was based on the increased access to information and knowledge facilitated by networked information technologies.

As Cohen and Fung argue in the opening chapter, thinking of the internet as contributing to the "public sphere" is important because of the role that public spaces and the free flow of communication and ideas play within democratic ideals.[25] A number of scholars assumed the internet had significant civic potential and would change the way that individuals accessed and consumed news and information, ushering in a new era of informed citizens and civic participation.[26] Individuals could bypass traditional media structures on platforms like YouTube and Facebook and thereby participate directly and shape public debate.[27] The internet also brought new opportunities for interacting directly with one's political representatives and offered more access to government-collected data.[28] It offered potential for many different publics to exist simultaneously, a way to surmount the limitations of the Habermasian ideal.[29] Despite this, there was no guarantee that the internet would usher in a new era of democracy. Early on in discussions about the internet's promise, Zizi Paparachissi noted that the utopian dreams of the internet were not yet earned and that "access to the internet does not guarantee increased political activity or enlightened political discourse," nor does it "ensure a more representative and more robust public sphere."[30] A decade and a half later, it is now apparent that Paparachissi's skepticism was well founded.

Early internet idealism was tempered by other concerns about the mass-mediated internet, including the core question about whether essentially privately owned online spaces could effectively constitute and steward the public sphere. We continue to have these discussions today when we consider whether companies like Facebook and Google are best suited to govern speech online. According to Cohen and Fung, platforms are "functionally public spaces" with the formal status of private entities and with much broader authority over what individuals say (all over the world) than governments.[31] This dynamic was noted early on by theorists like Lawrence Lessig, who called attention to the role that private entities, the companies and individuals writing the code for speech platforms, would have on standard setting, equating it with other forces, like markets, norms, and law, in its potential regulating power.[32] Other standards, like terms-of-service provisions, were also considered early on. Paul Schiff Berman questioned whether powerful private entities at the time, such as America Online, should be allowed to regulate speech on their platforms as "private standard-setting bodies—which now function so powerfully (yet so invisibly) to establish the code that regulates cyberspace."[33] In his view, these spaces which served as public but operated as private could, because of this dual status, evade many of discussions about constitutional values—such as the First Amendment—that he argued should ground the treatment of speech on these platforms. It is highly contestable as to whether the US-based constitutional value, the First Amendment, should guide normative discussions about the treatment of speech on what are now global communication platforms. And yet Schiff Berman's point—that conceiving of these public-private platforms as only private cuts off this kind of normative assessment—remains an important point as we consider the role social media companies and search engines, and their employees, have on speech today.

Platforms themselves are moving away from an embrace of the idea that they are the digital public sphere and toward an approach of the more private, "curated community," which acknowledges a common set of rules and standards by which users must abide to remain inside.[34] It is unclear however, whether this is a strategic, financial, or ideological shift (or all three). One platform representative has made the case that many social media platforms are intentionally shifting away from a "public square model"—encouraging users to embrace "openness and free speech and uploading everything"—to a contract theory approach that holds "we are a group, and we have X, Y, and Z ethical codes of how we treat each other." This rhetorical move away from openness to shared standards of a defined group can also be seen in Facebook's adoption of a new mission statement, "Building Global Community." This basic idea has been confirmed by Facebook's head of global policy management Monika Bickert, who noted that the company is not working to just "balance between safety and free speech" but must establish standards for speech "to create a safe community."[35]

It is difficult to argue against the need for (at least) minimum standards online. Disinformation serves to confuse citizens and hate speech can push individuals and groups off of networks, limiting the communicative power of some while protecting others unfairly. Such standards have long been in existence at most of the major platforms. Although they are only recently embracing the language of "community," platforms have been slowly establishing standards and limited restrictions for speech since they began hosting user-generated content. All platforms had content rules at their outset—for example, against illegal content, and also copyright protections were introduced early on—with most incorporating new rules over the past two decades in response to public concerns like terrorism and extremist content, revenge porn and harassment, and cyberbullying.[36] More recent rules against hate speech and disinformation (or "false news") are only the latest additions to the list of community guidelines users are expected to abide by when signing up for a service. And yet despite this new framing of "community," decision-making power over standards most often remains in the hands of a few employees at these platforms, with few, if any, checks to their power. In many cases, the conditions of community participation in these standards are structured by the platform in advance, through partnerships, and user agreements, that offer different patterns of influence and control for different members.[37] This can include in financial partnerships and advertiser-sharing agreements with media and amateur members (that influence the application of content standards like advertiser-friendly guidelines), in cooperation and agreements with government agencies as surveillance intermediaries,[38] or in granting access to user data through application programming interfaces, which can then be used to retarget content to members.

The upshot of all this activity is confusion among the public about what happens behind the scenes in private platforms. A lack of clarity over the distribution of rules and responsibilities in these privatized communities, and few formalized models to influence rule making in this structure, sidelines the potential influence of a variety of stakeholder groups—media industries, civil society organizations, governments, and users themselves. The remainder of this chapter identifies three distinct strategies of content moderation in order that a better understanding of private governance practices can lead to improved opportunities for wider influence in the future.

Strategies of Private Speech Governance

The majority of work on content moderation and speech governance has focused on the very large companies—Google (including YouTube), Facebook, and Twitter. This is for good reason. Facebook and Google are both operating on scales unmatched by any other platforms. As of 2019, Facebook properties (which includes WhatsApp and Instagram) boasted around 2.5 billion users.[39] YouTube has 1.8 billion monthly users.[40] Twitter, though much smaller, also has global impact because of its use by media and by government representatives and agencies. Because of their scale and broad influence, there has been considerable interest in how these three behemoths set standards for content. However, the focus on these companies and their specific mode of content moderation that is designed to operate at scale has eclipsed other content moderation models that may provide more desirable interactions between private companies and their publics.

In previous work, I have found that United States–based platforms differ extensively in their capacity to develop and enforce rules on their platforms, with three main strategies of content moderation that vary in terms of their governance structure, their implications for speech, and in their potential models for accountability or provision of checks and balances:

1. *Artisanal*, a case-by-case approach mostly used by platforms with smaller team sizes, such as Patreon, Discord, Medium, and Vimeo.
2. *Industrial*, a formalized structure with highly differentiated rules and responsibilities among team members, used by Facebook and Google (YouTube).
3. *Community-reliant*, wherein volunteer moderators, or users of the site, play an essential role in both developing speech rules and enforcing them amongst their communities, used by both Wikimedia Foundation (which owns Wikipedia) and Reddit.

In taking a comparative approach across platforms, I show how organizational dynamics play a key role in how data and content are generated by users, collected, and schematized. I also show, in these different strategies, how users transfer authority and governance of speech to platform companies, and the consequences for the digital public sphere as these companies continue to scale and take on greater responsibility for curating speech online.

Each strategy has a different approach toward content governance, rule making, enforcement, and trade-offs. These strategies should not be considered mutually exclusive, as evidence suggests not only that can one company move from one strategy type to another (e.g., from artisanal to industrial) but also, in some instances, platforms can deploy multiple strategies at a time. In each case, the strategy yields a different orientation to both users and employees responsible for making and enforcing content decisions. These strategies vary significantly in terms of these relationships, in some cases emphasizing one-to-one interactions between those making content standards and those violating the standards, or in structuring decisions locally, within communities, however they may be defined (e.g., region, language, interest, belief system, politics). The vast majority of users, however, are governed by an industrial model content moderation—a one-size-fits-all solution (mostly) that is used by companies like Facebook and Google to police the mass internet, using automated systems in addition to human labor in the form of outsourced workers responsible for reviewing hundreds of pieces of content per day in less-than-desirable workplace conditions.[41] Understanding the differences between these strategies can help us imagine different alternatives and frameworks for governing the digital public sphere.

ARTISANAL: "SMALL-BATCH" MODERATION

Artisanal strategies are often deployed at smaller companies, which have much smaller teams (as few as four employees), and typically house policy development and enforcement in one office, within the United States,[42] with policy and enforcement contained within one team, often in one room. Internal rules for decisions (which, for all company types, often differ substantially from the public-facing community guidelines available on websites) are typically not clearly formalized, with decisions about content violations usually discussed on a case-by-case basis. One platform representative using this approach compared it to a common-law system, based on precedent,[43] with another representative from a different platform (which has since industrialized) noting that rules often emerge from this process in the aggregate.[44] Decisions on difficult-to-classify content are often a group effort, with team members talking out hard cases together, occasionally using strategies like mock trials, or internal polls.[45] Although they have fewer team members, the user base of these companies is smaller, leading to fewer flags and allowing policy team members more time to decide on content violations. The result is that there can be multiple instances of outreach to users whose content is in review.[46] Platforms like Patreon view this as a process of "education" or creating awareness of the rules, with one employee saying "they hope to use more flexible and less defined policies as a way to start a conversation around a topic."[47]

Teams that follow this model rarely have a stable set of rules to adjudicate different content types. Although this means that these platforms may not be consistent with applications of the rules, this lack of formalization (along with a smaller amount of content that is flagged) allows for greater contextualization and greater attention to individual concerns. As a result of this approach, decisions about content often take much longer. One representative said that team members can spend "ten to twenty minutes" looking at violations and the context in which the violation occurs, as opposed to the mere seconds provided to workers at larger companies such as Facebook or YouTube.[48] This approach gives moderators more time to determine context and react accordingly, but other resource issues, such as limited language capacity outside of English, constrain the ability of these companies to attend to the cultural and political contingencies in many of the countries where they operate. This means they often have to be creative in how they respond to issues that require localized knowledge. Some companies will use translation tools (e.g., Google Translate) to mitigate these gaps, while others contract specific human-led translation services.[49] One company noted that it makes use of an informal network of experts—mostly academics—to solicit feedback and input for a potential decision.

Artisanal approaches benefit from fewer flags, and more time for review and outreach, and because of this process, rules often lack consistency, and open up the potential that bias of the moderators can influence individual decisions. However, in the embrace of the term *artisanal, small batch*, or *lean-and-mean*, it was evident that these companies saw value in embracing a hands-on, individualistic, and subjective approach to content moderation. The term *artisanal*, though deeply flawed in describing these practices, was used by a platform representative, and is used here, because of the juxtaposition of this model with the "industrial" version normally associated with content moderation today.

INDUSTRIAL: THE "DECISION FACTORY"

Industrial strategies emphasize consistency and limiting subjectivity, and thus potential bias. Used by Facebook, Google (including YouTube), and a number of other sites, the industrial strategy depends on a highly formalized structure of organization and rules, which is necessary to address content concerns at an immense scale (e.g., as of the end of 2019, Facebook had 2.27 billion monthly active users worldwide).[50] In this model, rule making is typically separated, both geographically and organizationally, from enforcement, with policy teams distributed across the United States and Europe while those doing the enforcement of rules are located in places like the Philippines, Turkey, or India.[51] Companies like YouTube also separate teams in terms of expertise and language fluency, so that content can be funneled to specialized moderators.[52] Part of the reason for this formalization and structure is because, as these companies scaled, moderators had to be on boarded en masse. One former team member noted that the goal for these companies is to create a "decision factory," which resembles more a "Toyota factory than it does a courtroom, in terms of the actual moderation."[53] He noted the approach as "trying to take a complex thing, and break it into extremely small parts, so that you can routinize doing it over, and over, and over again." This approach, and the tendency of companies like Facebook and Google, to use third-party contractors that in turn hire and mange moderators in areas like the Philippines and India with cheaper labor costs has led to significant concerns about the treatment of workers responsible for enforcing content rules.[54] These workers are also responsible for absorbing and moderating the most violent and prohibitive content, and they are expected to moderate with high degrees of accuracy and efficiency in stressful conditions, leading to additional concerns about trauma and emotional welfare within the industrial model.[55]

In this sense, industrial content moderation companies are large-scale bureaucracies, with highly specialized teams, and distributions of responsibilities and powers. Interaction with platform users is largely confined to the system of flagging and review, done through a platform user interface (rather than contact with an employee).[56] Flagging objectionable content is largely done by users themselves, though in the case of Facebook is increasingly performed by automated detection algorithms.[57] In rare cases, users can bypass the company altogether through gaining media attention for their cause or through interactions with civil society organizations that might have ties to the platform, or by deploying social connections to platform employees.[58]

When global platforms grow to be the size of a Facebook or a YouTube, they tend to prioritize consistency in decision making and to work to reduce variation and subjectivity.[59] One representative noted that contextualizing content moderation at the scale of a company like Facebook would be too complicated for an organization of that size. He noted that what constitutes whether content is discriminatory depends on the message, the speaker, and to whom the message is delivered, saying "who is historically disadvantaged with respect to whom is context dependent and situational." Perhaps because of this, platforms of this size tend to collapse contexts in favor of establishing global rules that make little sense in practice. The inability of large companies using an industrial model of content moderation to be sensitive to cultural concerns has generated controversy. On the one hand, this can mean that the Venus of Willendorf is accidentally censored for being too "pornographic."[60] In another, graver sense, such failures to address contextual considerations can lead to serious harms, as it did with hate speech that contributed to the persecution of Muslims in Myanmar, which was arguably fueled by disinformation and hate speech allowed to spread online by Facebook.[61]

Artisanal and industrial strategies differ in a number of ways—industrial teams are much larger, policy and enforcement are separate, and there is an emphasis on formalized and static rules. But they are similar in that they rely on users to flag content, and then employ (sometimes through contractors) moderators to develop and enforce rules, thus containing much of policy making within the private company itself.[62] In some cases, however, this is not the case. In addition to the tens of thousands of workers employed by Facebook (many as contractors hired via third parties), Facebook also depends on users to moderate—as in its Pages and Groups. For other companies, like Wikimedia and Reddit, this volunteer model accounts for the vast majority of moderation. These community-reliant strategies rely on volunteer labor but stress distributions of power, and they afford significant discretion to communities to set their own rules. Such methods offer other opportunities for interactions between platforms and their users.

COMMUNITY RELIANT

This third model of content moderation, "community reliant," is used by platforms such as Reddit and Wikimedia. The model offers more autonomy to community members while using their volunteer labor both to develop policy and to enforce content standards for their sites. Typically, this model distributes power between the parent organization (e.g., Wikimedia Foundation, Reddit Inc.) and its volunteer community, with the parent organization responsible for baseline rules and volunteers allowed to establish their own rules in their subcommunities. Representatives from these organizations noted that one way they separate their powers is through distinguishing between content and conduct, with the latter being under the purview of the organization and the former (for the most part) under the jurisdiction of subcommunities and volunteer editors.[63] For instance, Wikimedia will take action against harassment, defined as conduct. Content decisions in subcommunities can be done through edict (as is the case in Reddit) or "through a consensus process done over many years of building up policies" more akin to the Wikimedia model.[64]

Such a distribution of powers also finds an analogy within formal political structures of governance, with one representative for Reddit comparing its model to a "federal system" with baseline sitewide rules that must be obeyed but can be added to at the discretion of a moderator. Although this model creates some problems as communities with different norms interact on the site, policy managers from these companies tend to believe that enabling communities to make their own rules produces a greater sensitivity to potential cultural context concerns that the parent organization could not address on its own. The companies themselves tend to only get involved if users disobey the sitewide rules, leaving enforcement up to individual communities when it comes to additional rule sets.

In many ways, the community-reliant model is what was once imagined by scholars like Yochai Benkler in his embrace of "commons-based peer production."[65] The idea that volunteer labor was necessary to maintain the openness of online communities was also embraced by James Grimmelman on his early work on the "virtues of moderation" at Wikimedia.[66] But volunteer labor breeds its own concerns. There have been frequent controversies at Reddit, as subreddits come into conflict with one another, with subcommunities (like /r/pizzagate and others so heinous that they will not be named here) pushing the boundaries of what the wider community considers acceptable.[67] In these instances, Reddit has often been slow to respond, and when it has responded, corporate leadership has been attacked for new policies directed toward preventing serious concerns like harassment of marginalized groups.[68] Such disagreements get to the heart of problems with a distribution of responsibility of moderation between parent organizations and subcommunities. When combined with volunteer labor, there is a sense of ownership over decision making felt by users that makes broad, large-scale changes difficult. This can be problematic when, for instance, the reliance on volunteer labor produces inequalities in content moderation that the company must address. This has been the case with Wikimedia's reliance on volunteer labor, which has consistently shown to lead to gender bias in content because of the highly disproportionate number of male moderators.[69] In this sense, users play a much more active role in the daily governance of the sites, and because of this, rule making can often seem grounded in the context of the subgroup. However, due to reliance on volunteer labor and relative lack of oversight, inequalities and bias often emerge on these platforms, both in terms of a lack of diversity of editors and in terms of the kinds of communities (and rules) that are created.[70]

The Move to Automation and the Danger of One Global Rule

The three strategies described here are in no way static, and one platform that uses an artisanal model for content moderation can evolve to an industrial approach (though not necessarily community reliant, without a huge shift in outlook) or might deploy several strategies simultaneously. For a platform like Facebook, what has constituted content moderation has changed significantly over the site's history. An early Facebook employee recounted how, when "Site Integrity Operations" first began in 2008, primarily to deal with an influx of terrorist content, the entire team was only twelve people sitting in a single room.[71] The team quickly grew as the platform scaled, and problems multiplied. Twelve people became five hundred, and, according to statement made by Facebook, reached twenty thousand workers (full-time and outsourced laborers) by the end of 2018, with offices all over the world.[72] This is in response to rapid user growth over the same time period, and acquisitions like Instagram and WhatsApp that have greatly expanded the company's reach around the world. For platforms like Facebook, understanding this period of growth is essential for understanding how its content governance strategies have been shaped.

It is also essential to understand how the three strategies interact with the move toward automation. Policy makers and technologists currently view automation as the solution to labor (and cost) concerns in content moderation as well as the seemingly insurmountable problems of disinformation and hate speech online.[73] Although this may have been possible for other content types, like spam and copyrighted content, researchers are skeptical that automation and artificial intelligence can succeed when it comes to contextual and cultural content, and have questioned whether machines, which lack judgment and nuance, would lead to the overcensoring of global speech.[74] For smaller companies using artisanal strategies, automation is currently minimal (spam and trademarked content) or nonexistent. However, these companies are often existing at a scale where such an artisanal approach is still manageable. For community-reliant organizations, automation is used in the form of bots that can be customized to each rule system in each community.[75] Evidence suggests that industrial content moderation is actually geared toward eventual automation. The use of formal rule structures, the requirements for high degrees of accuracy, and the use of third-party contractors (and admissions by companies like Facebook that they are increasingly using automation in detection for a wide variety of content) are all indicators of this move. The move toward automation will mimic the issues of industrial moderation, requiring the reduction of complex cultural concepts into simple, easy-to-execute terms, which has led to problems like a lack of cultural nuance in content moderation and potential overcensorship.

How to increase accountability, oversight, and democratic participation into policies made by platforms differs depending on the content moderation strategy and the organizational dynamics at each company. If policy makers choose to regulate private platforms to preserve or protect the digital public sphere, then which content moderation model, and what relationship with users, should be prioritized? Each model enables a different set of interactions with users and the "community" that the platform purports to serve. In each case, the rules of content moderation depend not on what is uploaded by users but on the resources and constraints at each company. Allowing platforms to self-regulate thus means that individuals can exist within many different governance structures at one time, with different rules and policies, and systems of adjudication and review.

Making Platform Governance Participatory, Localized, and Democratic

So how do we make these platforms, despite their variations and their status as private entities, more participatory and democratic? And why does it remain important to make these private governments, more accountable to their various publics, and through what means?

The question remains as to whether social media companies and other online platforms—artisanal, industrial, or community reliant—are playing a fundamentally different role in moderating the public sphere than other traditional media companies in their choices about what to print or air, or whether moderation of users is a fundamental part of the new gatekeeping role played by platforms.[76] In the past, it has generally been accepted (though criticized), that media companies, mostly privately owned within the United States, will make decisions about what content to air on their networks. For news media, this process of "gatekeeping" and "agenda setting" has been widely accepted and acknowledged; social, political, and financial pressures exerted on entities like news media agencies, both internal and external to the organization, can shape decision making on what is reported and how.[77] In the case of social media networks and other platforms, content is amplified according to a variety of pressures, both internal and external. Although algorithms play a role in prioritizing some content over others, such approaches are not independent from the policies of the company and its economic goals.[78] Just as mass media became concentrated over the course of the twentieth century, so, too, has the digital public sphere become centralized, with big-tech platforms like Facebook and Google not only dominating their specific products (social media and search) but also taking a massive share of digital advertising.

Content moderation, and the conditions under which user accounts are reviewed, removed, or deprioritized, are dependent on similar factors. The notion that platforms are neutral spaces, free from the financial, cultural, and regulatory demands that shaped information flows in past communication eras,[79] has been in question as this process of decision making, and its outcomes, becomes more visible. And yet, strategically, platforms have distanced themselves from the media industry, claiming they are fundamentally different because they focus on distribution, rather than production of content (a claim that is both no longer true, with YouTube and Facebook both producing their own content, and not relevant, given that much of the media industry has focused on distribution in the past as well).[80] Yet the individuals who own and work at platforms, as businesses, technologies, and as designed spaces, make choices about the type of content that is prioritized or made visible over their networks, through algorithms, through partnerships with other companies and organizations, or through human-led curating, or through moderation. And such decisions have clear consequences for public discourse. For instance, as Zeynep Tufekci discovered, during a moment of massive civil unrest across the United States against police violence, Twitter's trending topics picked up on the importance of protest events in Ferguson, whereas Facebook's system had prioritized the "ice bucket challenge" due to how the affordances were structured on each social network.[81] Facebook has been known, at other points, to change how its News Feed prioritizes content in line with the company's policy and economic aims.[82] In a sense, the idea that social media platforms in particular are a form of public space is what has differentiated them from other media companies (whose content they also now host, establish partnerships with, and with whom they now compete).

It is because of the reliance on user-generated content that the role of the individual citizen, rather than the platform, the editor, or the journalist, has been central in discussions on how to improve the digital public sphere. In their chapter, Joshua Cohen and Archon Fung make the case that the "digitally-mediated public sphere" offers some clear advantages over the twentieth-century mass-mediated public sphere of the broadcast era, due to the reduced barriers for access for citizens to produce their own media, circulate their ideas, and gain access to a wider range of views.[83] At the same time, however, they acknowledge there are complications with declaring digital technologies, like social media platforms, an unmitigated success for democracy. Reduced barriers to access, they argue, has meant that users, without the training and norms associated with journalism to refrain from spreading false information and hate speech, spread such content, whether intentionally or unintentionally, and without consequence. Indeed, Cohen and Fung place much of the responsibility for the "excesses of the digital public sphere" on content generated by users, "who fail to abide by the norms of common-good orientation, high-regard for truth, and civility," arguing that any solution should involve citizens exercising more "self-restraint," behaving more like the gatekeepers of the mass-mediated public sphere of the twentieth century. This is less a call for moderation by platforms, than self-moderation by citizens, and a process of broader socialization and education that would teach citizens how to behave with more civility online. Cohen and Fung shift the normative demands of platforms onto citizens themselves while still carving out a responsibility for platforms to design their own tools and nudges for users in accordance with norms that "foster freedom of expression, equal access to means of expression, diversity of views, access to reliable information, and opportunities for communicative action in the design and operation of social media and other digital platforms."

Taking a more bottom-up approach, like that employed by the community-reliant platforms done by volunteer participants of a network, could also be one step in the right direction toward making content rules more responsive to local demands and more participatory. However, there are a number of barriers to this approach. First, as explained earlier, the community-reliant model has to address concerns about bias, power dynamics, and accessibility. Second, it is difficult, even for the most discerning and well-trained individuals (including journalists themselves), to determine the source and credibility of content spread online. Evidence continues to emerge that much of the intentionally spread disinformation has come from state actors, political campaigns, and other entities engaged in strategic political communication. These campaigns are often well organized and funded, and they use tactics like leveraging the feeds of prominent social media influencers to make disinformation appear organic while also achieving broad reach.[84] In other cases, disinformation blogs and news media websites mimic the appearances and strategies of established media outlets, which have also amended their own strategies and tactics to be more amenable to the clicks and shares driving the social media (and now media) economy.[85] What's worse, telling the difference between truths and falsehoods online is about to get much harder, with artificial intelligence–based learning now being applied to images and videos to produce content known colloquially as "deep fakes."[86] Finally, the community-reliant model depends on platforms having a well-formulated, trustworthy, and public interest–oriented approach to the kinds of content they would like to incentivize on their networks through a process of nudging and guiding. So far, many of the platforms discussed in this chapter have made huge strides in expanding their teams and crafting community guidelines to include prohibitions against disinformation and hate speech. However, research shows that establishing partnerships with fact-checkers and using design strategies such as trust markers or "disputed" tags may actually have increased misinformation or strengthened trust in content that is not yet tagged (fact-checking often works slower than the pace of social media).[87] Such concerns lead Facebook to abandon efforts to tag content in this way.[88] On the other hand, giving sole power to these social media companies to simply embed these values into how content is prioritized also presents concerns given that many social media platforms, particularly Facebook, have failed to be responsive and honest in relation to long-standing public concerns.[89]

Perhaps in response, a number of governments have sought to formalize incentives for platform companies to increase resources for content moderation and have urged them to consider the public interest perspective. As disinformation spread and hate speech become a greater concern, regulatory bodies, particularly outside of the United States, have been concerned about allowing US-based companies to define acceptable speech within their borders. In Germany, the passage of the Netzwerkdurchsetzungsgesetz (NetzDG), a law aimed at combating false news and hate speech over social media, places the same restrictions against hate speech over social media that have been placed on other media in the past.[90] In France, regulators, concerned about whether Facebook is capable of combatting hate speech, are being embedded within the company to study its efforts at self-governance (leading to potentially new concerns about preemptive regulatory capture).[91] Other countries across Europe are enforcing individual decisions made by courts to remove offensive content, such as the decision by an Austrian court requiring Facebook remove what the court perceived as "hateful" comments against the Green Party leader there.[92] Some governments are taking this opportunity to define the contours of acceptable debate, which may not always fall in their citizens' favor. In Malaysia, new laws against "fake news" are sparking concerns that such rules will be used to enact censorship and will be used by governments to punish detractors and critical individuals or media. According to reports from Reuters, the Malaysia Anti-Fake News Act has already been used to convict a Danish citizen over inaccurate criticism of police over social media.[93] Other governments, like that of Sri Lanka, concerned that social media is responsible for violence within their countries have blocked sites entirely or for short periods of time.[94] Whether this leads to a further balkanization of the internet or to the adoption of global or at least regional standards will depend on the resources available to content policy teams inside platform companies. As one incentive, regulations, particularly those emerging from Europe, are pushing platform companies like Facebook to be more transparent about their content moderation practices. As a model to increase accountability of these companies, however, these laws must be considered in the context of the health of the democracy (or lack thereof) within the countries making the rules.

Conclusion

I have identified three distinct approaches to content moderation in a general model of self-regulation by platform companies. These companies, which had previously declared themselves not to be the "arbiters of truth," are mediating cultural, philosophical, and political disagreements around the world about what constitutes truth, hate, and authoritative information. As companies seek to make more and more content decisions outside of legal requirements, there is increasingly concern about whether they have too much power to regulate speech or whether they are doing too little to combat hate speech and disinformation. The debate hinges on whether we consider the information environment created by platforms as public or private spaces. Platforms, recognizing a rising appetite for public oversight, have begun to reconfigure, at least in rhetoric, their spaces as private communities. However, though platforms often articulate these values as "shared," current models for content governance—artisanal, industrial, and community reliant—are largely one-sided, with different trade-offs for speech and safety of users. The organizational dynamics, and the interactions with users, play an important role in the classification of content and the enforcement of standards.

As we seek to understand how to make these platforms more participatory and democratic, we must consider a number of issues. Do platforms play a fundamentally different role from that of existing media companies as editors of content? The fact that social media platforms in particular host user-generated content requires special considerations in terms of increasing accountability regarding discussions of speech. Relying on users alone to change their behavior, or on platforms to nudge individuals toward better decisions, is limited in its potential to deter disinformation and hate speech. Even so, government regulations to incentivize platforms to address harms have been piecemeal and present additional concerns. However, given variations and significant limitations in developing and enforcing rules, the current model of private governance by platforms is perhaps the most problematic. And yet it has continually been put forward as the only option given legal, technical, and social constraints. Although it may be the most feasible, we are left with the question of how we make the various bureaucracies of content moderation more participatory and democratic, regardless of politics.

Notes

1. Alex Hern, "Facebook, Google, and Twitter to Testify in Congress over Extremist Content," *The Guardian*, January 10, 2018, https://www.theguardian.com/technology/2018/jan/10/facebook-google-twitter-testify-congress-extremist-content-russian-election-interference-information.

2. "Facebook CEO Mark Zuckerberg Hearing on Data Privacy and Protection," C-SPAN, April 10, 2018, https://www.c-span.org/video/?443543-1/facebook-ceo-mark-zuckerberg-testifies-data-protection.

3. Francesca Tripodi, *Searching for Alternative Facts: Analyzing Scriptural Inference in Conservative News Practices* (New York: Data & Society Research Institute, 2018).

4. Francesca Tripodi, "Alternative Facts, Alternative Truths," Data & Society Points, February 23, 2018, https://points.datasociety.net/alternative-facts-alternative-truths-ab9d446b06c.

5. Samuel C. Woolley and Philip N. Howard, "Computational Propaganda Worldwide: Executive Summary" (Working Paper No. 11, Oxford Internet Institute, Oxford, UK, 2017).

6. Robyn Caplan, *Content or Context Moderation: Artisanal, Community-Reliant, and Industrial Approaches* (New York: Data & Society Research Institute, 2018).

7. For an example, see positive coverage of Tucker Carlson on Breitbart, in "Google Should Be Regulated Like the Public Utility It Is," August 15, 2017, https://www.breitbart.com/video/2017/08/15/tucker-carlson-google-regulated-like-public-utility/. Elizabeth Warren, "Reigniting Competition in the American Economy" (Keynote Remarks, New America Open Markets Program Event, June 29, 2016, Washington, DC). This argument was also put forward by Mark Andrejevic in "Public Service Media Utilities: Rethinking Search Engine and Social Networking as Public Goods," *Media International AustralFia* 146, no. 1 (2013): 123–32.

8. Tarleton Gillespie, *Custodians of the Internet: Platforms, Content Moderation, and the Hidden Decisions That Shape Social Media* (New Haven, CT: Yale University Press, 2018).

9. For an example of ongoing investigative work in this area, see *The Guardian's* "The Facebook Files," available at https://www.theguardian.com/news/series/facebook-files.

10. Sarah T. Roberts, "Digital Detritus: 'Error' and the Logic of Opacity in Social Media Content Moderation," *First Monday* 23, nos. 3–5 (2018).

11. Facebook Transparency Report, https://transparency.facebook.com/ (n.d.).

12. Roberts, "Digital Detritus."

13. James Grimmelman, "The Virtues of Moderation," *Yale Journal of Law & Technology* 17 (2015): 44–109.

14. Sarah T. Roberts, "Content Moderation," in *Encyclopedia of Big Data* (New York: Springer, 2016).

15. Robyn Caplan and Tarleton Gillespie, "Tiered Governance, Demonetization, and the

Shifting Terms of Labor and Compensation in the Platform Economy," *Social Media + Society* (2020) [forthcoming]/

16. Grimmelman, "Virtues of Moderation."

17. Kate Klonick, "The New Governors of Speech: The People, Rules, and Processes Governing Online Speech," *Harvard Law Review* 131 (2017): 1598–1620.

18. Klonick.

19. Gillespie, *Custodians of the Internet*, 229.

20. The Communications Decency Act, 47 U.S.C. § 230 (1996).

21. Jürgen Habermas, "The Public Sphere: An Encyclopedia Article," *New German Critique* 3 (1974): 49–55.

22. Stuart Geiger, "Does Habermas Understand the Internet? The Algorithmic Construction of the Blog/Public Sphere," *Gnovis: A Journal of Communication, Culture, and Technology* 1, no. 10 (2009): http://www.gnovisjournal.org/2009/12/22/does-habermas-understand-internet-algorithmic-construction-blogopublic-sphere/.

23. Nancy Fraser, "Rethinking the Public Sphere: A Contribution to the Critique of Actually Existing Democracy," *Social Text* nos. 25–26 (1990): 56–80. And see Geiger, "Does Habermas Understand the Internet?"

24. Yochai Benkler, *The Wealth of Networks: How Social Production Transforms Markets and Freedom* (New Haven, CT: Yale University Press, 2006).

25. See also Zizi Papacharissi, "The Virtual Sphere: The Internet as a Public Sphere." *New Media & Society* 4, no. 1 (2006): 9–27.

26. M. Carpini and S. Keeter, "The Internet and an Informed Citizenry," in *The Civic Web* (Lanham, MD: Rowman & Littlefield, 2002), 129–53.

27. Jean Burgess and Joshua Green, *YouTube* (Cambridge, UK: Polity Press, 2009).

28. A. Mickoleit, *Social Media Use by Governments* (Working Papers on Public Governance No. 26, Organisation for Economic Co-operation and Development, 2014).

29. Fraser, "Rethinking the Public Sphere."

30. Papacharissi, "Virtual Sphere."

31. See Joshua Cohen and Archon Fung's contribution to this volume.

32. Paul Schiff Berman, "Cyberspace and the State Action Debate: The Cultural Value of Applying Constitutional Norms to Private Regulation," *University of Colorado Law Review* 71 (2000): 1263.

33. Berman.

34. Caplan, *Content or Context Moderation.*

35. Interview with Monika Bickert, first published in Caplan, *Content or Context Moderation.*

36. Gillespie, *Custodians of the Internet.*

37. Sherry R. Arnstein, "A Ladder of Citizen Participation." *Journal of the American Planning Association* 35, no. 4 (1969): 216–24.

38. Caplan and Gillespie, "Demonetized." See also *Harvard Law Review* 131 (2018): 1715–22.

39. Josh Constine, "2.5 Billion People Use at Least One of Facebook's Apps," *TechCrunch*, July 25, 2018, https://techcrunch.com/2018/07/25/facebook-q2–2018-earnings/.

40. Adi Robertson, "YouTube Has 1.8 Billion Logged-In Viewers Every Month," *The Verge, May 3, 2018,* https://www.theverge.com/2018/5/3/17317274/youtube-1-8-billion-logged-in-monthly-users-brandcast-2018.

41. Casey Newton, "The Trauma Floor: The Secret Lives of Facebook Moderators in America," *The Verge*, February 25, 2019, https://www.theverge.com/2019/2/25/18229714/cognizant

-facebook-content-moderator-interviews-trauma-working-conditions-arizona. See also Sarah T. Roberts, *Behind the Screen: Content Moderation in the Shadows of Social Media* (New Haven, CT: Yale University Press, 2019).

42. Caplan, *Content or Context Moderation.*

43. Caplan.

44. Caplan.

45. Caplan.

46. Caplan.

47. Interview with C, February 18, 2018, phone interview.

48. Interview with S, March 9, 2018, phone interview.

49. Interview with S, March 9, 2018, phone interview.

50. For a depiction of industrialized content moderation, see Ciaran Cassidy and Adrian Chen's film *The Moderators*, available at https://vimeo.com/213152344. See also J. Clement, "Number of Monthly Active Facebook Users Worldwide as of 3rd Quarter 2019," *Statista*, January 3, 2020, https://www.statista.com/statistics/264810/number-of-monthly-active-facebook-users-worldwide/.

51. Adrien Chen, "The Laborers Who Keep Dick Pics and Beheadings Out of Your Facebook Feed," *Wired*, October 2014, https://www.wired.com/2014/10/content-moderation.

52. Comments made by Nora Puckett, senior litigation counsel for Google, at the Content Moderation at Scale conference, May 7, 2018, Washington DC.

53. Caplan, *Content or Context Moderation.*

54. Mary L. Gray and Siddarth Suri, *Ghost Work: How to Stop Silicon Valley from Building a New Global Underclass* (New York: Eamon Dolan/Houghton Mifflin Harcourt, 2019).

55. Casey Newton, "The Trauma Floor: The Secret Lives of Facebook Moderators in America," *The Verge*, February 25, 2019, https://www.theverge.com/2019/2/25/18229714/cognizant-facebook-content-moderator-interviews-trauma-working-conditions-arizona.

56. Kate Crawford and Tarleton Gillespie, "What Is a Flag For? Social Media Reporting Tools and the Vocabulary of Complaint," *New Media & Society* 18, no. 4 (2014): 410–28.

57. "Hate Speech," Facebook Transparency, https://transparency.facebook.com/community-standards-enforcement#hate-speech.

58. Paul Mozur, "Groups in Myanmar Fire Back at Zuckerberg," *New York Times*, April 5, 2018, https://www.nytimes.com/2018/04/05/technology/zuckerberg-facebook-myanmar.html. See also Caplan and Gillespie, "Demonetized."

59. Klonick, "New Governors of Speech."

60. Aimee Dawson, "Facebook Censors 30,000 Year-Old Venus of Willendorf as 'Pornographic,'" *The Art Newspaper*, February 27, 2018, https://www.theartnewspaper.com/news/facebook-censors-famous-30-000-year-old-nude-statue-as-pornographic.

61. Anthony Kuhn, "Activists in Myanmar Say Facebook Needs To Do More to Quell Hate Speech," NPR, June 14, 2018, https://www.npr.org/2018/06/14/619488792/activists-in-myanmar-say-facebook-needs-to-do-more-to-quell-hate-speech.

62. Sarah T. Roberts, "Commercial Content Moderation: Digital laborers' Dirty Work," in *The Intersectional Internet: Race, Sex, Class and Culture Online*, ed. Safiya Umoja Noble and Brendesha M. Tynes (New York: Peter Lang, 2016), 147–59.

63. Interview with Jacob Rogers from Wikimedia. First published in Caplan, *Content or Context Moderation.*

64. Interview with Rogers.

65. Benkler, *Wealth of Networks.*

66. Grimmelman, "Virtues of Moderation."

67. Andrew Marantz, "Reddit and the Struggle to Detoxify the Internet," *New Yorker,* March 19, 2019, https://www.newyorker.com/magazine/2018/03/19/reddit-and-the-struggle-to-detoxify-the-internet.

68. Jason Koebler, "Right Now, Reddit's Top Posts Are Swastikas, Fat Shaming, and Ellen Pao Hate," *Vice Motherboard,* June 11, 2015, https://motherboard.vice.com/en_us/article/nzeygb/right-now-reddits-top-posts-are-swastikas-fat-shaming-and-ellen-pao-hate.

69. Emma Paling, "Wikipedia's Hostility to Women," *The Atlantic,* October 21, 2015, https://www.theatlantic.com/technology/archive/2015/10/how-wikipedia-is-hostile-to-women/411619/.

70. Joseph Reagle and Lauren Rhue, "Gender Bias in Wikipedia and Britannica." *International Journal of Communication* 5 (2015): 1138–58.

71. Caplan, *Content or Context Moderation.*

72. Anita Balakrishnan, "Facebook Pledges to Double Its 10,000-Person Safety and Security Staff by End of 2018," CNBC.com, October 31, 2017, https://www.cnbc.com/2017/10/31/facebook-senate-testimony-doubling-security-group-to-20000-in-2018.html.

73. Drew Harwell, "AI Will Solve Facebook's Most Vexing Problems, Mark Says: Just Don't Ask When or How," *Washington Post,* April 11, 2018, https://www.washingtonpost.com/news/the-switch/wp/2018/04/11/ai-will-solve-facebooks-most-vexing-problems-mark-zuckerberg-says-just-dont-ask-when-or-how/?utm_term=.0ad6e8c49135.

74. James Vincent, "AI Won't Relieve the Misery of Facebook's Human Moderators," *The Verge,* February 27, 2019, https://www.theverge.com/2019/2/27/18242724/facebook-moderation-ai-artificial-intelligence-platforms.

75. Reddit, "/r/Automoderator," https://www.reddit.com/r/AutoModerator/.

76. Zeynep Tufekci, "Algorithmic Harms beyond Facebook and Google: Emergent Challenges of Computational Agency," *University of Colorado Law Review* 13 (2015): 203–18.

77. Stephen D. Reese and Jane Ballinger, "The Roots of a Sociology of News: Remembering Mr. Gates and Social Control in the Newsroom," *Journalism and Mass Communication Quarterly* 76, no. 4 (2001): 641–58.

78. Robyn Caplan and danah boyd, "Isomorphism through Algorithms: Institutional Dependencies in the Case of Facebook," *Big Data & Society* (January 2018), https://journals.sagepub.com/doi/full/10.1177/2053951718757253.

79. Tarleton Gillespie, "The Politics of 'Platforms,'" *New Media & Society* 12, no. 3 (2012): 347–64.

80. Philip M. Napoli and Robyn Caplan, "When Media Companies Insist They're Not Media Companies: Why They're Wrong, and Why That Matters," *First Monday* 22, no. 5 (2017): https://firstmonday.org/ojs/index.php/fm/article/view/7051/6124.

81. Tufekci, "Algorithmic Harms beyond Facebook and Google."

82. Caplan and boyd, "Isomorphism through Algorithms."

83. See Cohen and Fung's chapter "Democracy and the Digital Public Sphere" in this volume.

84. Jonathan Corpus Ong and Jason Vincent A. Cabñes, "Architects of Networked Disinformation: Behind the Scenes of Troll Accounts and Fake News Production in the Philippines," 2018, 74, https://doi.org/10.7275/2cq4-5396.

85. Robyn Caplan, Lauren Hanson, and Joan Donovan, *Dead Reckoning: Navigating Content Moderation after Fake News* (New York: Data & Society Research Institute, 2018).

86. Robert Chesney and Danielle Citron, "Deep Fakes: A Looming Challenge for Privacy, Democracy, and National Security," *California Law Review* 107 (2019): 1753–1820.

87. Gordon Pennycook and David G. Rand, "The Implied Truth Effect: Attaching Warnings to a Subset of Fake News Stories Increases Perceived Accuracy of Stories without Warnings," SSRN, August 7, 2019, https://ssrn.com/abstract=3035384.

88. Kerry Flynn, "Facebook Abandons an Attempt to Label Fake New," *Mashable*, December 12, 2017, https://mashable.com/2017/12/21/facebook-fake-news-abandon-disputed-flag-related-articles.

89. Sheera Frankel, Nicholas Confessore, Cecilia Kang, Matthew Rosenberg, and Jack Nicas, "Delay, Deny and Deflect: How Facebook's Leaders Fought through Crisis," *New York Times*, November 14, 2018, https://www.nytimes.com/2018/11/14/technology/facebook-data-russia-election-racism.html.

90. Claudia Haupt, "Online Speech Regulation: A Comparative Perspective," American Political Science Association, August 30, 2018, Boston.

91. Mathieu Rosemain, Michel Rose, and Gwenaëlle Barzic, "France to 'Embed' Regulators at Facebook to Combat Hate Speech," Reuters, November 12, 2018, https://www.reuters.com/article/us-france-facebook-macron/france-to-embed-regulators-at-facebook-to-combat-hate-speech-idUSKCN1NH1UK.

92. Davey Alba, "A Court Order to Terminate Hate Speech Tests Facebook," *Wired*, May 9, 2017, https://www.wired.com/2017/05/court-order-terminate-hate-speech-tests-facebook/.

93. "First Person Convicted under Malaysia's Fake News Law," *The Guardian*, April 30, 2018, https://www.theguardian.com/world/2018/apr/30/first-person-convicted-under-malaysias-fake-news-law.

94. Zaheena Rasheed and Amantha Perera, "Did Sri Lanka's Facebook Ban Help Quell Anti-Muslim Violence?" *Al Jazeera*, March 13, 2018, https://www.aljazeera.com/news/2018/03/sri-lanka-facebook-ban-quell-anti-muslim-violence-180314010521978.html.

7

The Democratic Consequences of the New Public Sphere

Henry Farrell and Melissa Schwartzberg

The ignorance of democratic citizens and the poor quality of democratic decision making perennially concern political theorists and political scientists. Most recently, a variety of libertarian scholars—Jason Brennan (2016), Bryan Caplan (2007), and Ilya Somin (2016)—have written books claiming that citizen ignorance and bias fundamentally undermine standard arguments for democracy. Scholars like Chris Achen and Larry Bartels (2016) concur in large part with this judgment while holding out hope for a "realist" group-based account of democracy.

In separate work with Hugo Mercier, we find that this line of skepticism is partly misplaced, arguing that cognitive psychology provides strong foundations for a "no-bullshit" approach to understanding—and perhaps improving—democracy. In this chapter, we tackle a related but more particular set of worries. Both political science scholarship and popular debate suggest that the advent of social media exacerbates many of the problems that democratic skeptics have highlighted. Specifically, they claim that the new public sphere that search technology and social media is creating, contrary to initial enthusiasm, worsens ignorance by spreading myth and falsehoods. They believe that it accentuates the problems caused by cognitive bias, not only by making it easier to seek out bias-confirming information but also by deploying near-invisible algorithms that create "filter bubbles." These bubbles filter out information that disconfirms our priors while highlighting information that strengthens them.

In this chapter, we look first to critically assess these claims, assaying the evidence that social science research provides about the relationship of information, bias, and online technology. We note that many of these claims are likely overblown, although some are indeed troubling. We argue for an alternative way of thinking about how these new technologies are reshaping public and political knowledge by focusing on transformations in the informational environment, and its corresponding transformations to the structure of "epistemic" trust, changes in the sources on which citizens rely to form beliefs about the world, particularly politically relevant beliefs.

Our aim is to begin to investigate whether diminished trust in traditional news sources constitutes an important challenge to democratic decision making. To fully address that question, we would first need to have a thorough account of the sources of belief on which democratic citizens should rely. This would require us to address a set of thorny epistemological questions: What does it mean to trust sources of evidence? What makes certain sources of politically relevant information more reliable than others? When citizens disagree about the reliability of sources, how should they resolve these disagreements? We raise these questions to set aside most of them, as our project here is necessarily circumscribed. Drawing on the substantial literature in social epistemology on epistemic authority and epistemic trust, we consider possible transformations in how citizens determine that some source will be authoritative for them, whether in the strong sense of "preempting" their own reasons for belief (following the work of Linda Zagzebski), or—perhaps more plausibly—in the weaker sense of providing agents with additional reasons for holding existing beliefs or enabling them to update their beliefs through a Bayesian process, for instance.[1]

We take up these issues in the context of a discussion of how new forms of media have generated a change from a "strong gatekeeper" approach to public debate, in which access to media was heavily circumscribed, to a "weak gatekeeper" approach, where access is largely unregulated (although attention is shaped by algorithms and commercial decisions). Even as information flows have been increasingly channeled into the great rivers of content that flow through Google, Facebook and YouTube, they have not been dammed or diked (although they have been redirected). Efforts to control political content have been quite controversial (at least in the US context) and have aimed only at the most egregious and persistent purveyors of falsehoods. In brief, this transformation—from strong gatekeeper to weak gatekeeper—means that citizens no longer trust the same sources of information, and the reliability of the sources they do trust varies substantially.

The move from strong to weak gatekeeping may have salutary democratic benefits. It opens public discussion to new and different communities that would have had difficulty in articulating their shared problems and conditions, and organizing around them, in an earlier world. However, it also creates new vulnerabilities for democracy, exposing it not only to less attractive publics but also to forms of abuse and attack that involve "flooding" democratic debate with nonsense and deceit, to the point that some forms of democratic conversation might even become unviable (Farrell and Schneier 2018; Hasen 2017). By thinking more systematically about the changes wrought by new forms of communication, we can move away from both the uninformed adulation of a few years ago and the unmitigated pessimism of much current discussion to a better understanding of the trade-offs, and the strengths and weaknesses, of different media systems and their associated public spheres.

Trust and Epistemic Authority

The standard argument about facts and democracy runs something like this: without shared beliefs about vital facts, we cannot hope to have any rational basis for political decision making or justification for the exercise of political power. In Hannah Arendt's (1967, 88) oft-cited words, factual truth is the "ground on which we stand and the sky that stretches above us"; it constitutes the foundation of our shared political life. This raises an important question: when the communicative structures through which citizens learn political facts, and exchange perspectives about those facts, change, what are the consequences for democratic decision making?

Put synoptically, we argue that these transitions entail challenges to the structure of epistemic trust. We form many of our beliefs by relying on information supplied by others rather than through independent investigation. Unless we ourselves are scientists, economists, or sociologists, we typically rely on experts to help us answers of the following sort: Is the planet growing warmer? Is the economy improving? Is crime on the rise? Indeed, even the experts must rely on broader social systems of information (e.g., peer-reviewed publications). Yet whereas citizens once trusted traditional media sources to provide grounds for their politically relevant beliefs, challenges to the "mainstream media" have meant that citizens turn to diverse sources of information, which have uneven commitments to accuracy and to the value of expertise. Indeed, some such sources may deliberately seek to undermine claims to expertise on the part of scientists, for instance, arguing that their research is ideologically driven or motivated by the pursuit of research grants. But from the outset of our discussion, we want to emphasize that this is not solely a feature of the "new media"; indeed, the *Wall Street Journal*, as traditional a source as one might seek, has sought to attack climate science in their editorials. Yochai Benkler, Robert Faris, and Hal Roberts (2018) argue that the key epistemic problems can be found in Fox News rather than Facebook. So worrisome challenges to epistemic authority are not strictly a feature of the transition in the gatekeeping model, although the weak gatekeeping model may exacerbate these tendencies.

First, let us clarify what we mean when we discuss epistemic trust and authority in this context. As already noted, there is a substantial literature in philosophy on epistemic trust and authority, on the conditions under which we can claim that a person relies on an authority—an "epistemic superior," if not an "expert"—to possess relevant knowledge or competence with respect to some domain, and to provide testimony or information that gives that person reason to believe what the authority reports. Broadly, we construe trust as a three-part relationship: A trusts B with respect to X (Hardin 2002). What trust means, particularly in the epistemic context—whether it is a special form of reliance or dependence, or an attitude on the part of a truster toward a trustee—is a matter of disagreement (McCraw 2015). Most of these issues we set aside, however, as we do not intend to provide a distinctive contribution to the epistemology of trust or testimony, or even to the theory of belief formation in the face of disagreement among authorities. Rather, we seek to clarify the changes in the mechanisms by which people identify epistemic authorities in the form of media sources that they trust to provide politically relevant information, and on which they rely in forming politically relevant beliefs and judgments. One way to characterize our basic question is: why does A trust B to be reliable with respect to information X, whereas C trusts D to be reliable with respect to information X? After all, if we are epistemically responsible agents who wish to form true beliefs, we should tend to converge on the sources we trust to be reliable, including media sources. So why do we diverge?

A few examples highlight the basic structure of epistemic trust and potential sources of disagreement. Leah trusts Emily Nussbaum at the *New Yorker* to provide reliable information about which television shows are worth watching. Note immediately that this might mean that Leah trusts the way in which Emily Nussbaum forms her beliefs, or it might mean that she relies on the *New Yorker* to provide the wider institutional context in which Nussbaum and other writers form their beliefs. Imagine now that the difference between Leah and Emily Nussbaum in their competence with respect to their ability to form beliefs about which television shows are worth watching is not very significant; Emily Nussbaum's main advantage is her ability to allocate considerable time to the project that Leah lacks. As a result, Leah could without great difficulty independently determine whether Emily Nussbaum remains reliable when she makes claims on her personal Twitter feed rather than in the pages of the *New Yorker*. We might think that the more equal Leah and Emily Nussbaum are in their capacity to form true beliefs, the less Leah would seem to need to rely on Nussbaum; Leah could reasonably withdraw her dependence on Emily Nussbaum, or in the case in which she is untethered by the disciplining constraints of the *New Yorker*, without incurring significant epistemic harm, although her search costs would probably increase.

But consider another, politically significant context of epistemic trust. Izzy relies on the *New Yorker* for information about the anthropogenic sources of climate change. The *New Yorker* itself is not conducting scientific inquiry; rather, it is reporting the findings of scientists, as published in scientific journals such as *Nature*. So Izzy's confidence in a traditional news organization may in fact derive from his beliefs about the reliability of peer-reviewed journals in particular or about the reliability of scientific inquiry—conducted through the project of peer-verification—more generally. But the *New Yorker* also has a famously rigorous fact-checking process. Of course, one question is *whom* does Izzy trust, or *who* is the trustee? Does he trust the *New Yorker*, or is the *New Yorker* is merely a conduit to *Nature* or to peer-reviewed scientific research? Izzy knows that he is not competent to evaluate holistically the evidence in support of the claim that global warming has human causes; it is rational for him to rely on the beliefs of those who are epistemically better situated. Nor is he capable of comprehending the expert-level presentation of scientific data presented in *Nature*. Instead, he trusts the claims concerning climate change supported by scientific consensus, and he trusts in the accurate reporting of such claims (and the existence of such consensus) by the *New Yorker*. In sum, Izzy is willing to trust information that the *New Yorker* itself deems reliable. Of course, if the data in the *Nature* article is fraudulent, or if the *New Yorker* misrepresents the scientific arguments, Izzy may have misplaced his trust, but he will not be in a position to realize that absent some other credible (to him) source debunking the claims.

Again, these accounts presume that people are in fact highly motivated epistemic agents who seek to hold only true beliefs. Although this might be the case in many domains—people want to hold true beliefs about certain matters (e.g., how their local public school is performing relative to others, whether that suspicious-looking mole is malignant)—there is good evidence from cognitive psychology that individuals prefer to hold beliefs that are compatible with their prior judgments, enabling dubious epistemic mechanisms like confirmation bias and motivated reasoning to operate. In particular, if people are politically motivated and highly partisan, in the sense that they positively evaluate their party's own policy position and negatively evaluate those policies advocated by the opposing party, they may choose to trust sources that will provide further evidence for the political beliefs they already hold.[2] We might frame this as the following: people's trust in the epistemic reliability of sources may derive from their trust in their sources' political reliability, that is, the extent to which these sources affirm their preexisting partisan commitments. If correct, this is worrying; at the limit, it suggests that our desire to hold true beliefs is secondary to our desire to maintain and strengthen our partisan commitments.

We draw a somewhat different inference. We hold that people are not in fact indifferent to the reliability of the information provided by media sources, and that partisanship has not swamped their cognitive faculties so utterly that they can no longer distinguish reliable from unreliable reports. Rather, our argument is that this is a basically rational response on the part of individuals to the fragmentation of the media environment under the weak gatekeeping structure, albeit a response with troubling consequences.

To characterize this structure briefly, a 2018 article in *Science* holds that by lowering the cost of entry to new media competitors, many of which do not adhere to norms of objectivity and balance, the internet has undermined the "business models of traditional news sources that had enjoyed high levels of public trust and credibility" (Lazer et al. 2018, p. 1094).[3] What does it mean that people no longer "trust" the traditional news media?

Let's say, to begin, that it means that people no longer trust the traditional news media to report accurately concerning claims in some set *j*. Insofar as the claims in this set may be trivial or politically uncontroversial, or independently easy to verify, distrust of the news media just reduces to distrust of the reliability of reporters and editors. This would recommend remedies designed to improve in a public fashion the reliability of the information provided by such reporters. However, insofar as the claims in set *j* are in fact difficult to independently verify, or raise (even if indirectly) controversial moral or political questions, distrust in the news media may be epiphenomenal. Rather, what appears to be distrust in the news media may instead reflect a deeper distrust of scientific consensus, or of claims to moral or political expertise. Indeed, people may distrust both traditional epistemic authorities (e.g., scientific consensus) and the capacity of traditional news sources to accurately convey the (unreliable) consensus. We believe that this is the sort of worry that many have when they consider the breakdown in trust of traditional media: we are left in an epistemic vacuum which could easily be filled by fake news on social media or by divination. Although we are sympathetic to this worry, our aim is to make clear the underlying structure of the relationship between partisanship and epistemic trust so as to reveal potential harms and benefits for democracy.

Briefly, our argument is that under a weak gatekeeping structure, people require a heuristic by which they identify sites that will provide them with the information they seek. For many people, copartisanship constitutes this heuristic. In general, however, decline in trust across party lines has increased; indeed, interpartisan trust gaps in many democracies are wider than trust gaps across the identities that characterize underlying social cleavages (and to which parties correspond) (Carlin and Love 2018). Insofar as people trust their copartisans to provide them with more accurate information than those who oppose them, partisanship guides people in their selection of epistemic authorities.

Note—again—that this is a feature not merely of the online environment but of the weak gatekeeping model and the proliferation of media sources more generally. For instance, Kieran seeks reliable information as to whether citizens in the United States will benefit from, or be harmed by, the Affordable Care Act. There are countless sources to which Kieran might turn for this information, but because the answer is likely to require a great deal of information to which he lacks independent access, and because the answer is highly politically salient, he reasonably worries that sources are likely to disagree: they will weigh particular pieces of information differently, rely on different projections, and evaluate the normative benefits and harms from different vantage points (e.g., the prospect of coercion versus universal access). Because Kieran trusts Chris Hayes as a copartisan, believing that Hayes is likely to weigh and evaluate these pieces of information in a way that Kieran would approve of if he did have independent access to the information, Kieran trusts Hayes to be reliable with respect to information about the Affordable Care Act.

Again, Kieran has a range of sources available to him across the spectrum of partisan, and avowedly nonpartisan or independent, views. But in the new media landscape—which, as we shall argue, has only a weak gatekeeping structure—Kieran is not sure whom to trust. As such, he uses partisan valence as a guide for epistemic authority, at least concerning politically relevant information. To be sure, he likely does not use partisan valence (if he could identify it) in the selection of dermatologists for the evaluation of moles, although he may choose a dermatologist who attended the same college or otherwise has a shared cultural background. (Anecdotally, however, one of us knows parents on the Upper West Side of Manhattan who avoided seeing a pediatrician with Republican political leanings.) That is, in the context of a proliferation of prospective authorities, similar backgrounds may undergird our selection of whom to trust.

Here is a slightly different, more complicated, example. Fiona, a Republican, trusts former Fox commentator Bill O'Reilly to provide her with reliable information with respect to immigration.[4] That is, in a diffuse media environment, Fiona chooses to take Bill O'Reilly's information to be reliable because of Fiona's belief that O'Reilly shares Fiona's commitments *qua* Republican partisan. Again, information as to the economic costs and benefits of immigration is difficult for Fiona to access independently, and it likely entails countervailing, if not contradictory, pieces of evidence, yielding different and highly salient political judgments. Fiona believes that O'Reilly weighs this information in a way that Fiona herself would find convincing if she were to have unmediated access to the information.

However, when O'Reilly is fired from Fox News, Fiona must find a new source of information. She may choose to transfer his trust to Fox News more generally as an institutional source, although she may also blame Fox News for capitulating to pressure to fire him (thus undermining Fiona's relationship with Fox News). Or she may trust those alternative sources that Bill O'Reilly has endorsed, such as Breitbart, as she believes that Breitbart shares her conservative commitments (e.g., prioritizing the interests of compatriots over prospective members) and therefore trusts the evidence Breitbart provides (e.g., about the flow of violent immigrants from Mexico). Under this account, again, the factual accuracy of the information is not irrelevant; Fiona wants to hold true beliefs. But under the weak gatekeeping model, Fiona's decision to trust Breitbart derives primarily from her belief that Breitbart is engaged in a shared political project. Equally, her distrust of more traditional media may stem from her belief that these media sources are engaged in a competing political project. It is with these transformations in mind, we argue, that we should understand the issues that the shift from strong to weak gatekeeping models raise.

From Strong to Weak Gatekeeping

The standard story goes something along the following lines. In an earlier era (which was itself the product of a contingent meeting of market forces and technologies), news in the United States was dominated by large semi-oligopolistic organizations. Three dominant TV broadcast networks shaped television news and skewed it toward the center in pursuit of the median viewer (Prior 2007). An underdeveloped national market for newspapers (dominated by the *New York Times*, the *Wall Street Journal*, the *Washington Post*, and, in a somewhat different capacity, *USA Today*) was superimposed over a mostly local market for newspapers, which again was typically dominated in each locality by one or two publications. Radio was highly fragmented and localized because of technological limitations and ownership restrictions, but it also tended to steer away from political controversy because of the Federal Communications Commission's fairness doctrine.

What is notable about this period (itself the product of a temporary convergence of technologies and regulatory structures) is that it facilitated a strong gatekeeper model. The expense of print and broadcast technology implied a "few to many" information ecosystem, in which a relatively small number of organizations dominated the production of news. In general, market forces led news organizations to try to cater to the political center (with the exceptions of some local newspapers (e.g., the *Manchester Union-Leader*) and radio stations that took advantage of semi-monopolistic circumstances to put forward politically idiosyncratic views. Market forces, social norms, and regulatory instruments such as the fairness doctrine encouraged journalistic organizations to behave as informational gatekeepers, maintaining a standard of perceived factual reliability, and excluding politically controversial and unorthodox perspectives. Other aspects of the regime sometimes facilitated the distribution of nonorthodox opinions. Thus, for example, Clarence Manion's conservative talk show was widely distributed because it was cheap and because it satisfied public interest broadcasting requirements that were imposed on local radio stations (Hemmer 2018). However, this worked only to a point: when Manion made a controversial broadcast, he found that his access to the airwaves dwindled rapidly.

In describing this ecosystem, we emphatically do not want to extoll it as a lost golden age. To be sure—as John Dewey long ago noted—control over media can be used to stifle public opinion and the formation of publics whose interests may be averse to those of powerful actors. Many of the mainstream "facts" that passed through its gates appear today to be grossly inaccurate, based on unwarranted assumptions rooted in racism, sexism, and Cold War paranoia rather than on reliable evidence. Important factual information was often systematically filtered out or confined to small and heterodox forms of journalism, such as *I. F. Stone's Weekly*. From a Deweyan perspective, many possible publics found that their voices were excluded. African American perspectives, for instance, were sidelined to an insulated and subordinated sphere of communication. (We return to this issue in assessing the democratic implications of these transitions.)

A series of technological and political shocks disrupted the strong gatekeeping model. These shocks included the relaxation of rules on ownership of radio stations and abandonment of the fairness doctrine and the advent of cable television, as well as the creation of the internet, which both opened room for new amateur and for-profit publishers to challenge gatekeepers, and for new advertising models such as Craigslist that undercut newspapers' dominant share of local advertising. These tended on the whole to move away from a gatekeeping model based on a shared centrist consensus toward a model of looser control in which a wider variety of perspectives could find public expression.

However, the increasing nationalization of news markets provided a partial countervailing force. Market tendencies to economies of scale created a different set of pressures toward monoculture. The relaxation of ownership requirements allowed companies like Clear Channel to buy up swaths of local stations, replacing idiosyncratic local programming with automated playlists (sometimes disguised through prerecorded DJ patter referring to local events) and syndicated (typically conservative) opinion (Foege 2008). Cable TV allowed viewers to see any of hundreds of specialist channels but had strong "rich get richer" tendencies, so that large cable networks tended to prevail over smaller competitors while economies of scale pushed broadcasters to produce content at the national rather than the local level.

The conditions of news have changed yet again, thanks to services like Google, Facebook, and Twitter. These services do not operate as significant providers of news, and many operate under Section 230 of the Communications Decency Act, which provides them with safe harbor for disseminating content, and says that they should not be held responsible (e.g., for publishing readers' comments). Hence, in developing their own profit models, they have become general aggregators of news, collating and organizing news content together with other sources to make it findable through search and also circulating news items to their customers as part of their feed.

Replacing traditional gatekeeping, these online services feature a model of decentralized algorithmic curation that puts out key decisions as to what is newsworthy to readers. Under this model, sharing has become crucial. Stories live and die depending on how many clicks they receive, and hence how much online advertising revenue they generate. This model supplants one of news provision based on hierarchical status as a gatekeeper of reliable information, with one where gatekeeping involves a combination of personally chosen intermediaries (whether friends, family, influencers, celebrities, journalists or organizations on Twitter, Facebook and other media sites) and automated algorithmic curation. (The model may be changing again as companies like Facebook face political pressure over how their corporate strategies affect political news.)

Hence, at the national level, newspapers like the *New York Times* found that its conservative internal culture clashed with the new business imperatives of a social media–driven world (Usher 2014). New online services cannibalized existing markets for advertising, so that growth in advertising services in the United States is now nearly entirely captured by Google and Facebook. This obliged newspapers both to embrace the new economy and to look for alternative models of revenue generation. Journalists who saw themselves as authoritative dispensers of the truth felt obliged to create social media presences on Twitter and Facebook to drive interest in, and traffic to, their pieces. Lead articles at the *Times* and *Washington Post* suddenly had to jostle for position with pictures of cats and minor items of personal news in people's Facebook feeds.

Even worse, their editors and reporters found themselves competing against media operations that had grown up around entirely new business models. Upworthy, which was founded by Eli Pariser (whom we discuss further below) became notorious for crafting "clickbait" headlines that were designed to tease the reader into reading feel-good stories that were optimized for social media sharing. BuzzFeed and the *Huffington Post* thrived on quick, cheap content (although they both set up more substantial news operations as they grew), while Vox.com and other sites began with the mission of using new forms of presentation to make high quality content more attractive, often adopting a sharper political identity than traditional news organizations. New partisan websites such as Breitbart.com and the *Blaze* looked to challenge and outflank more traditional political publications. These new business models not only provided (sometimes fleeting) success to new businesses but also fundamentally challenged the approaches of traditional organizations.

For a period, traditional newspapers, which were already bleeding revenue, became more disinclined to engage in long-term investigative journalism with an uncertain payoff and more concerned with "winning" on social media and funding their operations from the associated advertising revenue. In many cases, they began to adapt or abandon traditional forms of dispassionately presenting information in favor of headlines and stories that were crafted to provide a more immediate response and to increase the likelihood that the story would be shared.

This arguably had some benefits, especially in bringing hitherto-unstated biases to the fore. When news sources became more overtly political (rather than treating their political assumptions as reliable background facts), it became easier for readers to evaluate the extent to which the news sources were balanced or unbalanced. Of course, this also may have increased distrust in news sources, which was sometimes warranted and sometimes less so. However, these new economies also provided the basis for new forms of news production that were parasitical on the reputation of preexisting gatekeeping structures. During the run-up to the 2016 election, for example, a small number of Macedonian websites became notorious for running websites with fake stories that were tailored precisely to cater to the political biases of

Donald Trump supporters. They did this not because the websites' owners were committed to Trump's victory, but because they perceived a profitable marketplace for such content. The combination of advertising (from Google and other sources) and cheap server space gave them a profit model based on the provision of active disinformation. Other businesses saw the same opportunity, building websites that made them appear to be "real" newspapers or broadcasters, including Paul Horner's notorious ABCnews.com.co (which had no relationship to ABC News), the Boston Tribune, and World News Report, all of which reported fake information for profit. Because there was no gatekeeper for the gatekeeping community itself, it was easy and cheap to set up websites that looked superficially like traditional publications.

The hitherto-slighted system of conservative radio and publishing became its own separate entity, defining itself in opposition to a "mainstream media" that it saw as inherently rotten and biased (Hemmer 2018; Benkler, Faris, and Roberts 2018). Talk radio created space for new forms of political organizing. The day that Newt Gingrich became Speaker of the House, he claimed on Rush Limbaugh's talk radio show: "Without C-SPAN, without talk radio shows, without all the alternative media, I don't think we'd have won. . . . The classic elite media would have distorted our message. Talk radio and C-SPAN have literally changed for millions of activists the way they get information."[5] Fox News deliberately took the talk radio format and adapted it to cable television (Sherman 2014). The advent of the internet provided new opportunities for conservatives: blogs (which rapidly became integrated into the existing conservative media system), websites like the *Drudge Report* which helped propagate conservative talking points to mainstream news, and eventually publishing websites such as Breitbart.com and the *Daily Caller*. The former was intended from the beginning as a conservative riposte to the *Huffington Post*. The latter was originally created as a right-wing rival to the quality journalism of the *New York Times* but rapidly moved toward a profit model based on publishing trash and innuendo. Both made money through advertising to a large conservative audience. Both became integral parts of a national disinformation ecosystem focused on Fox News that took claims, arguments, and perspectives from the fringes (e.g., Hillary Clinton was secretly dying; Clinton was the leader of a secret pedophile ring based at a DC pizzeria) and brought them into general circulation (Benkler, Faris, and Roberts 2018).

This converged with a nationalization of media markets around Facebook and Google, which had striking consequences for local newspapers. Again, previously, newspapers had an effective monopoly on certain forms of advertising aimed at local consumers, which they could use to support news services aimed at much the same population. Now, they find themselves outcompeted by automated algorithmic markets, which allow advertisers not only to target consumers in a particular region but also to show ads only to specific subdemographics within that region. The result is, as Case's chapter discusses, a sharp and general decrease in the quality of local news. There are also indirect consequences as national and international predatory actors are able to leverage the weaknesses in algorithms to undermine local economic knowledge, for example through creating websites that appear to focus on local news, but instead (like the radio stations of an earlier era) provide a mixture of canned and algorithmically generated content (Bengani 2019).

We present this potted history for a reason: these macro-level changes in media structures have important consequences for political knowledge. They mean that we have moved from a media model where there was relatively strong—and politically centrist—gatekeeping to a model featuring open gates. Under the old model, publishing was expensive and access was difficult. Now publishing is cheap, and access (or at least the ability to publish content, if not to draw attention to it) is easy. This, of course, has not led to a state of true openness. A publishing economy based on scarcity of access has been replaced by one based on scarcity of attention. However, there are important differences. An economy of broadcast networks and printing presses was one in which both economic and political forces made it easy to control the flow of opinion. Now, in contrast, even if it is hard to gain attention, it is hard to control it or use what control there is to enforce a shared consensus.

Micro-Level Changes in the Public Sphere

What are the consequences of this transition for the structure of epistemic trust underlying democratic decision making? Optimists point to the benefits of increased openness in encouraging democratic participation. Although some of the exuberant claims of the early internet are increasingly implausible, others still have purchase. Thus, for example, the decay of centrist gatekeeping means that the public has much greater access to a variety of voices, as well as a historically unprecedented ability to express their own voice (even if it is only irregularly heard by people outside their own close circles of intimates). Entire subcultures that had been hitherto denied the ability to understand and express their common interests can now become self-organized and self-aware publics. This serves to some degree as a corrective to the tendencies of elite media to focus on the problems and questions that are most likely to interest upper-middle-class people in urban environments. It would have been much harder for Black Lives Matter to become a self-aware and active force on the national level without some mobilizing technology, and it would have been especially hard in a previous America where the national channels of communication were dominated by a centrist consensus that only sporadically paid attention to the issues of the African American community. It is also true that elite media has tended until recently to downplay the problems of white rural America, in part because of the sorting processes identified by Case and Deaton (2020), where people of different classes and social backgrounds have come increasingly to occupy separate social universes. Of course, one should also note that some publics—for instance, those that are actively racist, fascist, or given to deranged conspiracy theory (e.g., QAnon)—are those that many of us might prefer to do without.

In contrast, pessimists draw on three closely linked sets of arguments, which we discuss here. However, as we note, these claims fail to fully acknowledge the larger structural change in epistemic trust: notably, the bifurcation of the public sphere into a left-through-center media ecosystem and a right-wing ecosystem that operate according to very different epistemic norms.

First, Cass Sunstein and others have claimed that the Internet was generating group polarization, because it made it easier for people to find other like-minded people to talk to and avoid people who disagreed, making all in the group more extreme. Sunstein (2017) suggested that increasing "homophily"—the tendency of likeminded people to cluster together and to actively seek out information that supported their existing prejudices—would lead to increased extremism. Changes in technology meant that people had access to a far greater variety of voices than before, including people who had extreme views. This in turn meant that people might gravitate toward online sources and interlocutors that shared their general opinions, and that interactions among all these like-minded people could lead them in an ever more extreme direction.

Second, Eli Pariser (2011) has argued that the rapid adoption of business models based on machine learning might lead to filter bubbles, in which people unwittingly live in an informational universe that more or less reflects their own biases back at them. This mechanism is driven by the interaction between individual preferences and machine learning. Both search engines (e.g., Google's) and social media (e.g., Facebook) rely on machine learning to provide results that the user will find "useful." These machine-learning processes also seek to classify us as individuals on the basis of, for example, which links or news-feed items we have clicked on.

This may lead to the algorithm systematically weeding out results or items that clash with individuals' priors. If, for example, a liberal is somewhat less likely to click on news stories shared in his feed by conservative friends than by his liberal friends, then the news feed is likely, over time, to provide less such stories. This may produce a self-reinforcing tendency, leading to a filter bubble in which the information that an individual finds uncongenial may be filtered out, so that he or she only sees information or perspectives that reinforce priors. Thus, an individual might be exposed only to biased news and, most perniciously, might not even know that she is being exposed only to biased news if she relies heavily on algorithmic sources to provide her with her diet of information.

Finally, Zeynep Tufekci, Guillaume Chaslot, and Maciej Ceglowski each point to how machine learning can lead to passive extremism in its efforts to maximize user engagement. Typically, algorithms are tailored to maximize user "engagement" with content so that people will spend as much time on Facebook, or YouTube, or the other service in question as possible, seeing advertisements and hence driving up revenue. However, the goal of maximizing engagement may produce worrying consequences. People tend to be more engaged by material that they find shocking, frightening, or disturbing. This may lead algorithms to systematically downgrade nonalarmist material and instead tend to direct people toward highly alarming and indeed conspiratorial content.

Chaslot, a former Google engineer who worked on the YouTube recommendation system, reports that "watch time was the priority. Everything else was considered a distraction." After being fired by Google, Chaslot carried out research, examining the chains of videos that YouTube recommends and plays via autoplay. He found systematic bias toward conspiracy videos (and in particular conspiracy videos spreading rightwing memes).[6] Zeynep Tufekci and Maciej Ceglowski report similar observations.[7] For roughly contiguous reasons, Facebook's advertising model gave Donald Trump far cheaper advertising rates than Hillary Clinton: his ads were more likely to be alarming and to be clicked on than Clinton's, hence increasing Facebook's profits.[8]

These various arguments are often conflated. However, they rely on quite different implicit assumptions about the nature and operation of gatekeeping. Sunstein's argument is based on the general weaknesses of gatekeeping: when the strong centralizing tendencies in media no longer operate, people will tend to cluster in their own tribes. Pariser's argument invokes almost the opposite mechanism: suggesting that the apparently free exchange of information in search and social media, like a cleverly constructed ant farm, is in fact structured by invisible glass barriers that separate us from one another and guide our activities. Chaslot, Tufekci, and Ceglowski concur with Pariser about the invisible forces but suggest that rather than separating us, they lead us ineluctably to the same destination—shocking and even radicalizing content.

How well do these claims match up to the empirical evidence? It is plausible that the shift that Sunstein identifies, while real, is not nearly as dramatic as he avers. Indeed, the main empirical finding of Gentzkow and Shapiro (2010) is that online media does better than offline media in exposing people to different news sources. The reason for this, in their argument, is that gatekeeping structures and market forces toward centrism still play an important role. In their account, news organizations that cater to both left and right viewers and readers are still central within the media system; as they pungently describe it, "firms that invest all that money in quality are not going to . . . cater to the neo-Nazi vegetarian tiny little corner of the market. They're going to position themselves in the center to appeal to a wide audience."[9] More recently, Boxell, Gentzkow, and Shapiro (2017) find that the "growth in polarization in recent years is largest for the demographic groups least likely to use the Internet and social media."

Second, an increasingly polarized electorate does not mean that people's social networks, especially online, are entirely homogeneous politically. Bakshy, Messing, and Adamic (2015) report that the ideological diversity (or lack of same) of friendship ties on Facebook have important consequences for which information people see, but also find that 20 percent of the average US user's Facebook friends have an opposing partisan disposition. This suggests that people are likely exposed to a significant amount of contrary political information (although it is also possible that the higher the degree of ideological variation in a given corner of the network, the less likely people are to express strong political views; see Mutz 2006). Yet under Pariser's argument about algorithms, even if one were a good Sunsteinian, who followed others with sharply different perspectives, one might still have an unreliable understanding of the range of perspectives and information they provide, unless one were perpetually vigilant in ensuring that one paid as much or more attention to clashing views as to congenial ones. Otherwise, algorithms and reading habits may reinforce each other in a feedback process that over time will heavily discount contrary viewpoints and information or even render it invisible.

What does this mean for the structure of epistemic trust we sketched earlier? Again, let us presuppose an epistemically responsible agent, Jack, who seeks to hold true beliefs. Jack knows that partisan valence should not necessarily guide his decision as to whom to trust as an epistemic authority, even in the context of political decision making. To have an accurate sense of the range of political perspectives in his society, and to seek out possible countervailing information to his own beliefs, Jack follows a wide variety of people on social media with different partisan commitments. Those people may help him to acquire true beliefs, to the extent that they are both trustworthy, in the sense that they operate using reliable mechanisms of belief formation, and have different partisan preferences from his own. Jack will be more likely to find his prior beliefs challenged in valuable, informative ways than by attending strictly to people who share his views. Both traditional Millian arguments for free political debate and recent argumentative accounts of reasoning (Mercier and Sperber 2017) point to the salutary benefits of this kind of inquiry. Yet if Jack also had a second set of ambitions for his acquisition of information in the politically relevant domain, in which he prioritized the construction of a set of action-guiding political commitments, he might adopt a different set of practices. In this context, he might value information from those whom he knows to share his allegiances—while they will expose him to less novel, surprising or discomfiting information, they may provide information that will better help him to build collective endeavors (Farrell and Schwartzberg 2008).

The filter-bubble problem, under this interpretation, arises from the fact that even if Jack seeks both clashing and reinforcing opinions, the machine-learning processes that social media companies currently employ will not distinguish between them and will instead treat both as a signal of what Jack finds interesting or engaging. For many, perhaps most, people in the current information architecture, this will mean that their interest in people who disagree with them will be swamped over time by their preference for agreeing perspectives. (The opposite problem—that they find themselves more often guided towards clashing perspectives than they might want—is unlikely to arise, except for politically quite unusual people.)[10]

Finally, and crucially, the difficulties that Tufekci, Chaslot, and Ceglowski point to arise from the combination of a different kind of algorithmic confusion (between stimulating and reliable content) and commonly held but often unwarranted beliefs about the reliability of algorithms in yielding trustworthy results. The problem, properly understood, is not so much that people may be led to conspiracy websites by algorithms but that their heuristics may lead them to trust these sites as authoritative (although see Munger and Philips 2019). That is, absent some additional mechanism—whether internal to the agent or through some additional filter—designed to parse out sites that, while partisan, provide basically reliable information from sites that traffic in conspiracies, there is a risk that the compounding effect of algorithms on partisan heuristics will produce deference to conspiracy theorists. Again, this is a problem of changes in the structure of epistemic trust on the weak gatekeeping model. People trust algorithms to provide them with reliable information to an unwarranted extent and, in addition to partisan heuristics, tend to adopt unjustified cognitive shortcuts to decide which online information is reliable and which unreliable, such as whether a given item of information is near the top or the bottom of a list and how much space it occupies (Burghardt et al. 2017). Tripodi (2018) describes a set of reading practices among religious conservatives that might be dubbed "Google inerrancy," where they trust that the top results for quite specific (and sometimes ideologically loaded) searches are reliable. Under some circumstances (Richey and Taylor 2018), such results can be accurate and helpful. Under others, algorithms (including Google's) may lead to grossly misleading information. In sum, the extent that people place a degree of epistemologically unwarranted trust in algorithms, they will sometimes be badly misled.[11]

Macro-Level Changes to the Public Sphere

Beyond the micro-level concerns, there are two important macro-level problems that require further inquiry. The first is that the new public sphere has developed in a lopsided fashion; these distortions are not symmetrically distributed across the partisan spectrum and have developed over decades rather than from the very recent advent of the internet and social media. As Benkler, Faris, and Roberts (2018) describe it, the major problem is Fox News rather than Facebook: distortions from the old model of strong gatekeeping have survived into the new world where gatekeeping is weak or algorithmically based. The history, as described in Hemmer's (2018) *Messengers of the Right: Conservative Media and the Transformation of American Politics*, is straightforward. Until the 1980s, conservatives had extraordinary difficulty in penetrating the existing public sphere. As conservative political entrepreneurs Richard Viguerie and David Franke (2004, 2) described it, "The general public didn't hear about our candidates, issues, causes, books, or magazines (except as hysterical warnings about the 'Far Right threat)." To survive, they had to rely on a variety of subsidies and artful exploitation of gaps in the existing system, which sometimes closed when conservatives were perceived as going too far.

Conservative politicians and media entrepreneurs created and mutually reinforced a narrative about mainstream media bias that persisted as they began to gain footholds, first in talk radio and then in cable television. This oppositional mind-set, which disparaged traditional reporting standards as a cloak for bias, helped shape the conservative media ecosystem so that it became a thing apart from traditional media and directly opposed to it. Stories were evaluated on whether they helped the cause rather than whether they were correct.

This has had large-scale structural consequences for the relationship between conservative and mainstream media. As Benkler, Faris, and Roberts (2018) find through mapping out online links and relationships, there are two quite distinct media ecosystems in the United States, operating according to very different logics. On the one side, the traditional media ecosystem links together liberal-to-left and mainstream publications into a common system of ties, on the other, a largely separate conservative ecosystem operates in isolation from, and opposition to, its mainstream counterpart. In each, there are stark differences in size and audience between a fringe of small and numerous sites, and a small number of much larger sites with large direct audiences (and the ability to spread information to other sites, and thus acquire a large indirect audience too).

Importantly, Benkler, Faris, and Roberts find no obvious difference in the propensity of people on the left and right to believe in salacious disinformation. The key difference is how it spreads. In the left-through-mainstream ecosystem, such rumors tend to circulate for a short period on the fringes and die from lack of oxygen. In the right-wing ecosystem in contrast, disinformation on the fringes is often taken up by the larger sites, which broadcast it to their audiences, and to other audiences too, so that it has far greater reach.

So fragmentations in epistemic trust are a function not merely of strong versus weak gatekeeping models but also of different institutionalized commitments on the left and on the right to providing reliable, well-supported information. The right-wing media ecosystem, rather than presenting truthful information in a compelling fashion, has a pronounced tendency to present false information. Notably, this is now the consequence of evolutionary pressures on the level of the media ecosystem rather than individual publications: aberrant voices are obliged to adapt, or selected out. When Tucker Carlson founded the *Daily Caller*, he said that he wanted a right wing alternative to the *New York Times*, deploring existing conservative media for its failure to build institutions that cared about "accuracy."[12] Carlson and the *Caller* abandoned this position with remarkable rapidity, as it became clear that there wasn't an ecological niche for accuracy but that there was one for active disinformation—those who are familiar with Carlson's recent career will find his earlier admonitions astonishing.[13]

The second macro-level issue is that the public sphere is potentially vulnerable to weaponized falsehood on a very broad scale because of the new ecosystem's reliance on algorithms. Legal scholars like Rick Hasen (2017) and Tim Wu (2017) point out that new threats are emerging in the era of "cheap speech."[14] It is possible rapidly and easily to create a large number of false identities on social media and to deploy either cheap human labor (China's "50 cent army") or automated techniques ("bots") to use these identities to manipulate and reshape public conversation. What Molly Roberts (2018) has identified as "flooding" attacks (see also Farrell and Schneier 2018) are not primarily aimed at convincing people of falsehoods. Instead, their effect is more pernicious—they undermine the kinds of exchange and debate of knowledge that allow for a genuine public or publics to form, "seeding doubt and paranoia, and destroying the possibility of using the Internet as a democratic space."[15] This has provided a boon for authoritarian regimes, which want to divide and disorient potential opponents (Roberts 2018). People looking to organize dissent find that they are immediately confronted by legions of voices disagreeing, distracting, and spouting nonsense, so that genuine agreement and disagreement becomes submerged under a torrent of disinformation. This may have second-order consequences when people realize that they do not know whether their interlocutors are real human beings or automated processes, provoking a generic sense of political helplessness and despair. As Peter Pomerantsev (2014, 67) describes its application in Russia: "The brilliance of this new type of authoritarianism is that instead of simply oppressing opposition, as had been the case with twentieth-century strains, it climbs inside all ideologies and movements, exploiting and rendering them absurd. . . . The Kremlin's idea is to own all forms of political discourse, to not let any independent movements develop outside of its walls. Its Moscow can feel like an oligarchy in the morning and a democracy in the afternoon, a monarchy for dinner and a totalitarian state by bedtime." The victims of these manipulations "never talk of human rights or democracy; the Kremlin has long learned to use this language and has eaten up all the space within which any opposition could articulate itself" (Pomerantsev 2014, 126).

Such attacks can be mounted in democracies as well. We do not know, and may never know, whether Russian disinformation attacks had any political consequences for the 2016 elections (Sides, Tesler, and Vavreck 2018). However, as new technologies develop, it will become increasingly easy to apply these attacks at scale. Generative adversarial networks make it easier to probe machine-learning systems at scale so as to discover vulnerabilities; similar techniques are making it trivially easy to generate realistic seeming fake video and audio content.

In sum, it is unclear that the transition to a weak gatekeeper system with algorithmic curating will produce or exacerbate the unfortunate micro-level effects that Sunstein, Pariser, and others identify. However, even if they do not, there are plausible macro-level problems that can operate even if people do not huddle together in clumps of confirmation bias, are not badly affected by filter bubbles, and are not always radicalized. The bifurcation of the American public sphere long predates the advent of social media, although it has become far easier to observe (and possibly more pernicious) now that the patterns of linkages and information spread have become more easily visible. But new media structures are systematically more vulnerable to flooding attacks than previous structures that had far fewer entry points and far more effective choke points.

Epistemic Trust and Democracy

Does this all spell doom for democratic decision making? Although we think that the problems are serious, we also believe that the sky is not falling. As we have argued, shifts in whom people regard as epistemically trustworthy may be welcome, a response to persistent epistemic injustice (Fricker 2007), in which less powerful social actors are ascribed diminished credibility. Rather than constructing a narrative of the Fall, we are better off trying to think systematically through the trade-offs.

Strong gatekeeper systems have the advantage of creating a relatively high degree of consensus within a given society over what is true and what is not true. When this gatekeeper system goes together with relatively reliable systems for exploring and discovering the truth, the outcome is likely to be salutary for democracy. People will have a high degree of trust in the institutional systems that provide them with news and information, and that trust will be largely warranted. Under such a system, people will have good reason to trust in the reliability of the information that they are being provided.

However, there is no necessary reason for believing that strong gatekeeper systems will work in this way all, or even most of the time. As contemporary histories relate (Perlstein 2001), societies with strong gatekeeping, such as the United States at the beginning of the 1960s, may also have very troubling yet pervasive beliefs that are reinforced by media gatekeepers. Thus, for example, even after the political eclipse of Joe McCarthy and the marginalization of the John Birch Society, there were relatively high levels of political paranoia and a nearly complete disinclination to know precisely what was happening in Vietnam. The social, economic, intellectual, and political deficiencies of African Americans were treated as nearly indisputable common knowledge, as were purportedly eternal verities about the role of women in society and the workplace.

Moreover, strong gatekeeper systems tend, by their very nature, to exclude divergent voices from political debate. It was extremely difficult for those with unorthodox understandings of politics on the right or on the left to break through into media attention. Advocates of racial equality had to adopt extraordinary measures to get public attention in a country where beatings, church burnings, and even lynchings were largely treated as part of the background noise of everyday politics.

Weak gatekeeper systems, in contrast, do not enforce strong consensus in society. Instead, they allow for a very large variety of people, with very different understandings of what is true and untrue, to participate actively in public debate, perhaps even creating their own media organizations or publishing platforms. When these systems go together with the necessary machineries for distinguishing between basic truths and basic falsehoods, they can combine diversity of opinion and disagreement with some agreement on the factual basis of what people are arguing about. In a world of weak gatekeepers, where everyone can become a publisher (and where, in a sense, everyone on social media is a publisher or republisher), many more categories of people are able to acquire epistemic authority. Some of this new authority will be warranted, a long-overdue remedy to structural epistemic injustice. As we have suggested, the breakdown of the traditional gatekeeping model may enable previously excluded or submerged voices to acquire credibility. Insofar as the barriers to entry into the media market become less responsive to traditional ascriptive markers of expertise, one might hope that new and more reliable sources emerge.

Again, however, there are no necessary reasons to believe that weak gatekeeper systems will work well all or most of the time: people may place unwarranted trust in unreliable authorities. People can use media (whether established channels such as Fox News or newer sources such as Breitbart.com) to construct parallel universes based around shared political illusions. The kinds of dynamics that Sunstein, Pariser, Tufekci, Chaslot, and Ceglowski worry about might metastasize into democratic breakdown, as people become incapable of even reaching the minimal degree of agreement that is necessary for a properly functioning democracy. At the limit, structures of epistemic trust grounded in partisanship may lead people to discount even obvious truths in favor of politically convenient ones.

This may also undermine some of the presumed virtues of a more open media system. On the individual level, we might worry that the breakdown of gatekeeping may lead a person to rely on her preconceptions of who constitutes a credible source of testimony: rather than remedying epistemic injustice, it might reproduce forms of injustice under homophilic or segregated networks. If we ascribe reliability to others only insofar as we share party or other political forms of identification with them, it is hard to be optimistic that we will successfully undermine patterns of testimonial injustice.

However, note that virtually any model of epistemic trust would be vulnerable to such forms of injustice. When Izzy decides that the *New Yorker* is no longer trustworthy with respect to climate change, he must seek out new sources of information. If he holds (however unwittingly) sexist views about scientific credibility—or if these new sources themselves reproduce sexism by failing to cite or highlight work of female scientists—what might nonetheless be warranted epistemic trust in the new source will nevertheless be epistemically unjust.

The challenge, then, is not to re-create the old media system (which in any event would be a Sisyphean task). Recognizing the trade-offs between the old and the new models allows us to see that the old model too had enormous flaws. Instead, we must think practically about how we can move forward from where we are, working within a larger ecosystem that is largely impossible for any individual to reshape (even Mark Zuckerberg is subject to market forces).

The first step is to recognize the possible advantages of the current system and how to build on them. For example, we may now think more clearly about how best to judge when new would-be authoritative sources warrant our trust. Linda Zagzebski (2012) has argued that the authority of another person or source is justified in part by the way that authority forms its belief; that is, I believe that the authority forms its belief in a way I would think deserves emulation (by identifying and weighing evidence).[16] As we have suggested, traditional sources tended to conceal their ideological commitments, rendering their belief-formation mechanisms less transparent. A distinct advantage of the new structure is that, at least in principle, supporting evidence and partisan biases may be more available for public scrutiny or subject to challenge, meaning that the grounds on which authority may be justified may be more secure. Yet insofar as we rely only on partisan or narrowly relational reasons for trusting the information sources provide, we may find ourselves in troubling epistemic cul-de-sacs.

One possible way out of these cul-de-sacs is by leveraging high levels of intrapartisan trust to break free. David Dagan and Steven Teles (2016), for example, document how conservatives became more open to information about the costs of mass incarceration after decades of "tough on crime" rhetoric. The key factor, in their account, was the willingness of trusted voices in the conservative community to talk about the problems (some conservatives had experienced these firsthand as a result of their own incarceration). Other voices, which were widely distrusted by conservatives and might have endangered this process of belief change (e.g., George Soros's Open Society) were willing to step back at key moments and let other institutions and actors take the lead. Such strategies are unlikely to work at the moment with US conservatives, who research suggests are committed to particularly problematic forms of partisan-based trust in grounding their choices of media sources at the moment (Faris et al. 2018; Boxell, Gentzkow, and Shapiro 2017). However, they may become more plausible again after Donald Trump leaves office, especially if he loses power in a general electoral disaster.

One can of course think about more systematic institutional solutions. We leave this for future work. However, we note that research on institutional possibilities should start from a more realistic set of assumptions about human cognition, assumptions that recognize both the ways in which cognitive biases can lead to the sorts of problems we have discussed in this chapter, and how those same biases can be harnessed for the benefit of social problem solving (Mercier and Sperber 2017; Farrell, Mercier, and Schwartzberg n.d.)

Conclusion

In conclusion, we wish to consider the wider implications of this view for liberal democratic theory. Although we have cautioned against panic concerning fake news, we do not deny that there are truly worrying implications of the dissemination of such information and the willingness of citizens to rely on those who seek to transmit or impart false beliefs. Social institutions facilitate the "formation, preservation, and transmission of true beliefs" (Goldman 1999), which can take the form of either the creation of new beliefs or the identification and correction of false beliefs (Buchanan 2004). Of course, there is a distinguished tradition of confidence in the capacity of liberal norms securing the free flow of information to remedy problems of false belief. To be sure, liberal institutions—effective institutional arrangements for freedom of thought, conscience, and association, and democratic decision making on terms of equal participation—may well serve as a means of reducing such risk by their capacity to correct erroneous judgments. But while liberal democratic institutions are necessary, they may well be insufficient.

Following Allen Buchanan (2004), we might worry about "unwarranted status trust": the trust (including epistemic trust) accorded to persons or groups simply because they have been identified as possessing a certain status or membership in a certain group. Although status trust may be warranted—a medical license justifies trust in the judgment of a doctor—it is likely to be corroded and become unwarranted in cases in which status tracks nonmerit considerations. Using partisanship as a heuristic to guide our selection of trustworthy sources yields similar problems, because such trust may often be unwarranted, placing us in moral and epistemic peril. Equally, partisanship may lead us to place much less trust than warranted in information structures that appear to be opposed to our partisan beliefs, or that we have been told by trusted partisan figures are opposed to those beliefs. Buchanan also highlights the important check placed on unwarranted epistemic trust by "epistemic egalitarianism," by which he means both equal moral status among members and, crucially, a commitment to viewing oneself and one's fellow citizens as competent to form and revise beliefs.[17] Insofar as the loss of shared epistemic trust under the strong gatekeeping system may tempt us to challenge the basic judgmental capacities of our fellow citizens—both to render decisions themselves and to identify reliable experts—we do believe this constitutes a serious concern for democracy.

To reiterate, our argument in this chapter is not that exposure to fake news renders citizens incompetent or that citizens are irrational in their selection of sources. Rather, we have sought to claim only that the transition from the strong to the weak gatekeeping model has produced widespread challenges to our capacity to readily identify trustworthy sources for reliable beliefs. But if this transition does reflect a fragmentation of citizens' wider relations of trust—not merely epistemic but also political and moral trust—then the consequences for democracy may be grave indeed.

Notes

1. On epistemic trust and authority, see especially Goldman 2001; Origgi 2004; Fuerstein 2012; Zagzebski 2012; McCraw 2015. For a Bayesian account of belief revision, see Mulligan n.d. For an account challenging Zagzebski's preemption view that enables an agent to base beliefs on as many reasons that she has, see Dormandy 2018. Other fields (notably psychotherapy) also investigate epistemic trust, but this would take us well beyond the scope of our chapter.

2. We take no stand on doxastic voluntarism more generally.

3. The authors define fake news as "fabricated information that mimics news media content in form but not in organizational process or intent."

4. Note that Stephen Colbert's satirical attachment to "Papa Bear" has its bite in its accuracy about the trust that many O'Reilly viewers had in him.

5. Quoted in Kurtz 1996, 290.

6. Paul Lewis, "'Fiction Is Outperforming Reality': How YouTube's Algorithm Distorts Truth," *The Guardian*, February 2, 2018, https://www.theguardian.com/technology/2018/feb/02/how-youtubes-algorithm-distorts-truth.

7. Zeynep Tufekci, "YouTube, the Great Radicalizer," *New York Times*, March 10, 2018, https://www.nytimes.com/2018/03/10/opinion/sunday/youtube-politics-radical.html, Maciej Ceglowski, "Build a Better Monster: Morality, Machine Learning, and Mass Surveillance," *Idle Words*, April 18, 2017, http://idlewords.com/talks/build_a_better_monster.htm/.

8. Antonio Garcia Martinez, "How Trump Conquered Facebook—Without Russian

Ads," *Wired*, February 23, 2018, https://www.wired.com/story/how-trump-conquered-facebookwithout-russian-ads/.

9. Caroline O'Donovan, "Q&A: Clark Medal Winner Matthew Gentzkow Says the Internet Hasn't Changed News as Much as We Think," Nieman Lab, April 19, 2014, http://www.niemanlab.org/2014/04/qa-clark-medal-winner-matthew-gentzkow-says-the-internet-hasnt-changed-news-as-much-as-we-think/.

10. Again, while the mechanism is plausible, it is difficult to have confidence in the size of its general impact. The best large-scale study (Bakshy, Messing, and Adamic 2015) finds that while filter-bubble effects have moderate consequences for the amount of information to which Facebook users are exposed, the ideological diversity of people's friendship networks explains substantially more of the information that they see or do not see. See https://www.cjr.org/analysis/fake-news-media-election-trump.php.

11. Again, however, the large-scale consequences of these mechanisms are an open question. On Tripodi's account, it would seem that these results are more likely to cause harm (when they do cause harm) by strengthening prior beliefs than by inculcating new ones. A wide body of evidence from political science suggests that people are highly resistant to information that clashes with their basic political values (Nyhan and Reifler 2018). This makes them less open to good new information, but also less open to bad new information.

12. Joel Mears, "The Great Right Hype: Tucker Carlson and His Daily Caller," *Columbia Journalism Review*, July–August 2011, https://archives.cjr.org/feature/the_great_right_hype.php.

13. Lyz Lenz, "The Mystery of Tucker Carlson," *Columbia Journalism Review*, September 5, 2018, https://www.cjr.org/the_profile/tucker-carlson.php.

14. https://knightcolumbia.org/content/tim-wu-first-amendment-obsolete.

15. Adrian Chen, "The Real Paranoia-Inducing Purpose of Russian Hacks," *New Yorker*, July 27, 2016, https://www.newyorker.com/news/news-desk/the-real-paranoia-inducing-purpose-of-russian-hacks.

16. As noted earlier, although we think this is a sensible account of the reasons we might trust a particular authority, we do not argue that a belief that *p* on epistemic authority must take the form of preempting other reasons for believing that *p*, as Zagzebski suggests. Rather, as noted, we assume that a belief that *p* on epistemic authority can take other forms of reason provision or revision.

17. For another account of the foundations and implications of epistemic egalitarianism, see Schwartzberg 2014.

References

Achen, Christopher H., and Larry M. Bartels. 2016. *Democracy for Realists: Why Elections Do Not Produce Responsive Government*. Princeton, NJ: Princeton University Press.

Arendt, Hannah. 1967. "Truth and Politics." *New Yorker*, February 25, 49–88.

Bakshy, Eytan, Solomon Messing, and Lada Adamic. 2015. "Exposure to Ideologically Diverse News and Opinion on Facebook." *Science*, May. https://doi.org/10.1126/science.aaa1160.

Bengani, Priyanjana. 2019. *Hundreds of "Pink Slime" Local News Outlets Are Distributing Algorithmic Stories and Conservative Talking Points*. New York: Tow Center for Journalism, Columbia University.

Benkler, Yochai, Robert Faris, and Hal Roberts. 2018. *Network Propaganda*. New York: Oxford University Press.

Boxell, Levi, Matthew Gentzkow, and Jesse M. Shapiro. 2017. "Is the Internet Causing Political Polarization? Evidence from Demographics." Working Paper No. 23258, National Bureau of Economic Research, Cambridge, MA.

Brennan, Jason. 2016. *Against Democracy*. Princeton, NJ: Princeton University Press.

Buchanan, Allen. 2004. "Political Liberalism and Social Epistemology." *Philosophy and Public Affairs* 32 (2): 95–130.

Burghardt, K., E. F. Alsina, M. Girvan, W. Rand, and K. Lerman. 2017. "The Myopia of Crowds: Cognitive Load and Collective Evaluation of Answers on Stack Exchange." *PLOS One* 12 (3): e0173610.

Caplan, Bryan. 2007. *The Myth of the Rational Voter*. Princeton, NJ: Princeton University Press.

Carlin, Ryan E., and Gregory J. Love. 2018. "Political Competition, Partisanship and Interpersonal Trust in Electoral Democracies." *British Journal of Political Science* 48 (1): 115–39. https://doi.org/10.1017/S0007123415000526.

Case, Ann, and Angus Deaton. 2020. *Deaths of Despair and the Future of Capitalism*. Princeton, NJ: Princeton University Press.

Dagan, David, and Steven Teles. 2016. *Prison Break: Why Conservatives Turned against Mass Incarceration*. New York: Oxford University Press.

Dormandy, Katherine. 2018. "Epistemic Authority: Preemption or Proper Basing?" *Erkenntnis* 83 (4): 773–91. https://doi.org/10.1007/s10670-017-9913-3.

Faris, Robert M., Hal Roberts, Bruce Etling, Nikki Bourassa, Ethan Zuckerman, and Yochai Benkler. 2018. "Partisanship, Propaganda, and Disinformation: Online Media and the 2016 US Presidential Election." Research paper, Berkman Klein Center for Internet and Society, Harvard University, Cambridge, MA.

Farrell, Henry, Hugo Mercier, and Melissa Schwartzberg. n.d. *No-Bullshit Democracy*. Unpublished manuscript.

Farrell, Henry, and Bruce Schneier. 2018. "Common-Knowledge Attacks on Democracy." Research Paper No. 2018-7, Berkman Klein Center for Internet and Society, Harvard University, Cambridge, MA.

Farrell, Henry, and Melissa Schwartzberg. 2008. "Norms, Minorities, and Collective Choice Online." *Ethics & International Affairs* 22 (4): 357–67. https://doi.org/10.1111/j.1747–7093.2008.00171.x.

Foege, Alec. 2008. *Right of the Dial: The Rise of Clear Channel and the Fall of Commercial Radio*. New York: Faber and Faber.

Fricker, Miranda. 2007. *Epistemic Injustice: Power and the Ethics of Knowing*. Oxford: Oxford University Press.

Fuerstein, Michael. 2012. "Epistemic Trust and Liberal Justification." *Journal of Political Philosophy* 21 (2): 179–99. https://doi.org/10.1111/j.1467-9760.2012.00415.x.

Gentzkow, Matthew, and Jesse M. Shapiro. 2010. "What Drives Media Slant? Evidence from US Daily Newspapers." *Econometrica* 78: 35–71.

Goldman, Alvin. 1999. *Knowledge in a Social World*. New York: Oxford University Press.

Hardin, Russell. 2002. *Trust and Trustworthiness*. New York: Russell Sage Foundation.

Hasen, Richard L. 2017. "Cheap Speech and What It Has Done (to American Democracy) Essays." *First Amendment Law Review* 16: 200–231.

Hemmer, Nicole. 2018. *Messengers of the Right: Conservative Media and the Transformation of American Politics*. Philadelphia: University of Pennsylvania Press.

Kurtz, Howard. 1996. *Hot Air: All Talk, All the Time*. New York: Times Books.

Lazer, David M. J., Matthew A. Baum, Yochai Benkler, Adam J. Berinsky, Kelly M. Greenhill, Filippo Menczer, Miriam J. Metzger, et al. 2018. "The Science of Fake News." *Science* 359 (6380): 1094–96. https://doi.org/10.1126/science.aao2998.

McCraw, Benjamin W. 2015. "The Nature of Epistemic Trust." *Social Epistemology* 29 (4): 413–30. https://doi.org/10.1080/02691728.2014.971907.

Mercier, Hugo, and Dan Sperber. 2017. *The Enigma of Reason*. Cambridge, MA: Harvard University Press.

Mulligan, Thomas. n.d. "The Epistemology of Disagreement: Why Not Bayesianism?" *Episteme*, 1–16. https://doi.org/10.1017/epi.2019.28.

Munger, Kevin, and Kevin Phillips. 2019. "A Supply and Demand Framework for YouTube Politics." Working paper, Penn State Political Science, State College, PA.

Mutz, Diana C. 2006. *Hearing the Other Side: Deliberative versus Participatory Democracy*. New York: Cambridge University Press.

Nyhan, Brendan, and Jason Reifler. 2018. "The Roles of Information Deficits and Identity Threat in the Prevalence of Misperceptions." *Journal of Elections, Public Opinion and Parties* (May): 1–23. https://doi.org/10.1080/17457289.2018.1465061.

Origgi, Gloria. 2004. "Is Trust an Epistemological Notion?" *Episteme* 1 (1): 61–72. https://doi.org/10.3366/epi.2004.1.1.61.

Pariser, Eli. 2011. *The Filter Bubble: What the Internet Is Hiding from You*. New York: Penguin.

Perlstein, Rick. 2001. *Before the Storm: Barry Goldwater and the Unmaking of the American Consensus*. New York: Hill & Wang.

Pomerantsev, Peter. 2014. *Nothing Is True and Everything Is Possible: The Surreal Heart of the New Russia*. New York: PublicAffairs.

Prior, Markus. 2007. *Post-Broadcast Democracy: How Media Choice Increases Inequality in Political Involvement and Polarizes Elections*. New York: Cambridge University Press.

Richey, Sean, and J. Benjamin Taylor. 2018. *Google and Democracy: Politics and the Power of the Internet*. New York: Routledge.

Roberts, Margaret. 2018. *Censored: Distraction and Diversion Inside China's Great Firewall*. Princeton, NJ: Princeton University Press.

Schwartzberg, Melissa. 2014. *Counting the Many: The Origins and Limits of Supermajority Rule*. New York: Cambridge University Press.

Sherman, Gabriel. 2014. *The Loudest Voice in the Room*. New York: Random House.

Sides, John, Michael Tesler, and Lynn Vavreck. 2018. *Identity Crisis: The 2016 Presidential Campaign and the Battle for the Meaning of American*. Princeton, NJ: Princeton University Press.

Somin, Ilya. 2016. *Democracy and Political Ignorance*. Stanford, CA: Stanford University Press.

Sunstein, Cass R. 2017. *#Republic: Divided Democracy in the Age of Social Media*. Princeton, NJ: Princeton University Press.

Tripodi, Francesca. 2018. "Searching for Alternative Facts: Analyzing Scriptural Inference in Conservative News Practices." Research paper, Data and Society Research Institute, New York.

Usher, Nikki. 2014. *Making News at the* New York Times. Ann Arbor: University of Michigan Press.

Viguerie, Richard A., and David Franke. 2004. *America's Right Turn: How Conservatives Used New and Alternative Media to Take Power*. Chicago: Bonus Books.

Zagzebski, Linda Trinkaus. 2012. *Epistemic Authority: A Theory of Trust, Authority, and Autonomy in Belief*. Oxford: Oxford University Press.

8

Democratic Societal Collaboration in a Whitewater World

David Lee, Margaret Levi, and John Seely Brown

We are in a world of continuous change, requiring flexibility in learning and updating of skills. It is a whitewater world that tests our capacity to read the currents and react quickly.[1] Advances in automation and the sharing economy have contributed to a rapidly shifting job market. Natural disasters such as pandemics, earthquakes, hurricanes, and fires are affecting communities at increasing rates. Political polarization, instability, and terrorism are constant sources of fear. These technological, natural, and political rapids generate radical contingencies, accompanied by sudden and regular shocks to society, making it difficult for individuals and communities to know how to adapt and protect themselves from the downstream consequences of change. We argue that a necessary step in preparing for the world we live in are societal collaborations in the service of democracy, collaborations using digital and other technologies not previously available.

In a whitewater world, the constant acquisition of new skills and dispositions is essential for survival at work, in society, and in politics. It is no longer just an issue of reskilling but an issue of constantly reskilling—such that reskilling becomes a state of being. In building societal collaborations, perhaps the most important reskilling lies in repurposing tools and organizational arrangements that worked in an earlier era—and inventing, as necessary, new tools and arrangements.

Our definition of democratic practice in digitally based societal collaborations has two major aspects. The first involves facilitating participation while ensuring relative equality of voice and decision power among the participants. The second is that the organizations are inclusive. Here is where we deviate from much earlier thinking. This is not simply (or only) a question of ensuring diversity, or of defining the common interest in terms of the actual members. Elinor Ostrom's *Governing the Commons*, for example, focuses only on how those who share a common-pool resource overcome the narrow forms of self-interest (which would imply free riding) to act on their longer-term interest (protecting the commons).[2] Our view is that it is equally important to define the community as those to whom there are obligations, even if they are not part of the specific membership group, be it a local community or a nation. We are committed to creating an "expanded community of fate," enabling individuals to perceive their interests as bound with strangers on behalf of whom they are willing to act in solidarity.[3] To provide mechanisms for deepening our sense of obligations is part of what we expect of democratic societal collaborations.

Digital connections and platforms create both challenges and opportunities for democratic practice. Smartphones and the internet of things are conduits and facilitators of information and social interaction across the globe, with significant implications for agency and action. The initial effect seems to have been the construction of echo chambers and the manipulation of votes and preferences, but the technology, if properly used, also offers means for communicating and learning across traditional divides and for supporting new organizational forms for problem solving.

But to provide a digital infrastructure for a democracy that encompasses an inclusive community of fate, we must take seriously the concerns raised about doing the hard work of building organizations beyond the mobilization of protest.[4] Although there is, as this literature suggests, a tendency for digital democratic practices to downskill, this need not be the case. In fact, as we hope to show, we can develop digital platforms that continually upskill.

In what follows, we build on thinking in three domains: pragmatism, arguments about building and sustaining the commons, and research on organizational cultures and institutional design that facilitate collective action that goes beyond the narrow self-interest of those engaged in the action. After elaborating the analytic tools we use, we consider a series of efforts to establish societal collaborations meant to enhance the public good. We then consider what we have learned and the next steps to take.

We primarily focus on technologies for facilitating cooperation and upskilling in societal collaborations, emphasizing the institutions, practices, and tools that make those collaborations internally democratic and that strengthen democracy generally. Some of those institutions, practices, and tools are borrowed from the past, but some emerge from the dynamics of collaboration in the networked age. It is the latter that are of particular interest. We are concerned with how organizations use technology to exercise democratic power and what obstructions they face but, equally, with the kinds of leadership, hierarchy, power, and obligations developed within the organizations themselves.

Democratic Theory and Societal Collaborations Using Digital Technologies

Societal collaborations for the whitewater world solve problems while inventing and reinventing the appropriate tool set for skilling and reskilling of participants in democratic civic practice; skilling and reskilling participants in multipurpose, transferable tools that prepare them for flexible work; and creating a public that has an expanded community of fate. Such skills are important not only in their own right but also for supplying societal collaborations with what is essential for generative deliberation and problem solving.

Our model derives from the findings of Ahlquist and Levi in their study of the longshore worker (dockworker) unions on the West Coast of the United States and in Australia.[5] These unions are mini–democratic governments and reveal processes and rules that might be generalizable and scalable.

The leadership of most American and Australian unions are economic rent seekers who expect considerable personal advancement in return for improving the pay, hours, and benefits of their members. The unions they manage are hierarchical with few expectations of members except paying dues and striking on command. Even voting on contracts and leadership is restricted, generally done by a representative rather than directly.

The two unions Ahlquist and Levi describe are exceptions to these rules as well as proof of concept of an alternative organizational design. Their leaders are political, not economic, rent seekers. While committed to—and required to—advance the material well-being of their members, they also wish to advance class and other forms of solidarity that extend beyond the boundaries of a particular union's membership. They want to expand the community of fate, those with whom the members' interests are bound and on whose behalf they are willing to take costly actions, even if there is little likelihood of direct reciprocity. The only way the leadership cohort can hope to change the beliefs of members about the nature of the world in which they live and change their willingness to engage in costly actions is through democratic institutions that, one, provide equal opportunity for political influence through votes and persuasive argument, and, two, ensure leadership transparency and accountability. Moreover, the organizational design must include education about the political, social, and economic context in which choices are made.

This particular democratic experiment flourished and survived multiple technological and legal transformations because it had two consequences.

First, it proved effective in improving the living standards and well-being of members; leadership understood that this was their primary responsibility. Second, the engagements in costly action—both strikes on behalf of their interests as longshore workers and port closures on behalf of far distant others—led to reinforcing beliefs in democratic practice and an extended sense of obligations. The members went from being "wharf rats to lords of the docks" while also experiencing the pleasure of agency.[6] They developed a sense of efficacy and power.

But what these two unions accomplished was in a predigital era and in a period when labor unions were relatively strong. Moreover, the expanded community of fate emerged from a group of people who were already tightly interconnected through their work and their unions. Achieving similar outcomes from a democratic structure that links people who have yet to—and may never—interact personally raises a series of new dilemmas for a pragmatic democracy.

Our model builds on the success of this small set of unions, organizations whose main mission was to serve the economic interests of members but that were able both to achieve that goal and to evoke from their members costly actions in the interest of others. With a combination of internal democratic institutions and learning opportunities, they were able to transform the narrow self-interest that unions generally encourage and create a broader sense of obligation They enabled the workers to perceive their interests as bound with strangers on behalf of whom they were willing to act in solidarity, thus building an expanded community of fate.

How do we transform this model, forged in a different era and without the digital tools of today, into societal collaborations that work for our time? Listening that is both generous and generative is necessary to forge common understanding of the problems to be addressed.[7] One means to ensure this quality of listening is a deliberative process, one in which individuals not only give reasons for their preferences and actions but also use the process to revise beliefs, practices, and solutions and to develop emotional interdependence. We also see deliberation as critical in enlarging the sense of obligation to those outside the group and thus expand the community of fate.

To facilitate and support this kind of deliberation also requires new organizational tools and institutions that enable individuals to recognize they are in that community and have solidarity even though their connections are digitally mediated and not personal. In contrast, the digital world may facilitate personal efficacy. In the longshore example, the workers were relatively unique in their capacity to engage in efficacious actions; they could close the ports. The physical world enables few such leverage points. The digital world offers far more. As experiments with new societal collaboration proceed, the criteria for their success will not only be their effectiveness in solving the identified problem. Equally important will be the extent to which they build new civic skills and devise appropriate institutional arrangements for facilitating a sense of community and efficacy.

To a large degree, societal collaboration is about realizing pragmatist John Dewey's vision for democratic experimentalism.[8] Dewey believed that actions build productive knowledge. When applied to education, this implies learning by doing and pedagogical methods such as experiential, problem-based, and situated learning that emphasize direct experience in learning. When applied to democratic theory, this implies civil society as a collective problem-solving endeavor and democracy as the form of self-governance that "affords the greatest possible scope to the social intelligence of problem solving and the flourishing of individual character as its condition and product." That is exactly how we envision societal collaboration.

Dewey chose not to offer concrete proposals. As Charles Sabel notes,[9] Dewey felt it was pointless to "set forth counsels as to advisable improvements in the political forms of democracy" until the problem of communication and improved collective self-understanding had been solved: "The prime difficulty . . . is that of discovering the means by which a scattered, mobile and manifold public may so recognize itself as to define and express its interests. This discovery is necessarily precedent to any fundamental change in the [political] machinery."[10] But the problem Sabel puts forth is solvable, and he, with Joshua Cohen, has offered one promising approach.[11] Another is that of Elinor Ostrom, who offers a schema of institutional arrangements that both reveal and facilitate common interests over a range of problems. Ostrom won the Nobel Prize in Economics for demonstrating that one could mitigate the tragedy of the commons without the state or the market, through community self-governance.[12] Left to their own devices, individuals tend to overuse common-pool resources, such as pastures, timber, or fish, in ways that deplete the shared resource forever. Economics has traditionally stated that there are only two solutions to this: either have the state control and enforce regulations on resource use or have the resource sold (privatized) and allow the market to regulate it. Both solutions lack sensitivity to context, the unique local conditions of a community. Ostrom showed that communities often do better than states and markets in monitoring and regulating resource use. Her extensive field studies led to generalized design principles for successful community management.

Ostrom's model of managing the commons not only depends on a relatively homogeneous population but also emphasizes monitoring exploitation of existing resources. It is not well suited to the instabilities and radical contingencies of a whitewater world. Nor does it apply where populations are heterogeneous and the resources are neither material nor geographically bounded. She and collaborators recognized this to some extent and have enlarged the framework to include other kinds of problems and institutional solutions, most notably in addressing knowledge commons.[13]

Yochai Benkler's commons-based peer production builds on Ostrom to advocate a model of socioeconomic production in which large numbers of people work cooperatively through nonmarket mechanisms facilitated by the internet.[14] For Benkler, the rise of the internet-based networks enables people to easily share and remix information resources, what he calls the networked information economy. In some sense, the internet meets the need Dewey identified: the problem of communication and improved collective self-understanding that he felt was a prerequisite to fundamental change in the political machinery. The internet facilitates political innovation in open licensing that enables people to designate digital resources or knowledge as nonproprietary. By doing so, these resources can in essence be added to a global digital commons that the community at large could mix and match and evolve. Benkler emphasizes that this evolution of the digital commons could happen in a radically decentralized and nonhierarchical way as a natural outcome of the open licensing that put digital resources into the hands of anyone who wants to use them.

Like Benkler, we see digital technologies and the networked information economy as a critical part of our vision of societal collaboration. However, we believe that large-scale deliberation, collaboration, and upskilling happen best when we mix more traditional, but perhaps computationally mediated, organizational forms with innovative forms of peer production. Some of the most exciting emphasize radical decentralization and flattened hierarchy.

Our approach to societal collaborations requires recognizing heterogeneity as a challenge but also a resource for democracy. We build on the Jack Knight and James Johnson version of pragmatist democracy, in which they emphasize institutional means to bring together the diverse voices, values, and commitments of the populace. They argue that "pragmatists see the social and natural world as fraught with contingency. As a result, even our most fundamental beliefs inevitably will be called into question and potentially proven false. What is important is how individuals and communities respond to the resulting tensions and strains, to the real doubt that unforeseen consequences generate."[15] For them the key features of a pragmatist philosophical position are fallibilism, antiskepticism, and consequentialism. Fallibilism implies a willingness to revise beliefs in the light of evidence and experience.

Antiskepticism limits the range of doubts and reduces relativism; doubts, as well as beliefs, require justification. Consequentialism involves a commitment to experimentation and assessment of the effects of a set of actions, including on our beliefs. The priority of democracy, for them, means creating a community of interest and then putting the right decisions with the right people and organizations and using the vote to determine outcomes.

Knight and Johnson argue that democracy is the best available means for facilitating collective choices at least as long as it affords equal opportunity in political influence, that is, votes and voice not distorted by money or status. We agree but go further. The condition of equal opportunity is a key feature of both the unions that produce expanded communities of fate and of the societal collaborations we advocate. Of equal import are the tools and institutions that permit the continual exploration and revision of beliefs, practices, and goals. Digitally grounded societal collaborations that incorporate generative deliberative processes at scale and continual upskilling create the possibility of new forms of democratic problem solving, participation, and obligations.

From Mobilization to Problem Solving

Numerous contemporary initiatives draw on digital technology to organize large-scale collaborative actions in support of democratic problem solving. These initiatives are of at least two kinds: those that mobilize people to engage in political action, and those that coordinate people to solve community problems or improve government responsiveness and efficacy. Both can potentially contribute to democracy by building civic skills, creating new channels for popular engagement in agenda setting, and generating expanded communities of fate that create shared responsibility and accountability among diverse populations. However, each initiative faces challenges that must be overcome to achieve these goals.

Certain features of digital technologies make them especially appealing for mobilization and advocacy. Most important is the capacity to scale action and power through large-scale distribution and personalized targeting of information and communication. However, when scalability is the sole focus, digitally based collaborations may reinforce echo chambers by microtargeting information. They may also contribute very little to skill building. When deliberation exists at all, it rarely involves generous and generative listening. The point is to get lots of people mobilized to achieve the goals of the organization, not to create a learning environment. The original variant of MoveOn, discussed below, succumbed to these limitations.

For initiatives centered on community problem-solving, such as Safecast and Code for America, their use of digital technologies goes beyond scalability and targeting of information to effective coordination of collaboration. We see such initiatives as particularly promising sites for using digital technologies to scale democratic participation in ways that embrace learning and deliberation. When digital technologies are applied to mobilization and advocacy for a predefined cause, they can short-circuit the richer engagement and deliberation that happens on the ground. But when digital technologies are applied to problem solving around a need and the stakeholders relevant for that need, then participation can support civic learning. Of course, such initiatives have their own challenges. The reliance on algorithms and complex software can empower certain people at the expense of others. Technologies for collaboration are also still in their infancy and still undeveloped or unproven, especially when it comes to mediating large-scale contestation and the productive friction crucial for democratic agenda setting, deliberation, and decision making.

This section describes three large-scale collaborative initiatives and the digital and other technologies that they use, and it interrogates the extent to which each successfully coordinates heterogeneous actors (who may also be strangers), upskills participants, and advances democratic practices. We first consider MoveOn and GetUp! as illustrating the benefits and limitations of digital technologies used only for mobilization. This is contrasted with Safecast, in which digital technologies are used to support collaborative problem solving. We see in this case a powerful illustration of how digital technologies can create new sites of engagement that bring people together around common needs, expand their community of fate, and upskill participants in democratic civic practice and technological tools. The mechanisms of Safecast, however, support scaling participation (and the democratic benefits of collaborative problem-solving) in only relatively simple domains. To explore more complex domains, we turn to Code for America, which illustrates the importance of organizational forms and ecosystems to make collaboration effective, especially in settings that require deep partnerships with government and deep levels of empathy around complex societal needs. This leads us to the following question: how can digital technologies support initiatives such as Code for America to further scale collaborative problem solving (and the opportunities they hold for civic learning) in complex domains? Crowdsourcing has mostly developed around scaling simple, parallelizable work. How can we develop technologies that support and scale organizational forms and ecosystems that are central for initiatives such as Code for America? We pick this question up, and its relationship to upskilling, in the final section.

MOVEON AND GETUP!

MoveOn started with the simplest of digital tools to mobilize people to use their signatures to better advance issues about which they cared.[16] It was the brainchild of tech entrepreneurs Joan Blades and Wes Boyd, who realized the potential of the internet for political campaigns when their 1998 online petition to "move on" from the Clinton impeachment went viral. Initially, it was largely an email call for signing petitions and contributing money. It used the internet to urge people to vote, but there was little actual personal interaction, at least at first. Although it did solve certain time and information problems and did encourage (and may have even had an effect) on voter turnout, it did not solicit full engagement. While MoveOn does coordinate people for elections, meetings, and other actions, it relies more on nudges and a bit of new information rather than the creation of new organizational forms or personal interactions. It costs less than door-to-door canvassing and may prove equally effective (jury is still out). It reaches more and different individuals than was possible with direct mailings. However, over time, MoveOn has become as much of an irritant as those multiple pieces of solicitation in our mailboxes.

The Australian version, GetUp, goes beyond the internet to create community-based organizations and grassroots leadership. In the section of the website called "How Do We Do It," GetUp states its mission as this: "Sometimes we gather in raucous protest, at other times we partner with policy experts to develop new solutions—and everything in between. Whatever we do, we do it with as many people as possible, using our hands, our hearts, our voices to fight for the issues that matter most."[17] The list of campaigns and the varieties of actions is significant, and GetUp has definitely developed an offline network of volunteers who engage in protests and voter solicitation but also lobbying and community-based organizing.

MoveOn and GetUp are examples of simple tools focused on mobilizing large-scale engagement in elections and political actions. They are good illustrations of open-call crowdsourcing, in which crowds are recruited en masse through targeted recruitment or social media to contribute to parallelizable collective action efforts. While this is useful for large-scale search or protests that can even topple regimes as in the Arab Spring, they tend to be limited in complexity, and participation is typically transient.[18] As described by Cebrian, Rahwan, and Pentland, "Social media has been much better at providing the fuel for unpredictable, bursty mobilization than at steady, thoughtful construction of sustainable social change."[19]

When there have been attempts at ongoing collaborations, it tends to be quite local. These organizations also tend to be associated with the more liberal or leftist side of the debate. This is an advantage for mobilization, but it may prove a limit to the challenge to beliefs and the development of the skill of persuasive argument, two hallmarks of a pragmatic democracy. They certainly build civic skills and, at least in the case of GetUp, are experimenting with democratic organization. However, their emphasis is on advocacy and mobilization rather than on collective and democratic problem solving or on significantly expanding the community of fate by reaching out to those with different ideological perspectives.

SAFECAST

On Friday, March 11, 2011, at 2:46 p.m., a magnitude 9.0 earthquake struck 70 kilometers east off the coast of Japan. It triggered tsunami waves reaching heights of up to 40.5 meters and traveling up to 10 kilometers inland. There were 15,895 deaths confirmed, 228,863 people displaced, and 402,699 buildings collapsed. The tsunami produced by the earthquake breached the walls of the Fukushima Daiichi Nuclear Power Plant and disabled its reactor cooling systems, leading to partial nuclear meltdowns and explosions resulting in releases of radioactivity.

This horrific series of disasters led to the Safecast story, a powerful illustration of rapid globally distributed community organized response to address a clear need: measuring radiation levels to determine whether living or traveling in certain areas is safe. It started with an email thread between Sean Bonner, Joi Ito, and Pieter Franken, individuals who each had authentic preexisting connections to the disaster area: "In the days following, the discussion moved from confirming safety of friends and family, to ensuring their continued well-being in part by getting Geiger counters into their hands. Commercially available supplies dried up almost instantly and the discussion changed from buying to building. A plan to distribute devices was developing."[20] It became clear that the government would not be able to supply this service. In addition, data sets existed, but they were often not shared, not standardized, or limited in scope. Commercial Geiger counters were also expensive and not tailored to the measurements needed. Bonner, Ito, and Franken all had strong global networks and expertise in rapid innovation. As they identified issues that needed to be solved to build and distribute Geiger counters, they pulled in relevant experts into the conversation. Bonner and Ito were the organizers for a conference in Japan, which they decided to repurpose to focus on crisis response specific to the earthquake. This meeting, just one month after the earthquake hit, became the first in-person meeting where they brainstormed the key idea for designing a measurement device, a Geiger counter that could be strapped to a car to log measurements in motion, and also where they decided to focus on collecting data and providing it in a publicly available online database and map.

Safecast was born. They set a one-week deadline for building out the initial prototype, which was created by a team at the Tokyo Hackerspace. Exactly one week later, they had a working device. In the following months and years, they produced several improvements to this device, added a new stationary device, and most importantly, built a network of volunteers using these devices to contribute to data collection. Safecast decided to be intentionally apolitical in the debate on nuclear energy, which enabled them to bring together governments, churches, academics, and businesses, all of which agreed on the importance of collecting and providing reliable data. Today, Safecast is continuing to provide measures for radiation levels but has broadened its mission to collecting reliable data for environmental measurements at large.

The Safecast story demonstrates the democratic power available to communities in a networked information economy that enables them to collaborate to meet a real need even when governments and the market fail to provide solutions or are even hindering it. It also illustrates how an expanded community of fate was created in the process of this effort. The effort was seeded by a small number of individuals who had authentic connections to the community in need, a strong network, and experience in innovation. However, this quickly expanded to a larger team through online and in-person interactions, and then to a large-scale data collection effort. This effort was able to bridge divides across typical communities of fate to encourage collaboration despite the sensitivity of the political debate on nuclear energy. As members of the community mobilized around data collection efforts, they learned technical skills related to radiation measurements and sensors, and they gained civic skills as they worked together with local community organizations. The sensor they created was open source and available as a "do-it-yourself" kit, enabling the community to self-organize in creating a large-scale and trusted source of data that was open and transparent to all.

However, this story also embodies some of the limitations of collaborative problem solving. One of these is the uniquely large network and expertise that Bonner, Ito, and Franken had, which enabled them to kick-start the process of forming a team and creating the device. This condition could be challenging to transfer and places disproportionate power in the hands of those with this ability. Another limitation is the relatively simple scope of collaboration involved in Safecast (putting together sensor kits and collecting data). In this scenario, this simple scope was enough for solving the need, and it naturally provided opportunities for upskilling and civic learning. Other problem-solving scenarios, however, may need sustained and interdependent expert participation. New mechanisms would be needed to support coordination and upskilling in more involved technical aspects of innovation as well as in leadership, deliberation, and other civic skills. In the absence of such mechanisms, it would be hard to take advantage of the opportunities for civic learning provided by collaborative problem solving.

Safecast is a good illustration of the many efforts around peer production and open source as well as in citizen science and crowd mapping.[21] These efforts provide tremendous agency to community and nonprofit initiatives that was not possible before, and they exemplify upskilling in both technical and civic dimensions. However, they often rely on a small number of long-term experts to drive the complex and interdependent parts of an effort. The community at large typically engages in learning and contributing along simpler parallelizable dimensions of the problem-solving effort. Open source is the closest example of large-scale, complex initiatives that also provide opportunities for learning. However, even in these settings, novices encounter many challenges to going beyond peripheral participation to more central tasks, and projects are often dependent on a few core contributors.[22] Longer-lasting success stories are often sponsored by a company and tend to center on frameworks, libraries, and other technical infrastructure as opposed to collaborative problem-solving initiatives.

CODE FOR AMERICA

In 2009, Jennifer Pahlka, who was then working at O'Reilly Media, realized that advances in technology could make government not only more effective and efficient but also more responsive to citizen needs and more user-friendly. She founded Code for America (CfA), with support from a wide group of technologists, government officials, and foundation sponsors. The organization tries to create an ecosystem supporting complex initiatives in a way that is consistent, replicable, and scalable. In many cases they develop apps that facilitate access to needed services (e.g., CalFresh, welfare benefits) and that remove obstacles that hinder flourishing (e.g., making it easier to pay off parking fines that must be cleared to obtain a job). Their team involves staff but also fellows (people who commit a full year to kick-start a project) and local brigades (volunteer midcareer professionals who partner with local governments on projects). The process upskills the fellows, the brigade members, the government officials, and often those they are serving. All become more engaged with one another, with projects that matter, and with the people whose needs they must listen to, address, and involve in improving their own lives. As the organization has grown, it has also evolved its strategy and its mission, now framed in terms of delivery-driven government.[23]

The primary purpose of Code for America is to improve government with technology by empowering teams of volunteers and trained staff to analyze and solve problems. It has demonstrably succeeded in improving the quality of service provision in response to actual demands and needs while also upskilling its team members. Code for America also fosters democratic practice. It is relatively nonhierarchical and holds itself and its teams accountable to the government agencies they are assisting. Most importantly from the perspective of pragmatic democracy, it solves problems through careful experimentation with possible solutions and with giving its staff and fellows considerable voice in finding and implementing those solutions. Through the process of creating the ecosystem necessary to support collaborative problem solving in complex domains, CfA generates significant civic learning and upskilling opportunities for participants. Moreover, as participants engaged with government officials and poorly served populations in the process of improving service, the team members necessarily develop empathy for those outside their existing communities of fate. This may well prove a mechanism for expanding the community of fate.

Code for America develops digital technologies that support governments, and it uses diverse technologies to coordinate their own teams and to help government officials and recipients coordinate with each other. However, developing digital technologies that can better support and scale such complex initiatives is still an open question in crowdsourcing. Such technologies would need to better support in-person relationships, organizational structures, and ecosystems, as is required for delivering much more complex, sustaining solutions in partnership with governments. Recent studies have begun to explore ways to enhance collaboration around complex goals, such as through computationally enhanced organizations.[24] These, coupled with analytics on the digital exhaust of cloud platforms for team collaboration, may prove useful for scaling initiatives like Code for America.

We believe that a whitewater world requires finding new ways to better support collaborative problem solving, for providing both immediate benefits and opportunities for engaging participants in democratic civic practice, upskilling them in multipurpose transferable tools for work, and fostering an increased sense of obligation to others and an expanded community of fate.

Towards Society as a One-Room Schoolhouse for Professional and Civic Upskilling

As individuals engage with others outside their typical communities and develop empathy for those affected, they begin to contribute to expanded communities of fate. We saw that the availability of such opportunities depended on the complexity of the domain and the nature of collaboration. When there are ways to carve out simple tasks like data collection, many people can get involved, and a large literature in crowdsourcing provides tools for supporting such large-scale participation. However, when objectives are complex, it is much harder to expand participation to those without prior expertise or with limited time.

In this section, we discuss ongoing projects in the Tech4Good Lab at University of California, Santa Cruz, to develop tools that more deeply integrate upskilling and reskilling into the collaborative problem-solving process.[25] These tools not only upskill novices in the required professional expertise but also create new ways to align an individual's time so that time spent in professional learning can also be time spent in civic upskilling—in developing empathy and the pleasure of agency that comes from engagement in real-world societal collaborations.

Our work builds on research on motivating participation at scale in volunteer crowdsourcing through methods that align engagement with other activities such as learning, play, or hobbies.[26] In the following cases, digital technologies are integrated into more traditional forms of organization and learning, where they help to harness learning activities to advance collective goals. While they are only prototypes, we hope that they will provide new angles to thinking about how societal collaborations might incorporate skilling and reskilling of participants and to realizing a pragmatist democracy that provides greater opportunities for diverse voices, values, and commitments of a populace to contribute to collective problem solving for society.

LEARNER-POWERED COLLABORATION

In early 2016, the Syrian refugee crisis was escalating, with millions of people being displaced, up to a third of the country's population. This motivated David Lee to think about how crowdsourcing could scale and coordinate volunteer support for nonprofits. Initially, he wanted to start with more independent tasks that could be learned and carried out in parallel, like those illustrated in the Safecast story. However, in conversations with refugee resettlement agencies, he found that most of the ideas they were interested in involved highly interdependent work such as designing and building web or mobile apps requiring longer-term skills-based efforts like those of Code for America. He started asking, "What would it take to organize volunteers, many of whom are short-term novices to support complex crowd work?"

Early attempts kept on failing, but they revealed a close connection between societal collaboration and upskilling. Even when participants had experience, they were experienced in different approaches and technologies. In order to successfully coordinate them, there needed to be a deeper integration of learning into the work process.

But more than that, it quickly became apparent that upskilling was also the key motivator for drawing initial engagement. Eighty-five percent of participants said they would be willing to participate even if they were just recreating existing websites. And most wanted to help only with web development and not with writing guides (for a cultural orientation app). But as they participated, their motivation evolved, and they began to ask for tasks that they previously did not want to do. They had begun to feel the pleasure of agency and to develop empathy for those affected. An expanded community of fate was growing.

There is an untapped opportunity for motivating societal collaboration through learning. In a whitewater world, individuals need to be continuously learning. In Heather McGowan's words, "In the past, we learned in order to work. Now, we must work so that we can continuously learn."[27] But finding opportunities to learn as you work, to learn experientially and in a situated real-world context, is really hard.[28] Even if you can commit an entire summer, for example, opportunities for internships are rare. It's a vicious cycle where you need experience to get experience. And there are certainly no opportunities to obtain the tacit real-world knowledge of the workplace in small amounts of time, through small weekend or evening internships.

If it were possible to support learners in contributing to complex real-world work, this would not only meet the needs of learners but also open up new levels of access and opportunity for individuals to contribute to their communities and to support a pragmatist democratic vision of societal collaboration.

After our many early failed attempts at organizing volunteers to build web applications for nonprofits, we finally succeeded (for a simple case), and we were able to organize a group of learners with little to no background in HTML and SCSS to successfully build out static pages of apps for nonprofits.[29] We designed a platform for learning web development. But instead of progressing through a sequence of topics, like in a classroom, we modeled learning after the workplace, with users moving through small experiential roles that are structured in an organization-like hierarchy. The hierarchy provides a learning pathway for developing professional skills. And as people move through this pathway, they contribute to real-world projects for non-profits and mentor those below them in the hierarchy.

This pilot study was just a simple case but is being extended to more complex aspects of web development and design. The Tech4Good Lab currently involves forty undergraduate students, most of whom do not have experience when they first join, to make real and significant contributions to projects on education, work, and community engagement. One student said: "When I first joined the lab, I was still fairly new to programming, but I knew I had an interest in web development, tech for social good, and UI/UX design. Within the lab I was able to learn and develop my skills as a front-end developer, as well as find a passion for design. Best of all, I feel personally connected to our projects and mission in a way that I don't in my core classes. Diving into HCI research and working on projects that help my community makes every day exciting." These students aren't just gaining technical skills, they are also getting the opportunity to develop a sense of empathy, experience the pleasure of agency in meeting needs beyond their own, and heighten their sense of obligation towards causes that might have previously only sparked a momentary pull. The learning of technical skills provides a pathway for them to do this and to align the work they need to do with the causes they want to contribute to. The result is a virtuous cycle where professional learning provides a back door to civic learning and where civic learning provides a powerful source of motivation to help individuals persist in the deliberate practice needed for professional learning.[30]

SMALL-GROUP NETWORKS

In their current form, the hierarchies described so far are still best suited to domains focused on execution of a determined direction as opposed to the (arguably more critical) other parts of problem solving such as design or agenda setting. More work still needs to be done to extend toward these other domains. In another early stage project, we are exploring learning pathways that are based not on micro-roles in organization-like hierarchies, but on small-group interactions, which are more suitable for brainstorming and deliberation. Participants progress through a sequence of different small group types that support learning while also contributing to real-world work. These groups include exploratory reading to develop intuition, deep dives for mapping out literature and honing an idea, and ideation groups for brainstorming around a theme.

One interesting aspect of these small groups is the relational component that develops through the in-person interactions over a monthlong period, and the opportunity this provides for facilitating large-scale deliberation. Digital technologies have created a new reality in which large-scale deliberation is possible but built on tweets and posts. In a heterogeneous population, societal collaborations—as we know from the longshore and other cases—require new building blocks for deliberation that enable debate and learning in the context of trusted, in-person relationships. These small groups might be useful from this perspective.

In our early prototypes, small groups spend the first fifteen minutes of each meeting on team-building activities. The simplest default activity is "get to know you" questions posed by members of the group; these have ranged from simple icebreaker questions to deeper questions like "Where do you see yourself in 10 years?" or even vulnerable questions like "When was the last time you cried?" or "What is the most politically incorrect thing you believe?" We've found that students unanimously value this component and the relationships it has helped to foster.

Other activities directly involve the group in a deliberative process independent of the group's primary purpose for meeting. For example, in one activity based on the six hats framework, the group gets a brainstorming prompt such as, "How can UC Santa Cruz better support the cost of living for graduate students?" or "How can the US best reform immigration?"[31] Group members engage in a facilitated discussion during which they put on different hats representing different ways of thinking (e.g., black for critical, yellow for positive, white for information). The activity provides an opportunity to learn one simple framework for discussing, disagreeing, and working towards consensus in a group. The results from their group discussion are uploaded to the platform and contribute to a larger discussion that other groups can build on. This project is still a work in progress, but it provides another example of how digital technologies can help to align learning with new forms of large-scale collaborative problem solving and to provide opportunities for participants to get exposed to and develop a greater sense of obligation for issues in society.

TOWARD SOCIETY AS A ONE-ROOM SCHOOLHOUSE

The upskilling and problem-solving properties of the described projects do not, by themselves, meet the demands of democratic societal collaboration. They have the potential to provide greater agency to individuals and communities in meeting unmet needs and to provide greater opportunities for individuals to engage in civic learning. However, the extent to which they succeed depends on how they are deployed in practice. For example, the expertise needed to create such structures would limit their power to those with the ability to create them, unless they are provided in ways that are publicly accessible and adaptable. The use of digital technologies can be a boon but can also limit access to certain populations.

Additionally, these structures need to be consciously embedded in a larger ecosystem, in settings like the brigades of Code of America, where digitally mediated interactions could be tightly integrated with in-person relationships, organizational structures, institutional structures, and social spaces. There also needs to be much more research on supporting contestation and productive friction at scale in ways that also engender generous and generative listening. We need to continue developing more means to enable individuals to recognize their common community of fate with strangers.

Most importantly, we need to learn how to create governance structures that give priority to democracy in the sense Knight and Johnson required. That means devising decision-making institutions that recognize both specialized knowledge and common purpose but that put the ultimate power in the hands of the public via the vote.

Luckily, there are many others experimenting at the intersection of all of these themes. A growing literature in "learnersourcing" considers how to align learner activities with other objectives and how to create ecosystems to support communities of practice.[32] Others have explored systems that support deliberation in civic settings.[33] Still others are experimenting with new forms of democratic practice.[34] We see these as each providing insights and small prototypes towards digital technologies that support democratic societal collaboration.

We return to the analogy we started from, our observation that we live in a whitewater world. How do we help people navigate whitewater rapids? They need to be provided with opportunities to train from lower-level rapids (with lower stakes and guardrails) before being thrust into higher level ones with real consequences. Our experiments with integrating professional and civic upskilling have so far been limited to small pilots, but we have begun to incorporate these ideas into coursework. What would it look like if a college student's four-year coursework not only provided real-world problem-solving capabilities but also engaged the student in supporting initiatives like Code for America where he or she could learn about local or national issues, develop a greater sense of ownership over the community, and develop an expanded community of fate? And if possible for universities, why not for

K–12, with younger students working within smaller societal collaborations at the level of a school or neighborhood?

There are sixteen million college students in the United States alone, but only 10 percent of them are able to obtain internships each year. Imagine if all of them had opportunities for real-world learning, and not just during the summer, but throughout the school year, and in the context of supporting social innovators and nonprofits working on needs like climate change or homelessness. Many of these needs will never be served through financial markets, but maybe that's an opportunity: an opportunity to redesign education away from exams in a classroom and toward a community-engaged experience with society itself as a one-room schoolhouse for real-world learning, where education is highly integrated with civic engagement and supporting societal collaborations. These experiments are a start toward our larger goals. If successful, we will have a far firmer grasp of how to support participants through democratic societal collaborations to view a whitewater world not from the perspective of fear but from the perspective of an adventure.

Concluding Thoughts

This chapter is a modest beginning at tackling the creation of democratic societal collaborations that upskill their participants and forge expanded communities of fate. We hope it provides some new perspective on the possible affordances that digital technology may provide in supporting flourishing in a whitewater world. The value of societal collaboration derives not only from the outcomes produced but also from the process of collaboration itself. Flourishing goes beyond basic survival needs to social belonging, esteem, self-actualization, and self-transcendence. One of the promises of societal collaboration lies in its potential to provide people with the opportunity to contribute to something greater than themselves and to build relationships, dignity, and meaning without compromising their ability to survive economically. In the spirit of Dewey and of Knight and Johnson, our hope is that from some of these experiments in collaboration might emerge new democratic institutional arrangements better suited to navigating a whitewater world.

Notes

1. Ann M. Pendleton-Jullian and John Seely Brown, *Designing for Emergence*, vol. 1 of *Design Unbound: Designing for Emergence in a Whitewater World* (Cambridge, MA: MIT Press, 2018); *Ecologies of Change*, vol. 2 of *Design Unbound: Designing for Emergence in a Whitewater World* (Cambridge, MA: MIT Press, 2018).

2. Elinor Ostrom, *Governing the Commons: The Evolution of Institutions for Collective Action* (New York: Cambridge University Press, 1990).

3. John S. Ahlquist and Margaret Levi, *In the Interests of Others: Leaders, Governance, and Political Activism in Membership Organizations* (Princeton, NJ: Princeton University Press, 2013).

4. See, e.g., Zeynep Tufekci, *Twitter and Tear Gas: The Power and Fragility of Networked Protest* (New Haven, CT: Yale University Press, 2017); Wael Ghonim, *Revolution 2.0: The Power of the People Is Greater Than the People in Power, a Memoir* (Boston: Houghton Mifflin Harcourt, 2012); Manuel Cebrian, Iyad Rahwan, and Alex "Sandy" Pentland, "Beyond Viral," *Communications of the ACM* 59, no. 4 (2016): 36–39. See also several of the chapters in this volume that consider more directly than we do the influence of new technologies on democracy as a system—particularly those by Farrell and Schwartzberg, Cohen and Fung, Gangadharan, and Ananny.

5. Ahlquist and Levi, *In the Interests of Others.*

6. Elisabeth Jean Wood, "The Emotional Benefits of Insurgency in El Salvador," in *Passionate Politics: Emotions and Social Movements*, ed. Jeff Goodwin, James M. Jasper, and Francesca Polletta (Chicago: University of Chicago Press, 2001), 267–81.

7. Pendleton-Jullian and Brown, *Designing for Emergence* and *Ecologies of Change.*

8. Thomas M. Alexander, "Pragmatic Imagination," *Transactions of the Charles S. Peirce Society* 26, no. 3 (1990): 325–48; Charles F. Sabel, "Dewey, Democracy, and Democratic Experimentalism," *Contemporary Pragmatism* 9, no. 2 (2012): 35–55.

9. Sabel, "Dewey."

10. John Dewey, *The Public and Its Problems* (Athens: Ohio University Press, 1954), 140.

11. Joshua Cohen and Charles F. Sabel, "Directly-Deliberative Polyarchy," *European Law Journal* 3 (2002): 313–42.

12. Ostrom, *Governing the Commons.*

13. Charlotte Hess and Elinor Ostrom, eds., *Understanding Knowledge as a Commons: From Theory to Practice* (Cambridge, MA: MIT Press, 2011), esp. chap. 3, "A Framework for Analyzing the Knowledge Commons," by Ostrom and Hess.

14. Yochai Benkler, *The Wealth of Networks: How Social Production Transforms Markets and Freedom* (New Haven, CT: Yale University Press, 2006); Yochai Benkler, Aaron Shaw, and Mako Benjamin Hill, "Peer Production: A Form of Collective Intelligence," in *Handbook of Collective Intelligence*, ed. Thomas W. Malone and Michael S. Bernstein (Cambridge, MA: MIT Press, 2015), 175–204.

15. Jack Knight and James Johnson, *The Priority of Democracy: A Pragmatist Argument* (Princeton, NJ: Princeton University Press; New York: Russell Sage Foundation Press, 2011), 28. See also the excellent symposium on the book in *Crooked Timber* (February 2013), available at http://crookedtimber.org/category/knight-johnson-seminar/.

16. David Karpf, *The MoveOn Effect: The Unexpected Transformation of American Political Advocacy* (New York: Oxford University Press, 2012); Ariadne Vromen, "Campaign Entrepreneurs in Online Collective Action: Getup! in Australia," *Social Movement Studies* 14, no. 2 (2015). 195–213.

17. https://www.getup.org.au/about.

18. On search, see Joseph M. Hellerstein and David L. Tennenhouse, "Searching for Jim Gray: A Technical Overview," *Communications of the ACM* 54, no. 7 (2011): https://cacm.acm.org/magazines/2011/7/109892-searching-for-jim-gray/fulltext. On protests, see Sandra González-Bailón, Javier Borge-Holthoefer, Alejandro Rivero, and Yamir Moreno, "The Dynamics of Protest Recruitment through an Online Network," *Scientific Reports* 1, no. 197 (December 2011): 1–7.

19. Cebrian, Rahwan, and Pentland, "Beyond Viral."

20. See http://iopscience.iop.org/article/10.1088/0952-4746/36/2/S82 and https://blog.safecast.org/history/. See Pendleton-Jullian and Brown, *Designing for Emergence*, 207n3.

21. See Benkler, Shaw, and Hill, "Peer Production"; Robert Simpson, Kevin R. Page, and David De Roure, "Zooniverse: Observing the World's Largest Citizen Science Platform," in *Proceedings of the 23rd International Conference on World Wide Web* (2014): 1049–54; Brian L. Sullivan, Christopher L. Wood, J. Iliff Marshall, Rick E. Bonney, Daniel Fink, and Steve Kelling, "Ebird: A Citizen-Based Bird Observation Network in the Biological Sciences," *Biological Conservation* 142, no. 10 (October 2009): 2282–92; Matthew Zook, Mark Graham, Taylor Shelton, and Sean Gorman, "Volunteered Geographic Information and Crowdsourcing Disaster Relief: A Case Study of the Haitian Earthquake," *World Medical & Health Policy* 2, no. 2 (July 2010): 7–33.

22. Christoph Hannebauer, "Contribution Barriers to Open Source Projects" (PhD diss., University of Duisburg-Essen, 2016); Georg von Krogh, Sebastian Spaeth, and Karim R. Lakhani, "Community, Joining, and Specialization in Open Source Software Innovation: A Case Study," *Research Policy* 32, no. 7 (July 2003): 1217–41; Igor Steinmacher, Marco Aurélio Graciotto Silva, and Marco Aurélio Gerosa, "Barriers Faced by Newcomers to Open Source Projects: A Systematic Review" paper at the International Federation for Information Processing Conference on Open Source Systems, May 6–9, 2014, San José, Costa Rica, 153–63.

23. https://medium.com/code-for-america/delivery-driven-government-67e698c57c7b.

24. Daniela Retelny, Sébastien Robaszkiewicz, Alexandra To, Walter S. Lasecki, Jay Patel, Negar Rahmati, Tulsee Doshi, Melissa Valentine, and Michael S. Bernstein, "Expert Crowdsourcing with Flash Teams," *Proceedings of the 27th Annual ACM Symposium on User Interface Software and Technology*, October 5–8, 2014, Honolulu, 75–85; Melissa Valentine, Daniela Retelny, Alexandra To, Negar Rahmati, Tulsee Doshi, and Michael S. Bernstein, "Flash Organizations: Crowdsourcing Complex Work by Structuring Crowds as Organizations," *Proceedings of the 2017 Conference on Human Factors in Computing Systems*, May 6–11, 2017, Denver, CO, 3523–37.

25. https://tech4good.soe.ucsc.edu.

26. On learning, see Juho Kim, "Learnersourcing: Improving Learning with Collective Learner Activity" (PhD diss., MIT, 2015). On play, see Seth Cooper, Firas Khatib, Adrien Treuille, Janos Barbero, Jeehyung Lee, Michael Beenen, Andrew Leaver-Fay, et al., "Predicting Protein Structures with a Multiplayer Online Game," *Nature* 466, no. 7307 (2010): 756–60. On hobbies, see Sullivan et al., "Ebird."

27. Heather McGowan, "Work to Learn," *Computing Research Association Summit on Technology and Jobs* (2017), YouTube video, 30:35, posted by Computing Research Association, January 5, 2018, https://www.youtube.com/watch?v=5x4zqUi2Nc0.

28. John Dewey, *Democracy and Education: An Introduction to the Philosophy of Education* (New York: Macmillan, 1916).

29. David Lee, Sina Hamedian, Greg Wolff, and Amy Liu, "Causeway: Scaling Situated Learning with Micro-Role Hierarchies," *Proceedings of the 2019 CHI Conference on Human Factors in Computing Systems*, May 4–9, 2019, Glasgow, Scotland. Paper No. 74, https://doi.org/10.1145/3290605.3300304.

30. K. Anders Ericsson, Ralf T. Krampe, and Clemens Tesch-Römer, "The Role of Deliberate Practice in the Acquisition of Expert Performance," *Psychological Review* 100, no. 3 (1993): 363–406.

31. E. De Bono, *Six Thinking Hats* (London: Penguin, 2017).

32. Kim, "Learnersourcing." See also Haoqi Zhang, Matthew W. Easterday, Elizabeth M. Gerber, Daniel Rees Lewis, and Leesha Maliakal. "Agile Research Studios: Orchestrating Communities of Practice to Advance Research Training," *Proceedings of the 2017 ACM Conference on Computer Supported Cooperative Work and Social Computing*, February 25–March 1, 2017, Portland, OR, 220–32, https://doi.org/10.1145/2998181.2998199; Rajan Vaish, Snehalkumar (Neil) S. Gaikwad, Geza Kovacs, Andreas Veit, Ranjay Krishna, Imanol Arrieta Ibarra, Camelia Simoiu, et al. "Crowd Research: Open and Scalable University Laboratories," *Proceedings of the 30th Annual ACM Symposium on User Interface Software and Technology*, October 22–25, 2017, Quebec City, QC, 829–43, https://doi.org/10.1145/3126594.3126648.

33. Travis Kriplean, Jonathan Morgan, Deen Freelon, Alan Borning, and Lance Bennett, "Supporting Reflective Public Thought with ConsiderIt," *Proceedings of the 2012 ACM Conference on Computer Supported Cooperative Work*, February 11–15, 2012, Seattle, 265–74, https://doi.org/10.1145/2145204.2145249.

34. See, for example, the list of those presenting their experiments with blockchain and quadratic voting at the RadicalxChange meetings (https://radicalxchange.org). The work of Santiago Siri is one of many instances. Andrew Leonard, "Meet the Man with a Radical Plan for Blockchain Voting," *Wired*, August 16, 2018, https://www.wired.com/story/santiago-siri-radical-plan-for-blockchain-voting/.

9

From Philanthropy to Democracy: Rethinking Governance and Funding of High-Quality News in the Digital Age

Julia Cagé

The modern media industry is in a state of crisis. Digitalization has changed the nature of competition in media markets and the range of products provided. There is growing concern over news quality and the effectiveness of the media as a check on power. Furthermore, the number of journalists is plummeting in all developed countries, a major social change that may reflect media outlets' declining incentives to invest in quality. An open question—with important consequences for journalists who are facing social mutations that threaten their profession and more generally for the quality of the democratic debate—is whether news still has a commercial value and what kind of new business models and legal status need to be developed for media organizations. In this chapter, I first explain why quality news should be considered a public good and why this public good is essential in well-working democratic societies. I then investigate why this public good is underprovided—and underconsumed—in contemporary democracies. Finally, I discuss a number of solutions that could be implemented to provide adequate long-term financing and capitalization of news media while preserving media independence. In particular, I stand up for a new model—nonprofit media organization—that could leverage digital technologies to remove direct links between the philanthropists and the media they fund, and raise investment funds through crowdfunding.

I gratefully acknowledge the many helpful comments and suggestions from the participants at the "Digital Technology and Democratic Theory" workshop. All errors remain my own.

Will Foundations Be the Future of the Media Industry?

Philanthropy is booming in our democracies, particularly in the United States, where we see a growing role of private funders in the provision of public goods as the government retrenches. This phenomenon is not specific to the media industry—as highlighted in Reich, Cordelli, and Bernholz (2016), "in the United States and most other countries, we see philanthropy in all areas of modern life" (see also Reich, 2018)—but philanthropy also increasingly supports the provision of information.

The growing role of philanthropy in media funding has been well documented. The *Growth in Foundation Support for Media in the United States* report published in 2013 by Media Impact Funders reports that $1.86 billion was awarded in media-related grants from 2009 to 2011. The investigative website ProPublica, created in 2008, is funded entirely through philanthropy; its French counterpart, Disclose, launched in November 2018, is similarly raising money through crowdfunding and larger donations, including from US foundations (e.g., Open Society). As of today, there are about 150 nonprofit centers doing investigative journalism in the United States, and for-profit newspapers like the *New York Times*, much like foundation-owned newspapers like the *Guardian* in the United Kingdom, have recently set up nonprofit ventures to support their journalism. In France, Le Monde Afrique website, launched in 2015 by the daily broadsheet *Le Monde*, has received financial support from the Bill and Melinda Gates Foundation. Overall, philanthropy is becoming a very large part of the revenue streams of a growing number of news companies. In a series of articles published in the *Columbia Journalism Review*, David Westphal defines philanthropy as "journalism's new patrons."

Concurrently, during the past decade, we have also observed an increasing tendency of out-of-market billionaires to acquire media outlets, often at a very low cost but with even lower profit expectations. Jeff Bezos (Amazon) and the *Washington Post*, Patrick Soon-Shiong (a biotech billionaire entrepreneur) and the *Los Angeles Times*, Marc Benioff (Salesforce) and *Time Magazine*, are but a few examples of the way tech entrepreneurs with deep pockets are showing a new "taste" for the media industry. While these new media moguls publicly claim that they are acting as philanthropists,[1] it is more accurate to call them by their name: the new media patrons. The development—or more precisely the revival—of this patronage model is by no means specific to the United States, as is apparent from the recent entry of telecommunications billionaires on the French media market (e.g., Xavier Niel and *Le Monde*; Patrick Drahi and *Libération*, BFMTV, and RMC), and most recently of the Czech billionaire Daniel Kretinski (who made a fortune in the energy sector and is now buying shares in *Le Monde* and other media outlets).

It is important to distinguish between philanthropic funding (via charitable donations) of the media on the one hand and the patronage model on the other hand. But in the end, as we will see in this chapter, these two models pose similar problems with regard to journalists' independence and the disproportionate weight given to the preferences of the wealthy. Furthermore, an additional threat comes from the growing lack of transparency regarding media ownership and more broadly the funding of news (I discuss, among others, the case of Project Veritas and Robert Mercer's investment in right-wing media). This lack of transparency relies on the use of foundations—protecting donors' identity—to fund journalism.

Overall, (why) do we need nonprofit models to finance the production of information? And is the growing role of philanthropy—as it is broadly understood—due to digital technologies? In this chapter, I argue that there are at least two different reasons—with very different implications—we need to consider the nonprofit model when thinking about the future of the media in the online world. We will consider these two motives in turn. First, we need to take the profit motive out of information production because information is a public good; hence, it cannot be left at the mercy of the market. I will indeed make the case that democracy, both in the voting booth and beyond, requires informed citizens. Therefore, we need to rethink the media industry along democratic lines, which also entails a necessary distinction between the development of nonprofit media organizations with a democratic governance and philanthropic funding of the media in the absence of any countervailing powers.

Second, the need for philanthropic support to the media comes from the fact that the commercial value of information has plummeted in the digital era. But information still has a very high social value. Hence providing high-quality independent news requires coming up with alternative sources of funding. In this chapter, I defend the idea of a new corporate model for the media, adapted to the twenty-first century: the "nonprofit media organization" (Cagé, 2016).

Democracy, Informed Citizens, and Quality News as a Public Good

What is democracy? The fundamental principle of our modern political system is "one person, one vote."[2] In this chapter, I argue that it should rather be "one informed person, one vote," or more precisely "one person, one vote in an information-rich environment." An uninformed vote should indeed be considered an undemocratic vote; in the best-case scenario, it is a "random" vote, and most probably, it is a "captured" vote. Hence free, unbiased, high-quality information is indispensable to the democratic debate, both in the voting booth and beyond.

ONE PERSON, ONE VOTE?

In my opinion, democracy should be approached as "one person in an information-rich environment, one vote." Although this view of democracy may be perceived as overly narrow in light of the numerous discussions in the literature, I think it is nonetheless important to highlight the central role played by information in our modern democracies.

Democracy can be broached from a number of different points of view. In its simplest approach, democracy, considered as electoral democracy, is just "one person, one vote." As long as elections take place on a regular basis, with the appearance of fairness and freedom, a regime could be considered "democratic." Obviously, such an approach is insufficient and does not entirely correspond to reality (see e.g., Achen & Bartels, 2016, who argue that even well-informed voters do not make "rational" political decisions based on the information they acquire regarding the actual performance of incumbent political leaders).[3] But the objective of this chapter is not to discuss the relative importance of predictable biases versus information in the voting decisions made by citizens. I do agree that it is important to be well aware of the existence of such biases, but it does not mean that information is anecdotal, just that we may have overestimated its role in existing models of democracy. As highlighted by Goldman (1999), "although it takes full core knowledge to *guarantee* a democratically good result, high levels of core knowledge can make a democratically good result highly probable" (p. 327). In other words, while voters may not be perfectly informed, "core knowledge promotes democratically "successful" results, and greater core knowledge is always better."

More importantly, as I will discuss, it is has been shown empirically that when citizens are better informed, governments tend to be more accountable. Hence, among the range of factors that should be considered to judge the quality of democratic governance in any society, Gilens (2012) includes the following: "do citizens have access to the information necessary to evaluate their political leaders and competing candidates?"

In the end, for electoral competition to work, citizens need to have at least some political information, even if they do not always rely on this information when they vote, and the news media should provide them with such information. Furthermore, democracy goes well beyond electoral competition and the voting booth. Democratic societies indeed require a robust civil society in which citizens enjoy freedom of speech and an informal public sphere in which information circulates and where citizens communicate and deliberate. The quality of civic deliberation depends on the quality of the information citizens have. Such an idea is also defended by Joshua Cohen and Archon Fung who, in their chapter "Democracy and the Digital Public Sphere," describe the informal public sphere as "an arena for communicative action, easy access to diverse views and perspectives, and high quality, low cost information on issues of common concern."

In other words, both electoral competition and the public sphere require informed citizens, that is, the production of high-quality information and the consumption of such information by the citizens who need to be able to (or be willing to) access it.[4] Information is a public good—valuable for the realization of democratic ends—and the private market alone would tend to underproduce this public good.[5] Hence the need to improve our understanding of the business model of the media and the way digital technologies have affected this model, an issue I discuss in this chapter (in particular the collapse in advertising revenues and the increasing difficulty to monetize news in the digital world).

MEDIA AND PARTICIPATION

It is well documented that the media affects political participation as well as social norms and beliefs, and that more informed citizens increase government accountability. Using data on US daily newspapers from 1869 to 2004, Gentzkow, Shapiro, and Sinkinson (2011) find that the entry of the first newspaper in a county has a positive effect on political participation. Why? Because newspapers provide information; to begin with, they inform citizens that elections are taking place. They also provide people with information about the issues at stake: better-informed individuals are more likely to vote. Similarly, the introduction of the radio in the United States in the 1930s increased voter turnout; it also directly increased politicians' accountability (Stromberg, 2004).[6] In a country like Brazil, when radio is used to expose corrupt politicians, it continues to influence the selection of good politicians today and promotes political accountability (Ferraz & Finan, 2008).

In contrast, when the media becomes less informative (e.g., when newspapers produce less information or lower-quality information), people vote less. Using French data since World War II, I show that increased newspaper competition (the entry of the second or third newspaper in the market) has a negative effect on political participation, due to a decrease in the quality of news and in particular in the size of the newsroom (Cagé, 2020). This result is driven by a business-stealing effect. The entry of a competitor in a local news market reduces the circulation of incumbent newspapers by nearly 20 percent, their revenues by around 40 percent and leads to a decrease of between 19 percent and 35 percent in the number of journalists working for them. As a consequence, we observe a 0.3-percentage-point drop in turnout at local elections.[7]

Similarly, in the digital world, in the absence of traditional news media, people vote less, and when they do, they may choose to support more extreme parties. For example, according to *Politico*, Donald Trump outperformed in 2016—compared to Mitt Romney in 2012—in counties with the lowest numbers of news subscribers.[8] However, he did worse than Romney in areas with higher subscription rates. At the local level, the internet is not providing adequate substitute for local journalism: consistently, Gao, Lee, and Murphy (2018) have shown that local newspaper closures between 1996 and 2015 in the United States led to higher borrowing costs of municipalities in the long-run, even in localities with high internet usage.

This latest point raises an interesting question that we need to address to improve our understanding of the world of online media: why is it that the impact of traditional news outlets differs from that of digital-only media?

DIGITAL TECHNOLOGIES AND DISINFORMATION

An increasing number of studies have highlighted what may be considered a paradox: while digital technologies should facilitate access to information—in particular because a lot of information is now available for free online—it turns out that areas with higher Internet penetration tend to have lower turnout. Why? First of all, because the Internet—just like television a few decades ago (Angelucci, Cagé, & Sinkinson, 2020; Gentzkow, 2006)—has displaced other media with greater news content, starting with newspapers. In both the United Kingdom and Germany, increased internet access has negatively affected political participation, in particular among less educated and younger citizens (Falck, Gold, & Heblich, 2014; Gavazza, Nardotto, & Valletti, 2019).

Furthermore, by increasing media choice, the internet may have increased inequality in political involvement, an argument first highlighted by Prior (2007) in the context of the introduction of cable television in the United States. Determining whether internet has increased segregation (compared to offline news consumption) is an ongoing debate and a fascinating empirical question (see e.g., Bakshy, Messing, & Adamic, 2015; Gentzkow & Shapiro,

2011; Halberstam & Knight, 2016; Pariser, 2011, for opposing views). While the focus of this chapter is on the impact of digital technologies on the production of information rather than its consumption, it is nonetheless of interest to highlight the pro and cons of these new technologies. Moreover, at the end of the chapter, I question the extent to which the necessary reinvention of the business model of journalism may be enough to increase the actual consumption of high-quality news.

Similarly, the current debate on the regulation of fake news—and the willingness to prevent abuses in elections—reminds us of the importance of providing citizens with high-quality independent news that they can identify as such.[9] The public's declining trust in the media—as well as in a number of other institutions, starting with political parties and government—has been well documented. Let me first highlight that such a decline is not necessarily negative; it may reflect an increase in citizens' critical judgment, which is an important virtue of any democracy (Henry Farrell and Melissa Schwartzberg make this argument in chapter 9).

But such a claim, as acknowledged by Farrell and Schwartzberg, rests on the idea that citizens' distrust for evidence is rooted in critical judgment rather than in less salutary forms of thinking. One of the main issues today when we consider the public's declining trust in the news media comes from the fact that when citizens encounter news on social media, how much they trust the content is determined less by who creates the news (a reputable media source or an unknown media source) than by who shares it (a person they trust or not). This has been documented in a 2017 study of the "Media Insight Project."[10] In other words, the challenge nowadays for reputable news media is not only to provide information as a public good but also to convince citizens that the information they produce is trustworthy. This is particularly demanding, as Farrell and Schwartzberg point out (chapter 9), given that the transition from the strong to the weak gatekeeping model has changed the structural basis of epistemic trust from a reliability- to a relationship-based model.

WHAT IS HIGH-QUALITY NEWS?

Before tackling the key issue of the demonetization of high-quality news, it is important to address a final question: what is high-quality news? In other words, how is news quality measured? As a reader, you might think that when you read a news article, it is easy for you to get a sense of whether the quality of the article is high or low. But would you be able to define a number of objective criteria that have led you to such a conclusion? And how could you be sure that what is for you a high-quality news article (e.g., an informative news article) will also be informative for another reader with, for example, a different educational background?[11]

In other words, if I ask this question, it is primarily to emphasize how difficult it seems to answer it, particularly to measure news quality empirically. In what follows, I consider the originality of a news article as one of the dimensions of its quality, but I am well aware that there are several additional dimensions (in particular, in the last part of this chapter, I introduce the central issue of independence as one of the core dimensions of journalism quality). Should we consider the length and the structure of the article? Or the quality of the language (e.g., complexity of the sentences, use of adverbs, lexical density)? Or how well people are informed after reading a particular piece of news? And yet, what is the direction of the correlation between how well citizens are informed and the complexity of the language used?

As researchers in social sciences, we are able to measure "inputs" (e.g., the size of the newsroom) but also the structure of the ownership. Hence we can argue that the quality of the news produced by a media outlet is positively correlated with the number of journalists working in this outlet. But quality is much more than that. For example, should we not also consider the fact of bringing new voices, the diversity of the points of view, and so on? Answering these questions should be and will be the objective of future research.

The Demonetization of High-Quality Information

High-quality news is a public good, one of the central features of well-functioning democracies, whether we consider electoral competition or the informal public sphere. But it is an underprovided public good in contemporary democracies. Although some optimists will not agree with this alarming observation, claiming that digital tech makes citizens better informed than in the past—and overlooking the important issue of "information overload"—the fact remains that not only are jobs for journalists at legacy media outlets continuing to disappear,[12] but also the number of journalists hired by digital native websites represents only a very small percentage of those laid off in the traditional media industry. Overall, the total number of journalists—whether we consider journalists working for the newspaper industry, radio and television stations, or digital news outlets—has been falling continuously in Western democracies for nearly fifteen years (Cagé, 2016). In the face of such a collapse, there is growing concern over the industry's ability to produce high-quality information (Hamilton, 2016; Henry, 2007; Starkman, 2013).

FIGURE 9.1. Number of journalists working for daily (local and national) newspapers as a share of the active population, France and the United States, 1978–2015.

It is thus important to understand the determinants of such a collapse. Is the internet to blame for this decrease? If the so-called Great Recession and the growing competition from online and social media are part of the explanation, then focusing on the recent period misses the big picture. Figure 9.1 plots the evolution of the number of journalists working for daily newspapers in France and in the United States since 1978. As the figure clearly shows, neither the financial crisis of 2008 nor the advent of the Internet can bear all the blame for the drop in journalistic employment. Other factors were also at work. Admittedly, the crisis of the print media has worsened since the late 2000s. But the internet only amplified a broader phenomenon: increased competition in the media market owing first to radio, then television, and today the web.

This increased competition explains both the revolution in journalism—the decrease in the number of journalists working for print media as opposed to other media—and the economic crisis of the press, which has seen its advertising revenues collapse as competition has increased. Needless to say, the number of journalists I consider here is the number of "professional journalists"—those employed by the media (whether part-time, full-time or freelance journalists)—not a broader figure that would also take into account bloggers or so-called citizen journalists. There is indeed a key difference between professional and nonprofessional journalists, which matters when studying the provision of high-quality journalism: professional journalists are subject to codes of ethic and professional journalistic standards and are accountable to the media organization(s) they work for. On the contrary, bloggers are not accountable to anyone, which raises a number of issues in the current context of fake news and disinformation. Obviously, it does not mean than citizen journalists are useless and that we should overlook their contribution to the public debate and overall provision of information. In a number of contexts—including the current Syrian conflict—citizen journalism is playing a crucial role; it also played a key role during the Arab Spring, changing the way information was disseminated, a phenomenon that has been very well documented (Howard, 2010; Howard & Hussain, 2013; Tufekci, 2017). But we still need gatekeepers—to filter, fact-check, contextualize, organize information, and so on—that is, professional journalists.

COMPETITION AND THE COLLAPSE OF ADVERTISING REVENUES

Technological innovations in news delivery, such as the advent of the Internet, threaten the basic economic model of news operations. While my focus in this chapter is on the internet and associated digital technologies, I first step back and examine the consequences of the entry of what is considered today a traditional media platform, namely television. Studying what happened at the time of the introduction of television indeed allows us to better understand the mechanisms at play nowadays. The entry of television, much like that of the internet, was indeed for print media a shock on both sides of the market: a shock on the reader side, with citizens switching from reading print media to watching television and then consuming news on the internet, and a shock on the advertising side, with the television and the Internet offering more effective advertising platforms. My recent research (with a number of coauthors) has focused on understanding the consequences of increased media competition, and in particular the collapse in the commercial value of news.

In Angelucci and Cagé (2019), we build a new data set on daily local newspapers, national newspapers, and television in France between 1960 and 1974, and study the effect of the introduction of television advertising in 1967–1968.[13] The introduction of television advertising affected national daily newspapers more severely than local daily newspapers; national newspapers indeed rely to a greater extent on advertisements for brands, whose owners may also wish to advertise on television, while advertisements in local newspapers tend to feature classified ads or promote local establishments. We show that this introduction led to a 24 percent decrease in the advertising revenues of national newspapers compared to those of local newspapers and, as a consequence, to a 21 percent decrease in the size of the newsroom and a drop in the amount of journalistic-intensive content produced. (Interestingly, while some argue that media staff cuts must surely be offset by productivity gains, the evidence confirms that this is indeed not the case. High-quality journalism requires journalists to begin with—not news bots.)

Furthermore, on top of this competition on the advertising side, the entry of new platforms—such as television and the Internet—leads to increased competition for print media on the reader side of the market. In the United States, unlike in France, commercial advertising was allowed on television at around the same time television was introduced. In Angelucci, Cagé, and Sinkinson (2017), we examine how the entry of television affected local newspapers as well as consumer media diets in this context. To do so, we construct a unique data set of US newspapers' economic performance and product characteristics covering over 1,500 local news markets from 1945 and 1964, and exploit quasi-random variation in the timing of the entry of television in different markets, caused by a "freeze" in the Federal Communications Commission licensing. (Television took off in the United States after World War II, but the rollout of television was interrupted on September 30, 1948, when the FCC announced a freeze on the granting of new television licenses, mostly due to interference problems. While the freeze was originally planned to last only six months, it ended up lasting nearly four years, in part because of the outbreak of the Korean War, and did not end until April 14, 1952.[14]) We show that the entry of television led to a 3 percent drop in the circulation of evening newspapers—suggesting consumer substitution away from newspapers following the introduction of the new technology—and to a 4.6 percent decrease in their advertising revenues, entirely driven by the quantity of advertising sold to national advertisers (we observe no significant change in the quantity sold to local advertisers). In other words, television has been a substitute for newspapers in both the advertising and the consumer markets.

Furthermore, the introduction of television led to a drop in the total quantity of news produced—and most probably also in the quality of news, even if this cannot be measured empirically—with, in particular, an observed drop in the total number of stories printed in the newspapers. Interestingly, we also see a decrease in the quantity of national wire stories. Although such a decline may sound surprising given that national wire stories are more economical for print media than the production of original local news, it reflects the fact that, with increased competition, it has become more difficult for traditional local news providers to raise their profitability by bundling their in-house coverage of local events with nonoriginal—typically wire—content. The intuition is as follows: while "bundling" content, that is, providing both local and national news, maximizes profit and quality for a print media in the absence of a competing platform, then entry of a national news specialist (e.g., the internet) reduces the incumbent's incentives to raise its provision of news. The incumbent decreases the quantity of both local and national news it produces.

LESSONS FROM HISTORY

Overall, what do we learn from these past experiences? They may help rationalize current industry trends. First, a reduction in advertising revenues lowers media outlets' incentives to produce journalistic-intensive content.[15] In figure 9.2, I plot the evolution of daily newspaper advertising revenues and the number of journalists in the United States between 1980 and 2015. Obviously, correlation is not causality, but it is striking to note that the size of newspaper newsrooms has closely followed the evolution of advertising revenues. Note that this does not have to be the case; it is driven by the fact that the majority of American newspapers are for-profit newspapers focused on maximizing their profits and paying dividends to stockholders. Furthermore, in the United States, a number of newspapers are publicly held companies and thus have a fiduciary responsibility to their stockholders to maximize profits. Cranberg, Bezanson, and Soloski (2001) have documented that in the United States newspaper companies have seen significant growth in their cash flow after becoming publicly traded companies, despite a modest growth in revenues, reflecting a reduction in expenses. In a nonprofit setting, we might have instead observed a decrease in the cash flow as a response to the drop in advertising revenues.

While the drop in advertising revenues has been mainly driven historically by the introduction of new platforms on which companies are more willing to advertise—and by the increasing competition between platforms that have led to a decrease in advertising costs—an additional cause nowadays is consumers' desire to avoid advertising. We indeed observe an increasing use of ad blocking: according to the 2017 Digital News Report published by Reuters, among a quarter of internet users use ad-blocking software (31 percent in France and 23 percent in the United States).[16] Not surprisingly, the use of ad blockers leads to a decline in websites' advertising revenues; Shiller, Wald-

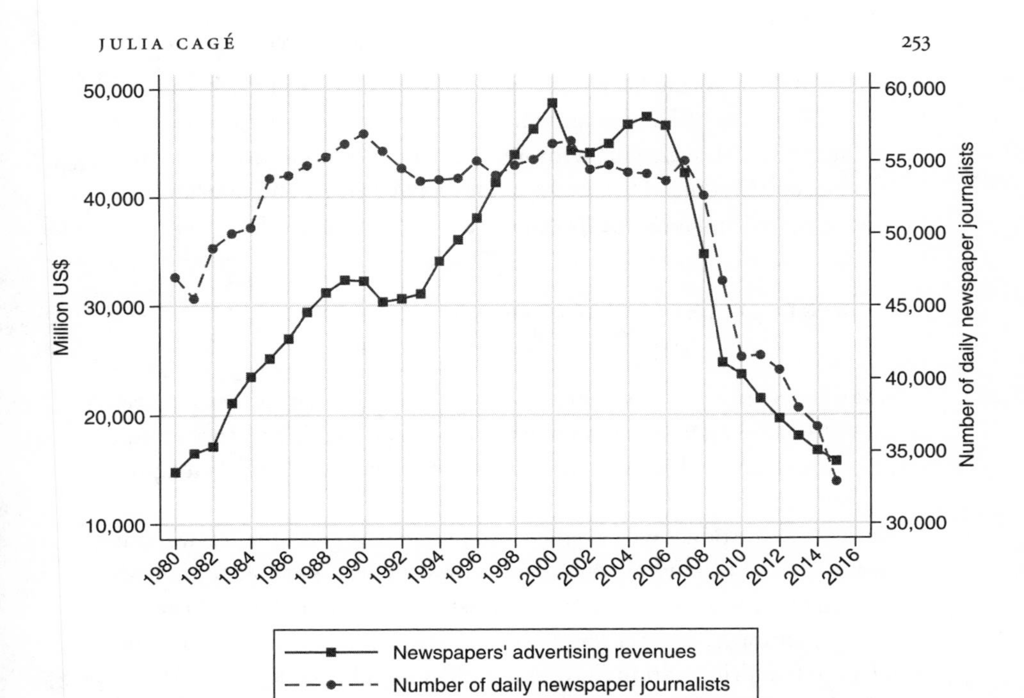

FIGURE 9.2. Newspaper advertising revenues (in dollars) and number of journalists in the United States, 1980–2015.

fogel, and Ryan (2017) have shown that it consequently implies a decrease in the quality of the sites whose users use ad blockers.

Second, what we learn from history is that newspapers are no longer the only channel through which to reach consumers: third parties are now often direct competitors. Classified ads have moved to specialized online outlets such as Craigslist—whose negative impact on US newspapers has been well documented by Seamans and Zhu (2014)—and soft news about local communities is now provided free of charge on social networks such as Facebook. As a result, media are becoming niche media, which raises questions regarding the consequences for the provision of general information and in particular of local journalism. The important question is then whether local newspaper companies can be profitable by providing local journalism only.

This negative overall picture is indeed especially true for the local newspaper market. After a number of very tough years, recently national newspapers have done better economically, and in the second quarter of 2018, the New York Times Company reported a $24 million profit. Recent reports by the Reuters Institute and the Pew Research Center have shown that more young people are willing to pay for news online, which gives cause for optimism as to the future of the news media. However, because of their small market size and the impossibility of bundling content, local newspapers are struggling much more than national ones. Furthermore, the need to rethink the governance, ownership, and funding of the media in the digital world is an important issue for both national and local media outlets, as I show clearly below.

DIGITAL TECHNOLOGIES AND THE INCENTIVES TO PRODUCE ORIGINAL INFORMATION

In addition to reducing advertisers' willingness-to-pay for newspaper readers' attention, the internet has altered the media industry's structure in other ways, in particular with an increasing ability for rival news outlets to appropriate stories.

Before the introduction of digital technologies, a print media, when investigating in the (costly) production of original news, was able to monetize the information it produced during at least a twenty-four-hour period: the time required for its competitors to reproduce, print, and sell this information (given that print editions have simultaneous daily updates; even in areas with both morning and evening newspapers, the news breaker had at least a half-day advantage). This is no longer the case in the internet era because of the combination of real-time updates, high speed of information propagation, and increasing use of copy and paste.[17] In today's online world, utilizing other people's work has become instantaneous. This makes it extremely difficult for news content providers to distinguish, protect, and reap the benefits of the news stories they produce.

In Cagé, Hervé, and Viaud (2020), we documented the extent of copying and estimated the returns to originality in online news production, using a unique data set we built that combines all the online content produced by French news media (newspaper, television, radio, digital news media, and two news agencies—Agence France-Presse and Reuters) during the year 2013 with new microaudience data. Our data set contains 2.5 million documents, that is, the entire content produced by the main French news media online, whatever their offline format. We show that a quarter of the news stories are reproduced online in under four minutes.[18] This is not due to journalists' enhanced productivity thanks to the internet but to the fact that two-thirds of the online content is simply copy and paste—only 32.6 percent of the online content is original. In other words, in the digital era, covering a news story does not necessarily imply providing original reporting on this story; verbatim copying is also often used as an alternative.

A good illustration of this phenomenon is the propagation of the breaking news published on Monday, October 21, 2013, by the national daily newspaper *Le Monde* regarding a US National Security Agency (NSA) surveillance program. On Monday October 21, at 6:01:13 a.m., the newspaper published on its website a worldwide exclusive revealing that the NSA had accessed more than seventy million phone records of French citizens from December 10, 2012, to January 8, 2013 ("France in the NSA's Crosshair"), followed almost simultaneously by a second article titled "Inside the NSA's Web of Surveillance" published at 6:01:23am. These two articles—costly and time-consuming to produce—were around 5,500 characters long. Thirty seconds after the publication of the first article by *Le Monde*, at 06:01:43am, the French news agency (the AFP) published a first short dispatch on the same topic (there was a news embargo before the publication by *Le Monde*), followed by a second, longer dispatch. Forty percent of the first dispatch was copied and pasted from *Le Monde*'s original article, as was 75 percent of the second one. Half an hour later, the first non-news-agency media outlet to report online on this extensive electronic eavesdropping was a radio station, RTL. Eighty-one percent of this article was simple copy of the longest AFP dispatch; similarly, 89 percent of the article published at 6:40:58 a.m. by the weekly newspaper *Le Nouvel Observateur* was copied from the AFP, and so on. Overall, within three hours after the publication of the first article by *Le Monde*, fifty-three articles related to the event had already been published online by different websites, and these articles were broadly similar to the original pieces published on the website of *Le Monde*. The bottom line is that for any citizens waking up around 7 a.m. in the morning, it made no difference whatsoever whether they consumed the information about the NSA surveillance program on the website of *Le Monde*—the media that bore all the cost of producing this information—or on any other website that had simply reproduced this information on their own platform. To put it another way, while the cost of the investigative reporting was borne solely by *Le Monde*, the benefits of this reporting—that is, the number of citizens who clicked on the story thereby generating advertising revenues for the other websites—was shared between a number of different outlets. All of which further reduced the financial benefits for *Le Monde*.

Such a phenomenon may at least partly explain why high-quality news is underprovided in contemporary democracies: the scale of copying online might negatively affect media outlets' news-gathering incentives. In the event that online audience was distributed randomly across the different websites and regardless of the originality of the articles, our results regarding the prevalence of copy and paste online would imply that the original news producer captures only 33 percent of the audience and of the economic returns to original news production (which as a first approximation can be assumed to be proportional to audience, e.g., via online advertising revenues). However, the original news producer incurs 100 percent of the cost of original news production; and one-third of the returns may well be insufficient to cover the costs.

Fortunately, despite increasing consumer switching online—and despite that news brands are struggling to cut through on distributed platforms[19]—online audience is not distributed randomly. In our study, we show that reputation mechanisms and the behavior of internet viewers make it possible to mitigate a significant part of what we could call a "copyright violation problem" (even if, on the one hand, copy from the news agency dispatches falls within the bounds of the copyright law and therefore should not be considered a violation, and, on the other hand, national copyright laws make it difficult to distinguish between what is, or is not, protected by copyright).

First, using article-level variations (with event, date, and media fixed effects), we show that a fifty-percentage-point increase in the originality rate of an article leads to a 39 percent increase in the number of times it is shared on Facebook. If we instead consider the number of times the article is shared on Twitter, we find that an increase of one thousand in the number of original characters leads to an 11 percent increase in the number of Tweets. In other words, the more original an article is, the more it is shared on social networks, which can be considered a reasonable proxy for its "popularity" (unfortunately, we do not have data on the number of times each article is viewed online).

Second, we combine media-level daily audience data and article-level social media statistics (number of Facebook and Twitter shares) to obtain an audience measure (number of views) at the article level. We first assume a simple linear relationship between the number of shares on social media and the number of article views. We then use a unique data set on the number of views and Facebook shares at the article level from the newspaper *Le Monde* (covering the period April–August 2017) to characterize the joint distribution of the number of Facebook shares and the number of visitors. We use these different estimates to obtain a lower and an upper bound of the number of times each article is viewed. Depending on the specification we use, we find that original content represents between 54 percent and 62 percent of online news consumption, that is, much more than its relative production (32.6 percent). Reputation mechanisms actually appear to solve about 30 percent to 40 percent of the copyright violation problem.

In other words, reputation effects partly counterbalance the negative impact of plagiarism on news-gathering incentives: while they can find the same content for free everywhere, consumers partly seem to favor original content. Why is this the case? First, in Cagé, Hervé, and Viaud (2020), we show that the copying media tend to offer lower-quality coverage than the original news producer. Second, this may also be partly due to the way search engines work. For example, while the exact algorithm behind Google News is not public, it is well known that Google uses "freshness" and original content as a ranking signal. If consumers mainly consume news through search engines—rather than going directly on media outlets' websites—they will first see the most original articles that are ranked first by these engines. They therefore click more on more original content independently of their perception of the reputation of the media. As highlighted by Boczkowski and Mitchelstein (2013), consumption choices are often made at the story level. Hence, regardless of the reputation of a given media, consumers wishing to learn about a news event may "decide" to read the most original piece, simply because this is the first article published within an event (and so the first they happen to see) or the one ranked first by search engines.

While this technological aspect is part of the story, in Cagé, Hervé, and Viaud (2020) we show that reputation also matters per se. Overall in 2013, media outlets with a larger fraction of original content tend to receive more audience. Furthermore, the originality of an article matters even when we control for its publication rank, meaning that, for a given publication rank (or reaction time), consumers do value originality per se. This is a very important point because it means that reputation can help mitigate the negative effect that digital technologies have on the production of high-quality news (knowing that such technologies also have a lot of positive effects for investigative journalism; the Panama Papers recently showed the value of using big data skills: without computers and algorithms, the journalists in charge of the investigation would have been unable to manage nearly three terabytes of leaked information). But this conclusion requires the media outlets to realize this and allocate their effort and journalist time accordingly. If media outlets only took into account media-level audience, and did not realize that more original articles generate more audience, then they would misallocate some of their resources (i.e., more time copying and pasting and less time in investigation and original content production).

Unfortunately, the observed drop in the average size of the newsrooms and the resources devoted to investigative journalism seems to indicate that media outlets underestimate the reputation effects. Moreover, digital technologies—and in particular the social media platforms—also have a negative effect on how and what kind of news is consumed by citizens, an issue we will address at the end of the chapter.

Rethinking Governance and Funding of High-Quality News

The main lesson from the second part of this chapter is that, due both to the collapse in advertising revenues and to the rapid propagation of online information, news no longer has a large commercial value. This does not imply that there cannot be commercial news outlets but that we need to find new media funding models that are not based on profit maximization. This is especially true because, as we have seen in the first part, information is a public good, and successful democracy depends on citizens acquiring such information. With that in mind, this last section is also based on the idea that the view that the new "media barons" such as Jeff Bezos and Pierre Omidyar will usher in a brighter future for the internet era is misleading. Moreover, it may well be dangerous not only for the future of the media but also, more importantly, for the future of democracy.

FROM PHILANTHROPY TO PATRONAGE

When people talk about "foundations" nowadays in relation to the media, what they most often have in mind is foundations—such as the Bill and Melinda Gates Foundation—that fund media outlets or specific investigative projects. From this point of view, in the future one may well consider Google's Digital News Initiative (initially launched for three years in 2015 with a $150 million fund and relaunched in 2018 with $300 million to be spent over the following three years) as a foundation-type initiative to support the media. But philanthropic funding suffers from a number of downsides: while philanthropy may offer one resource with the potential to fund the production of high-quality journalism, media outlets must resist potential hidden agendas. This is not specific to the media, however, and the risks of philanthropist funding have already been highlighted in the context of the funding of education, with questions about the power of donors to set research agendas. As highlighted by Reich (2018), we must consider philanthropy "as an act with political dimensions, in the sense that philanthropy can be an expression of political power. . . . Wealthy elites can pose problems for democratic politics, even—and perhaps especially—when elites direct their wealth toward the public sphere" (p. 64) (on the capture of democracy and on the privatization of the provision of public goods by the wealthy, see also Cagé, 2018).

As highlighted in the introduction, it is important to distinguish between two different kinds of philanthropic mechanisms at work in the media industry nowadays—even if, as we will see, they partly suffer from the same downsides: the philanthropic model and the patronage model. The philanthropic model consists in creating nonprofit news organizations that are then funded via charitable donations. Such a model is well illustrated by the examples of ProPublica and the billionaires Herbert and Marion Sandler; Pierre Omidyar, who created First Look Media and *The Intercept*; and recently the Markup, a news site that investigates big tech, subsidized by Craig Newmark (the Craigslist founder).

The patronage model, while it also claims to be philanthropic in spirit (in particular in view of the low profitability of the sector), involves billionaires such as Jeff Bezos, Patrick Soon-Shiong, and Marc Benioff buying and controlling news media organizations, keeping them as for-profit entities but willing to run them at a loss. This patronage model has a historical precedent. In the nineteenth century, before the appearance of the penny papers and the development of the mass media, "out of their own funds, wealthy political leaders sometimes provided start-up capital for newspapers" (Starr, 1982, p. 92).[20] The main difference with today's situation is that while historically the patronage was political inasmuch as these newspapers were endorsing political parties, nowadays a large share of the new media moguls seem to care much less about politics (ensuring that a Republican or a Democrat candidate is elected) but much more about regulation. Or, more specifically, about guaranteeing a low level of regulation (particularly in the case of e-commerce and the telecommunications sector).

Both the philanthropic and the patronage models raise the same issue: power resides where the money is. The media have all too often served as the playthings of billionaires seeking to influence an agenda. From this point of view, there is no difference between private ownership of for-profit entities and the funding of foundations. Most often, donors indeed retain control over the governance and the purpose of the foundation, and in particular over how the funds are spent. To provide an example that illustrates the complexity of the situation, given that without this external funding, the newspaper in question would have shut down its newsroom, if being owned by the founder of Amazon creates problems relating to independence and self-censorship for journalists working at the *Washington Post*, who may have more difficulty covering issues linked to e-commerce, then what is the difference between being owned directly by Jeff Bezos (which is the case today) or being funded on a daily basis by, let's say, a "Bezos foundation for the media" created, funded, and governed by the very same man? The answer is simple: there is no difference.

In other words, philanthropy is not without risks.

PHILANTHROPIC FUNDING OR NONPROFIT MODELS? A MATTER OF GOVERNANCE

The limits of the foundation model for the media have been well described, in particular by Rodney Benson. In the article "Are Foundations the Solution to the American Journalistic Crisis," he has documented that "foundation donations are not "free" but rather constitute a redirection of public resources . . . to nontransparent and unaccountable foundations that have assumed media policy responsibilities."[21] Moreover, foundations prefer to fund specific projects rather than general operations, which creates the possibility of conflicts of interest. Obviously, founders will always claim that they never impose changes to the content of the investigations they have funded—and this may well be the case—but do we really expect media outlets to apply for funding to investigate the funders?

Underlining the limits of the foundation model does not mean, however, that we do not need nonprofit journalism; on the contrary. The central question is not the corporate form of the news organization (for profit or not-for-profit) but rather its governance. We need nonprofit media outlets, that is, commercial media outlets that reinvest their profits—and do not pay dividends—and whose aim is to maximize the quality of the information they produce rather than their own profitability.[22] But the governance and ownership of these outlets have to be democratic. It is necessary to clarify the public debate around these lines and to draw a clear distinction between nonprofit journalism on the one hand and foundation-funded journalism on the other hand, highlighting the need to avoid dependence on a patron or on a small number of plutocrats. Many small donors are preferable to one large donor, precisely to avoid such dependence. Yet many small donors is definitely what digital technologies may allow new models to achieve.

PHILANTHROPY, LACK OF TRANSPARENCY AND THE FUNDING OF FAKE NEWS

So far, we have highlighted the downsides of both the patronage model and the philanthropic funding model of the media. Before turning to the solutions—and in particular to the nonprofit media organization model—it is important to highlight an additional threat to the media: the growing lack of transparency regarding who owns the media and more generally who funds the news. Funding the news—much like funding political campaigns—can indeed be used as an efficient tool to shape public discourse and policies (Cagé, 2018).

Perhaps unsurprisingly, the same deep-pocketed billionaires often invest in media while making campaign contributions. Such a pattern is particularly worrying given that it creates a model of partisanship in journalism that mirrors campaign contributions.

In the United States, the most striking example of this is that of Rebekah Mercer (the daughter of Robert Mercer, former chief executive officer of Renaissance Technologies, involved in the Cambridge Analytica scandal), who both generously invested in right-wing media and funded Donald Trump's presidential campaign. In particular, Rebekah Mercer bankrolled the conservative fake-news website Breitbart News. Moreover, her media investment does not stop at America's borders. For example, the Mercers finance the Gatestone Institute, an anti-Muslim think tank that spreads fake news in Europe, as well as the Canadian ultraconservative Rebel Media group.[23]

The Mercers are but one example; many others come to mind, including the Koch brothers (Mayer, 2016), known for financing Americans for Prosperity, who bought *Time Magazine* in 2018 (the news magazine was then sold a few months later to Marc Benioff, another billionaire). According to the news website Sludge, the Charles Koch Foundation and the Charles Koch Institute donated over $2 million to conservative media operations in 2017, funding, for example, the *Daily Caller*'s website.[24]

An additional example of the growing spread of fake news financed by billionaires is Project Veritas, an organization run by James O'Keefe that specializes in operations against the media (e.g., recently against the *Washington Post* and the *New York Times*). According to the *Washington Post*, relying on documents fielded with the International Revenue Service, Project Veritas received $1.7 million in 2017 from charity associated with the Koch brothers.[25] Furthermore, other contributors to Project Veritas in recent years include Gravitas Maximus LLC, an organization controlled by the Mercer family.

What is interesting here—and, I believe, needs to change in the future for the sake of the quality of the public debate—is that the lack of transparency regarding the identity of individuals who fund the news stems from the use of the foundation status to finance the media. This has been very well documented by Mayer (2016), who highlights that the Koch brothers' activism "was cloaked in secrecy and presented as philanthropy, leaving almost no money trail that the public could trace" (p. 5). In other words, philanthropy guarantees anonymity. This is increasingly true in the United States, with the development of new tax status such as the 501(c)(4) for "social welfare" groups to which donations can be made in secret and the 501(c)(6) for "business leagues" that similarly guarantee donor anonymity.[26]

DEMOCRATIZING GOVERNANCE

In order to produce high-quality news, one needs to set up a newsroom, and the production of nonspecialized information—rather than the creation of a niche publication—requires several hundred journalists; as a result, several million dollars are needed to create a new media company. Yet raising funds for an independent media start-up may prove very difficult. Unless one leverages the potential of digital technologies.

Philanthropies and foundations are one of the possible funding streams. Another possibility is to raise money with business entrepreneurs (broadly defined), which may include existing media tycoons. What is important first of all is to guarantee full transparency regarding the identity of these funders and, second, to ensure that these initial funders, despite providing the capital, have limited "power" to make decisions (basically a limited number of seats or voices on the board). In other words, big investors must give up some of their decision-making power. Why would they do so? What we need is an innovative solution adapted to today's realities: a new form of ownership participation with shared control rights and decision-making power, or, in other words, a new form of shareholder democracy adapted to the media and perhaps to other enterprises as well.

This idea is at the core of the nonprofit media organization (NMO) model I present in my book *Saving the Media: Capital, Crowdfunding and Democracy* (Cagé, 2016). The NMO is a new corporate model for the media, a model adapted to the twenty-first century that leverages digital technologies. Inspired in part by successful experiments in the media sector in recent years, as well as by major universities that combine commercial and nonprofit activities in a single entity, it is intermediate in status between a foundation and a corporation. One goal is to secure permanent financing for the media by freezing their capital. A second goal is to limit the decision-making power of outside shareholders with constraining bylaws. A third goal is to rely on many small donors—including readers and journalists—rather that on only one large donor, an objective that can be achieved through new digital tools such as crowdfunding campaigns.

The basic idea is as follows: first, the NMO, as a nonprofit corporation, pays no dividends, and shareholders cannot recover their investments (as in the case of a foundation). Like a foundation, it can accept unlimited gifts. Any physical or moral person can contribute—in particular readers and employees. Such gifts are to be tax deductible, as gifts to foundations currently are. Second, in the NMO, some additional power is granted to small contributors, who are regarded as participants in the management of the firm and not mere donors. More precisely, the law should specify that any investment above a certain threshold of capital share of an NMO would bring a less-than-proportional share of voting rights. In *Saving the Media*, I propose that investments above 10 percent of capital might yield only an additional one-third of a vote per share. Conversely, small investors, who contribute less than 10 percent of the company's capital, would receive a proportionate boost in their voting rights so that the total is always 100 percent. Obviously, one may defend different thresholds. What is important is that, de facto, the biggest shareholders would not wield power in proportion to their investment. The tax benefits would compensate for the decreased voting rights.

The NMO model differs from cooperatives, which often rely on some form of "one worker, one vote" voting scheme, which lacks flexibility. With NMOs, journalists or readers bringing more funds to the media will have more votes, provided this does not result in an excessive concentration of power with a small number of individuals. The NMO also differs from the proposals that were recently made regarding "quadratic voting" (Posner & Weyl, 2018). The chief objective of quadratic voting is to allow individuals to express the intensity of their preferences by casting more votes in some elections but not in others. The main application is related to settings with multiple votes, such as a large number of referenda or repeated elections, where voters would be given multiple credits, which they could allocate to the decisions they care most about. In the conclusion of their book, Posner and Weyl clearly pave the way for a full-fledged market for votes, where wealthy voters could buy off the votes of less wealthy voters, which would fly in the face of my proposal, namely limiting the power of billionaires. In another paper, Posner and Weyl (2014) describe an application of quadratic voting in corporate governance whereby shareholders would buy off differentiated voting rights from the company, the proceeds of which would be distributed back to shareholders in proportion to their shares, but with a cap on reimbursements to large shareholders (with more than 1 percent of stock). This could potentially be a way to limit the power of large shareholders, but this is more indirect and less explicit than in the NMO model and does not deal with the nonprofit dimension.

LEVERAGING DIGITAL TECHNOLOGIES AND PROTECTING INDEPENDENCE

The NMO model aims to limit the power of large investors in order to protect the independence of journalists. While we have so far mostly focused on the "originality" of the news articles to proxy for their quality, it is also vital to emphasize the importance of journalism independence for the quality of the information environment. Independence is indeed one of the core values of quality journalism, together with authority, integrity, and impartiality. From this point of view, one can draw a parallel between the production of high-quality journalism and the production of high-quality research; in both cases, the professional process is key and independence is essential.

In the digital era, there is the option to take advantage of all the new opportunities that arise with digital tech, and in particular the possibility to raise investment funds through crowdfunding (or participatory financing). The main advantage of crowdfunding is that it only involves small investors. A few media organizations have availed themselves of large-scale crowdfunding in recent years, in particular in the United States, such as the *Voice of San Diego* in 2015 with the "Bigger Voice Fund," the *American Bystander* since 2015, but also in the United Kingdom with the *Positive News* "Own The Media" crowdfunding campaign in 2015 and in France (e.g., *Les Jours, Causette,* Explicite, *Terra Eco, Mediapart*).

Other innovative business models in recent years include that of the *Correspondent,* which is entirely member funded. Originally launched in 2013 as a Dutch news website funded through a successful crowdfunding campaign, in the summer of 2019 the *Correspondent* recently finished a successful $2.5 million campaign to launch an English-language "unbreaking news" platform.

Pluralism is key in the digital world in order to guarantee media independence and the production of high-quality news: pluralism of investors and pluralism of funding sources. Philanthropic funding may be a good solution as long as there is a multiplicity of donors. An interesting parallel can be drawn here with advertising. Some argue that the greatest danger the media faces is pressure from advertisers. But advertising has long been viewed as a guarantee of media independence. For many years advertising money kept newspapers financially afloat and afforded them a degree of independence vis-à-vis governments in both Europe and the United States (Schudson, 1981). The issue of editorial independence from the influence of advertisers is significant nowadays because as ad revenues dry up, the media must compete hard for what little remains. Recent events may serve as a reminder: in 2015, Peter Oborne resigned with much fanfare from the *Daily Telegraph* in protest at the paper's failure to cover the Swiss Leaks scandal in order to avoid offending HSBC bank, one of its chief advertisers. If the *Daily Telegraph* had relied at the time on a multiplicity of advertisers, the controversy would never have happened. The same is true with respect to foundations: a media outlet whose activity depends solely on funding from one foundation may bias its reporting to please the foundation; however, this form of bias would not arise if it was receiving funding from a multiplicity of foundations and thus did not rely financially on a sole donor.

The key is reducing the power of donors and protecting the media from their agenda. One interesting solution would be to remove direct links between philanthropists and the media they fund, such as by setting up a media fund to which all philanthropists may contribute if they wish and that then allocates the funds to different media outlets. The allocation decision could be made by a group of experts and should be international to avoid any direct conflicts of interest. In other words, the main issue with today's philanthropic funding of the media is that this funding mechanism is direct. We need to establish indirect funding mechanisms to preserve independence. But what would be the incentives for large donors to pool their money in a common civic fund for journalism? Apart from the tax incentives, there may be none. That is why public funding of the media may have an important role to play here. While crowdfunding campaigns—funded by small donors—may be insufficient, a matching system could be introduced where the government matches (e.g., on the basis of three dollars for each dollar contributed) the small donations, so as to provide the media outlets with enough funding.[27]

However, it should be noted that, while pluralism of investors and funding sources is a necessary condition to guarantee the production of high-quality independent information, it is by no means enough. Even in the presence of sufficient funding and pluralism of funding sources, more is needed for high-quality information to be produced. Quality indeed also depends on strong professional norms in journalism, on the existence of talented editors and fact-checkers, and so on. This is why in recent years a number of countries have adopted journalistic codes of ethics and created press ethics council, such as the United Kingdom, which established the Independent Press Standards Organization, a voluntary regulatory body.[28] The introduction of such a regulatory body is currently under discussion in France, and in November 2018, Reporters without Borders published an International Declaration on Information and Democracy providing a number of principles for the global information and communication space.

Note finally that while independence from newspaper owners is the main dimension of independence, it does not guarantee the provision of accurate news by the media. A key factor in the nonprofit media organization model is the idea that both readers and journalists should have some decision-making power. Indeed, it may not be ideal to leave all the normative power in the hands of journalists in the news field. While Boczkowski and Mitchelstein (2013) have documented that journalists are willing to produce high-quality content, even though their audience may prefer non–public affairs news over more newsworthy stories, journalists are mainly white-collar workers whose preferences may differ from that of blue-collar citizens.

FROM THE PRODUCTION TO THE CONSUMPTION OF HIGH-QUALITY NEWS

Overall, would the reinvention of the business model of journalism I propose in this chapter—in particular through the NMO model—be sufficient to increase the actual consumption of high-quality news? I already mentioned the issue of news consumption in the internet era. Given that the digital world is a "high-choice media environment," to quote Prior (2007), even if a high-enough number of media outlets were producing independent high-quality journalism, how could it be ensured that this information would be consumed by citizens who may prefer to consume less informative but more entertaining news?

This problem is compounded by the increasing role played by the algorithmic curation of news. It is well documented that nowadays the vast majority of consumers prefer to access news through a side door rather than going directly to a news website (see, e.g., Reuters Institute, 2018). However, given that these access points are limited—Facebook and Google are in a nearly duopolistic situation—this gives them huge market power vis-à-vis the online news media and, more generally, all the newsrooms. Digital platforms have become publishers in a short space of time, and this leaves "news organizations confused about their own future" (Bell and Owen, 2017). In particular, news publishers have lost control over distribution; the news is increasingly filtered through algorithms and platforms that are opaque and unpredictable. In particular, each change in Facebook's algorithm has a huge effect on the size of a news website's audience. This has been particularly striking in recent years when Facebook has decided to reduce exposure to news, instead prioritizing interactions with family and friends, leading to a huge drop in the traffic from Facebook to news publishers' websites.

But what can be done to ensure that Google or Facebook algorithms better rank high-quality content? It seems urgent to reform the distribution of news online, which has shifted in recent years to digital platforms, such as Google and Facebook, which are now in a monopolistic situation. As highlighted by Robyn Caplan in her chapter in this volume, democratic discourse is becoming increasingly governed by platforms that are using content moderation policies to regulate speech online. While the focus of Caplan's chapter is mainly on hate speech and false information, I have similar concerns related to the policies and practices used by these companies with respect to curating and access to quality information.

In particular, I think it is necessary to regulate these platforms given that they have an impact on the curating, production, and agenda-setting capability of the traditional news media, and that, contrary to legacy outlets, they do not assume the so-called serendipitous historical role of news media. Concretely, what could be done to make sure citizens are exposed to and consume the high-quality independent journalism that is produced? In her chapter, Caplan notes that for a number of platforms, the switch to the subscription model was a much clearer signal than ads "for what constituted quality content." Of course, it seems complicated to introduce regulations that will force platforms to switch to subscription models. However, it may be advisable to change their incentives, for example, by strongly taxing their advertising revenues. Note, moreover, that the tax money thus raised could then be used to finance the legacy news media whose business models have being affected by the platforms.

More generally, even if the core business model for effective financing of publishers' websites still centers on advertising, I believe that the subscription model should be considered one of the solutions to the media crisis.[29] A number of successful media outlets have made the choice to rely solely on subscriptions in recent years. One of the best illustrations of this successful strategy is the French pure online media Mediapart. This media outlet, specialized in investigative journalism, was created in 2008 with a hard paywall. In recent years it has played a key role in uncovering several corruption scandals involving politicians from both the left and right. As of today, Mediapart has more than 140,000 subscribers who provide a revenue of €13.7 million (in 2017). With its 4,700,000 unique visitors per month and eighty-five staff members, the media outlet is highly profitable (and has been making a profit since 2012). The subscription model alone should not be considered an alternative to counter the collapse in advertising revenues. But media outlets such as Mediapart that operate behind a paywall are much less dependent on digital platforms for their traffic. Under the subscription model their traffic is indeed mostly direct. This lower reliance on platforms means that when Facebook decides to modify its algorithm—as it did, for example, in the summer of 2018—a media outlet such as Mediapart, contrary to the majority of the French media, was barely affected. In other words, it is easier for media outlets relying on a subscription model to ensure that their subscribers will be exposed to the high-quality content they produce.

Conclusion

In *Democracy's Detectives*, Hamilton (2016) makes a cost-benefit analysis of a number of investigative stories. As he highlights, "the results of reporting do not come cheaply, but they are a bargain for society." Accountability stories may indeed "generate millions in net benefits to society because of changes arising from new policies triggered by the reporting." The main issue today is that we are observing a shortage of investigative reporting, clearly reflected in the collapse in the number of journalists and the closing down of numerous media outlets (McChesney & Nichols, 2010; McChesney & Pickard, 2011). This shortage comes from the fact that while the benefits of investigative journalism are public, the costs are private. From this point of view, the solution is simple: the costs should be borne by society as a whole, and the news media must transcend the laws of the market. In practice, the implementation is more difficult.

Things are moving fast, and there is no perfect remedy. A number of different—and sometimes conflicting—aspects need to be considered. For example, I have shown that a reduction in advertising revenues lowers media outlets' incentives to produce journalistic-intensive content. Does this mean that we should aim to increase advertising revenues and that media outlets' reliance on advertising revenues is a good thing per se? We have seen that an increasing number of citizens are using ad blockers: they want to avoid advertising. Hence it would be better to find a model in which media can produce high-quality journalism without imposing advertising on consumers who do not want it. Furthermore, reliance on advertising may raise issues regarding editorial independence. Similarly, philanthropy cannot be considered *the* answer, but it may be part of a more complex answer, through the reliance on a multiplicity of small donors and the leveraging of digital technologies (e.g., through the NMO model). As we have seen, an important limitation of philanthropy is that it may introduce potential conflicts of interest. But this does not imply that we cannot innovate and ensure philanthropic funding (which could be indirect) and media independence at the same time.

For democracy, digital technologies are both an opportunity and a threat. The fact that they are an opportunity has been well documented in this handbook. In the chapter by Margaret Levi, David Lee, and John Seely, for example, the authors show that government can be improved through technologies. They also represent an opportunity now that media can use crowdfunding to raise funds, and digital technologies can help to develop new business models in the media. They also pose a threat, however: they have decreased the commercial value of news, and so are partly responsible for the current lack of high-quality journalism. They also sometimes act as an obstacle blocking consumers' access to high-quality news. The key issue that needs to be addressed for the future of the media and democracy is how the financing structure of the production and dissemination of information and influence affects democratic outcomes.

Notes

1. From this point of view, it is very interesting to watch Jeff Bezos's interview on why he bought the *Washington Post.* The founder of Amazon argues that he decided to buy the newspaper when he realized that the *Washington Post* was an important institution: "It is the newspaper in the capital city of the most important country in the world. The *Washington Post* has an incredibly important role to play in this democracy. There's no doubt in my mind about that." So why save this struggling institution? "When I'm 90, it's going to be one of the things I'm most proud of, that I took on the *Washington Post* and helped them through a very rough transition."

2. Alas, a principle flouted by the growing role played by money in politics. Political power is indeed increasingly conditioned on wealth (Kuhner, 2014), not only in the United States (Bartels, 2008; Gilens, 2012; Page & Gilens, 2017) but also in France (Bekkouche & Cagé, 2018), in the United Kingdom (Cagé & Dewitte, 2018), and in other Western countries (see Cagé, 2018, for a comparative approach).

3. See also Goldman (1999), who highlights that "ordinary American citizens have a minimal, even abysmal, knowledge of textbook facts about the structure of American government, the identity of their elected officials, and fundamental facts about contemporaneous foreign policy" (p. 317).

4. Cohen and Fung point out that one of the central features of a well-functioning democratic public sphere is access: "Each person has good and equal access to reliable and instructive sources of information on issues that are or may become matters of public concern."

5. A point also made by Avin Goldman (1999) in *Knowledge in a Social World*: "Messages in an open forum that enunciate truths are plausible examples of public goods. . . . There is reason to expect . . . that a private market would tend to underallocate the resources necessary to the discovery and transmission of true messages" (p. 203).

6. Stromberg (2004) investigates the effect of the introduction of the radio on the distribution of funds in one of the New Deal programs. He finds that this mass media has not only a direct effect on spending—politicians target voters using the radio—but also an indirect effect: citizens using the radio are better informed and vote more as a result; this increase in turnout in turn increases spending.

7. The empirical analysis follows Gentzkow et al. (2011): I estimate the effect of newspaper entry by comparing counties that experience an entry to similar counties in the same years that do not. The identifying assumption is that newspapers in these other counties form a valid counterfactual for the incumbent newspapers in counties that experience an entry, after conditioning for differences in preexisting trends, newspaper fixed effects, year fixed effects, and a large set of demographic covariates. Hence the effects I obtain can be interpreted as causal.

8. See Musgrave and Nussbaum (2018). Note that these findings should be interpreted with caution given that they highlight a correlation and cannot be interpreted as causal.

9. On the impact of fake news during the last US presidential election, see in particular Allcott and Gentzkow (2017).

10. http://www.mediainsight.org/Pages/%27Who-Shared-It%27-How-Americans-Decide-What-News-to-Trust-on-Social-Media.aspx.

11. Answering this question is particularly challenging. Some tech (over)optimists tend to think that, in the future, deep-learning algorithms will be the way to measure news quality in real time. But to work, deep-learning algorithms need to be trained, which means that ground truth is necessary with the imposition of a number of criteria. The issue is this: what will be informative for a researcher, a student, or a journalist may be considerably too complex for less educated citizens who nonetheless also need to acquire information through the media.

12. In the United States, the average number of journalists per newspaper decreased sharply from 39 in 2001 to 23.5 in 2015. Revealingly, the American Society of News Editors decided in 2016 that it would no longer provide information on the size of the newspaper workforce.

13. France was one of the last Western democracies to allow commercial advertising on television. In comparison, advertising was introduced on television in 1941 in the United States, in 1955 in the United Kingdom, in 1956 in Germany, and in 1957 in Italy and Spain (Parasie, 2010).

14. On the introduction of television in the United States, see Boddy (1993); de Leon (2015); Noll, Peck, and McGowan (1991); Starr (1982).

15. The introduction of the internet—just like that of television—decreases advertisers' willingness to pay for readers' attention, which in turn reduces the media companies' incentives to invest in news quality.

16. Their survey covers thirty-six countries, with a sample size of around two thousand individuals per country.

17. Also because of increased consumer switching, which is an essential distinguishing feature of online news consumption (see e.g., Athey, Calvano, & Gans, 2013).

18. On average, news is delivered to readers of different outlets 172 minutes after having been published first on the website of the news breaker. Although these empirical findings are based on the case of the French news media, they would most probably also hold in other Western democracies where the functioning of the media market is relatively similar.

19. In the Reuters 2017 Digital News Report, an experiment tracking more than two thousand respondents in the United Kingdom was performed to measure the extent to which consumers remember news brands. Fewer than half of consumers could recall the name of the news brand after consuming a news story, whereas most could remember the search or social network through which they found the story.

20. Hamilton (2004) similarly highlights that, before the emergence of nonpartisan reporting as a commercial product in the American newspaper markets in the 1870s, the type and amount of information provided depended on both the value of the readers and the value derived from political patronage.

21. Benson (2016).

22. Although it is not the topic of this chapter, let me highlight here that we also need public media that offer information to citizens for free. On average, individuals are better informed in countries that have well-funded independent public broadcasters than in countries that do not.

23. This *BuzzFeed* investigation is particularly informative: https://www.buzzfeednews.com/article/ishmaeldaro/anti-muslim-anti-immigrant-news-europe-to-america. See also https://www.lemonde.fr/pixels/article/2019/03/07/des-milliardaires-americains-financent-discretement-des-campagnes-de-desinformation-en-europe_5432486_4408996.html.

24. https://readsludge.com/2018/11/21/charles-koch-is-funding-rightwing-pro-trump-media-new-disclosure-reveals/.

25. https://www.washingtonpost.com/investigations/project-veritas-received-17-million-last-year-from-koch-backed-charity/2017/12/01/143e13ca-d6d3–11e7–9461-ba77d604373d_story.html?utm_term=.2f66cd6afd51.

26. In the United States, the "social welfare" group (under the IRS Code sec. 501(c)(4)) can participate in politics so long as it is not the organization's primary focus.

27. In a recent report for the Stigler Center for the Study of the Economy and the State, together with J. Gans, E. Goodman, B. Knight, A. Prat, G. Rolnik, and A. Schiffrin, we proposed a novel idea for saving high-quality journalism: give every adult American $50, via an income-tax deduction, to donate to a favorite news outlet. https://research.chicagobooth.edu/-/media/research/stigler/pdfs/media---report.pdf.

28. It corresponds to the former "Press Complaints Commission."

29. Even if pay models are becoming an important part of the business of digital news, in most countries there is still only a minority of news lovers who pay for online news (Cornia, Sehl, Simon, & Nielsen, 2017).

References

Achen, C. H., & Bartels, L. M. (2016). *Democracy for realists: Why elections do not produce responsive government.* Princeton, NJ: Princeton University Press.

Allcott, H., & Gentzkow, M. (2017). Social media and fake news in the 2016 election. *Journal of Economic Perspectives, 31*(2), 211–236.

Angelucci, C., & Cagé, J. (2019). Newspapers in times of low advertising revenues. *American Economic Journal: Microeconomics, 11*(3), 319–364.

Angelucci, C., Cagé, J., & Sinkinson, M. (2020). *Media competition and news diets* (Working paper). Cambridge, MA: National Bureau of Economic Research.

Athey, S., Calvano, E., & Gans, J. (2013). *The impact of the internet on advertising markets for news media* (Working Paper No. 19419). Cambridge, MA: National Bureau of Economic Research.

Bakshy, E., Messing, S., & Adamic, L. A. (2015). Exposure to ideologically diverse news and opinion on Facebook. *Science, 348*(6239), 1130–1132.

Bartels, L. M. (2008). *Unequal democracy: The political economy of the new gilded age.* New York, NY: Russell Sage Foundation.

Bekkouche, Y., & Cagé, J. (2018). *The price of a vote: Evidence from France, 1993–2014* (Discussion Paper No. 12614). Washington, DC: Center for Economic Policy Research.

Bell, E., & Owen, T. (2017). *The platform press: How Silicon Valley reengineered journalism.* New York, NY: Tow Center for Digital Journalism.

Benson, R. (2016). Are foundations the solution to the American journalistic crisis (Working Paper No. 2016-001). New York, NY: NYU Department of Media, Culture, and Communication Media Ownership Project.

Boczkowski, P. J., & Mitchelstein, E. (2013). *The news gap: When the information preferences of the media and the public diverge.* Cambridge, MA: MIT Press.

Boddy, W. (1993). *Fifties television: The industry and its critics.* Urbana, IL: University of Illinois Press.

Cagé, J. (2016). *Saving the media: Capitalism, crowdfunding and democracy.* Cambridge, MA: Harvard University Press.

Cagé, J. (2018). *Le prix de la démocratie.* Paris, France: Fayard. Published in 2020 in English translation as *The Price of Democracy.* Cambridge, MA: Harvard University Press.

Cagé, J. (2020). Media competition, information provision and political participation: Evidence from French local newspapers and elections, 1944–2014. *Journal of Public Economics.*

Cagé, J., & Dewitte, E. (2018). *It Takes Money to Make MPs : New Evidence from 150 Years of British Campaign Spending* (Working paper). Paris, France: Sciences Po.

Cagé, J., Hervé, N., & Viaud, M.-L. (2020). The Production of Information in an Online World. *The Review of Economic Studies.*

Cornia, A., Sehl, A., Simon, F., & Nielsen, R. K. (2017). *Pay models in European news* (Working paper). Oxford, UK: Reuters Institute.

Cranberg, G., Bezanson, R. P., & Soloski, J. (2001). *Taking stock: Journalism and the publicly traded newspaper company.* Mahwah, NJ: Wiley.

de Leon, C. L. P. (2015). *That's the way it is: A history of television news in America.* Chicago, IL: University of Chicago Press.

Falck, O., Gold, R., & Heblich, S. (2014). E-lections: Voting behavior and the internet. *American Economic Review, 104*(7), 2238–2265.

Ferraz, C., & Finan, F. (2008). Exposing corrupt politicians: The effects of Brazil's publicly released audits on electoral outcomes. *Quarterly Journal of Economics, 123*(2), 703–746.

Gao, P., Lee, C., & Murphy, D. (2018). *Financing dies in darkness? The impact of newspaper closures on public finance* (Working Paper No. 44). Washington, DC: Hutchins Center, Brookings Institution.

Gavazza, A., Nardotto, M., & Valletti, T. (2019). Internet and politics: Evidence from U.K. local elections and local government policies. *Review of Economic Studies.*

Gentzkow, M. (2006). Television and voter turnout. *Quarterly Journal of Economics, 121*(3), 931–972.

Gentzkow, M., & Shapiro, J. M. (2011). Ideological segregation online and offline. *Quarterly Journal of Economics, 126*(4), 1799–1839.

Gentzkow, M., Shapiro, J. M., & Sinkinson, M. (2011). The effect of newspaper entry and exit on electoral politics. *American Economic Review, 101*(7), 2980–3018.

Gilens, M. (2012). *Affluence and influence: Economic inequality and political power in America.* Princeton, NJ: Princeton University Press.

Goldman, Alvin I., 1999. *Knowledge in a social world.* New York, NY: Oxford University Press.

Halberstam, Y., & Knight, B. (2016). Homophily, group size, and the diffusion of political information in social networks: Evidence from Twitter. *Journal of Public Economics, 143*, 73–88.

Hamilton, J. T. (2004). *All the news that's fit to shell: How the market transforms information into news.* Princeton, NJ: Princeton University Press.

Hamilton, J. T. (2016). *Democracy's detectives: The economics of investigative journalism.* Cambridge, MA: Harvard University Press.

Henry, N. (2007). *American carnival: Journalism under siege in an age of new media.* Berkeley, CA: University of California Press.

Howard, P. N. (2010). *The digital origins of dictatorship and democracy: Information technology and political Islam.* New York, NY: Oxford University Press.

Howard, P. N., & Hussain, M. M. (2013). *Democracy's fourth wave? Digital media and the Arab Spring.* New York, NY: Oxford University Press.

Kuhner, T. (2014). *Capitalism v. democracy: Money in politics and the free market constitution.* Stanford, CA: Stanford University Press.

Mayer, J. (2016). *Dark money: The hidden history of the billionaires behind the rise of the radical right.* New York, NY: Doubleday.

McChesney, R., & Nichols, J. (2010). *The death and life of American journalism: The media revolution that will begin the world again.* New York, NY: Nation Books.

McChesney, R. W., & Pickard, V. W. (2011). *Will the last reporter please turn out the lights: The collapse of journalism and what can be done to fix it.* New York, NY: New Press.

Musgrave, S., & Nussbaum, M. (2018, April 8). Trump thrives in areas that lack traditional news outlets. *Politico.* Retrieved from https://www.politico.com/story/2018/04/08/news-subscriptions-decline-donald-trump-voters-505605

Noll, R. G., Peck, M. J., & McGowan, J. J. (1973). *Economic aspects of television regulation.* Washington, DC: Brookings Institution.

Page, B. I., & Gilens, M. (2017). *Democracy in America? What has gone wrong and what we can do about it.* Chicago, IL: University of Chicago Press.

Parasie, S. (2010). *Et maintenant, une page de pub: Une histoire morale de la publicité à la télévision française, 1968–2008.* Bry-sur-Marne, France: Institut National de l'Audiovisuel.

Pariser, E. (2011). *The filter bubble: How the new personalized web is changing what we read and how we think.* New York, NY: Penguin.

Posner, E. A., & Weyl, E. G. (2018). *Radical markets: Uprooting capitalism and democracy for a just society.* Princeton, NJ: Princeton University Press.

Posner, E. A., & Weyl, E. G. (2014). *Quadratic voting as efficient corporate governance* (Working Paper No. 643). Chicago, IL: University of Chicago Law School.

Prior, M. (2007). *Post-broadcast democracy: How media choice increases inequality in political involvement and polarizes elections.* Cambridge, UK: Cambridge University Press.

Reich, R. (2018). *Just giving: Why philanthropy is failing democracy and how it can do better.* Princeton, NJ: Princeton University Press.

Reich, R., Cordelli, C., & Bernholz, L. (2016). *Philanthropy in democratic societies: History, institutions, values.* Chicago, IL: University of Chicago Press.

Reuters Institute. (2018). *Digital News Report 2018.* Oxford, UK: Author.

Schudson, M. (1981). *Discovering the news: A social history of American newspapers.* New York, NY: Basic Books.

Seamans, R., & Zhu, F. (2014). Responses to entry in multi-sided markets: The impact of Craigslist on local newspapers. *Management Science, 60*(2), 476–493.

Shiller, B., Waldfogel, J., & Ryan, J. (2017). *Will ad blocking break the internet?* (Working Paper No. 23058). Cambridge, MA: National Bureau of Economic Research.

Starkman, D. (2013). *The watchdog that didn't bark: The financial crisis and the disappearance of investigative journalism.* New York, NY: Columbia University Press.

Starr, P. (1982). *The creation of the media: Political origins of modern communications.* New York, NY: Basic Books.

Stromberg, D. (2004). Radio's impact on public spending. *Quarterly Journal of Economics, 119*(1), 189–221.

Tufekci, Z. (2017). *Twitter and tear gas: The power and fragility of networked protest.* New Haven, CT: Yale University Press.

10

Technologizing Democracy or Democratizing Technology? A Layered-Architecture Perspective on Potentials and Challenges

Bryan Ford

Democracy is in the midst of a credibility crisis. Some of the most well-established Western democracies have become increasingly polarized to the point of tribalism and authoritarianism.[1] The information sources that voters use to understand the world and make their decisions are increasingly suspect.[2] While democracy preaches a gospel of treating all citizens as equal, established democracies fail in numerous ways to protect the equality of citizens' influence at the ballot box.[3]

Outside the ballot booth, people in real democracies depend on government to protect not only their physical safety but also their economic and social equality and human rights. Here, too, established democracies fail to protect their citizens from private coercion or feudal rent-seeking structures.[4] They fail to ensure equal access to equal economic opportunity by accelerating transfers of public wealth to the already rich in the face of skyrocketing economic inequality.[5] They fail to offer an adequate social safety net to protect the ability of the unlucky or disadvantaged to participate in society as equals with dignity, and they even fail even to protect many people from effective slavery.[6] As Robert Dahl asked: "In a political system where nearly every adult may vote but where knowledge, wealth, social position, access to officials, and other resources are unequally distributed, who actually governs?"[7]

Many perceive tremendous potential for technology to improve democracy: for example, by making it more convenient (vote from home with your laptop or smartphone), more participatory (express your opinion more than once every few years), or more inclusive (even in the developing world smartphones have become ubiquitous). But this somewhat "techno-utopian" view, common among the denizens of the early internet, has gradually been overshadowed by our realization of the many ways technology can undermine democracy, either by accident or by design.

Technologists have often talked about technology as somehow inherently "democratizing"—using that term simplistically to refer to technological capabilities becoming inexpensive and widely available. The unstated and evidence-free implication embedded in this use of the term *democratizing*, however, is that any inexpensive and widely available technological gadget somehow makes society automatically more democratic. Our actual experience in practice seems to suggest the opposite. The evolution of "democratized" social networking capabilities into advertising-driven instruments of mass surveillance; the weaponization of "democratized" free expression capabilities into instruments of fear, chaos, and polarization; the transformation of "democratized" financial technologies like Bitcoin into shiny objects mainly attracting money launderers and financial scammers: all offer abundant experiential evidence of how antidemocratic a "democratizing" technology can be.

But we have also seen how technology is almost infinitely flexible and adaptable. Technology is what we design it to be. Can we design technology to be genuinely democratic—to support and facilitate democracy reliably rather than undermining it? This chapter explores several ways in which democracy in today's digital world increasingly depends on technology for better or worse, ways that technology is currently failing democracy, and potential ways in which technology could be fixed to support democracy more effectively and securely.

Because effective democracy depends on far more than the occasional act of voting, we explore technology's interaction with democracy "top to bottom," across multiple levels at which the ability of people to self-govern depends on behavioral practices that are heavily affected by technology. Yes, effective democracy requires people to have both the right and the ability to vote. When they do vote, they need effective choice, not just a choice "between Tweedledum and Tweedledee."[8] Technologies such as e-voting, online deliberation, and liquid democracy show promise in expanding the convenience and effectiveness of democratic choice, but each brings associated risks and major unsolved challenges that we outline.

Effective democracy also requires that people live in a social and economic environment satisfying the conditions for intelligent, informed, and effective democratic choice. People need reliable information sources protected from both subversion through "fake news" and polarization through automated overpersonalization. People need free expression and free association to discuss ideas and organize effectively—but they also need protection from trolls and other abusers seeking to amplify their voices via sock puppets (multiple fake identities orchestrated by one person) or via fully-automated, anonymous bot armies. People need an economic environment offering them the empowerment and leisure time needed to become informed and participate deeply in the deliberative phases of democracy, and not just in the final vote. Finally, people need the digital ecosystem to be able to recognize and identify them *as people*—that is, as formal "digital citizens"—and to be able to distinguish these real people from the millions of fake accounts of bot farmers inhabiting the internet,[9] without undermining effective participation through exclusionary and abuse-ridden digital identity systems.

Having examined some of the promises, failures, and unsolved challenges at each of these levels, I attempt to sketch briefly a long-term vision of a potential architecture for effective digital democracy, layered in the classic fashion followed in network protocol architecture.[10] The following sections outline, from top to bottom, such a layered architecture for digital democracy.

The top layer, which I address first, represents the highest-level functionality that I consider the primary end goal: namely effective technology-supported self-governance through democratic deliberation and social choice. Subsequent sections address critical "building block" layers for effective technology-supported democracy: an information layer ensuring that participants have manageable feeds of high-quality, accurate, and unbiased information as an adequate basis for deliberation and decisions; an economic foundation layer to help ensure that citizens have the baseline means and freedoms to invest the time and attention required for genuine democracy; and finally, a digital citizenship layer ensuring that technology can securely but inclusively protect the rights and resources of real people from being abused, undermined, and diluted by online fakery. Finally, in the last two sections I briefly recap this architecture and summarize how appropriate technologies for each layer could eventually fit together into a fundamentally more solid foundation for digital democracy than exists today.

Democratic Deliberation and Choice

As networked computing technology was just emerging, visionaries immediately recognized its potential use to involve people more richly in the democratic process.[11] Instead of trekking to a physical polling place every few years to make a nearly binary choice between candidates that voters have at best heard about on TV, technology promised the possibility of "virtual town halls" in which millions could observe and participate continuously in democratic deliberation processes. Bringing that online democracy vision into reality has been far more slow and fitful, however.

As a starting point, e-voting systems promise the convenience of voting from the comforts of one's own home, or remotely from outside the country of one's citizenship, without fundamentally changing the nature or frequency of democratic choice.[12] Switzerland's long-held and extensive practice of direct democracy results in citizens being asked to vote typically four or more times per year. This participatory approach in part motivated Switzerland's pervasive adoption of voting by mail, followed by its early adoption of e-voting.[13]

Any technology that permits voting outside the controlled environment of the ballot booth, however, may increase risks of undetected voting fraud such as coercion or vote-buying attacks. The McCrae Dowless mail-in-ballot fraud incident in North Carolina recently highlighted these risks.[14] E-voting systems present particularly critical security concerns, however, due to the risks they may present of scalable electronic attacks, such as by an attacker anywhere in the world successfully compromising the vote-counting servers or exploiting a security bug common to many end-user devices.[15] Further, the same efficiency and scalability that makes e-voting attractive could potentially enable attackers to coordinate large-scale voter fraud, through dark *decentralized autonomous organizations*, or DAOs, for example.[16] Some of these challenges are likely to be solvable only in coordination with "lower layers" of the technology stack, such as communications and identity layers, discussed later.

There have been many attempts—with varying success—to get citizens involved not just in one-off votes or polls but in true deliberative processes where participants learn about and discuss issues in depth, often with the help of domain experts.[17] Selecting participants by sortition or randomly sampling a target community can keep costs manageable while ensuring that the deliberative body is diverse and representative.[18] Because the size and cost of each such representative group is small, governments and other organizations can in principle launch and run many such deliberative groups in parallel on different topics, making the process efficient and scalable. Recent computer science efforts to scale automated decentralized systems, such as blockchain and smart contract systems, have relied on essentially the same principle of running many small representatively sampled groups in parallel.[19]

Some of the key benefits of democratic deliberation, however, are embodied not so much in the outcome of deliberation (i.e., the lessons learned or the report written at the conclusion), but in the impact of deliberation on the participants themselves—such as giving the participants deeper understanding of issues that affect them, a sense of participating actively in their community, and (hopefully) a feeling of having their voices heard. Deliberation in small sampled groups, however representative and scalable, has the key limitation of awarding the latter class of benefits only to the few lucky winners of the lottery. The larger population benefits at best indirectly, from the participants' reports about their experiences and/or from the effects of better policy decisions hopefully being made.

The idea of delegative or liquid democracy pursues the goal of enabling everyone to participate in regular or even continuous online deliberative processes while recognizing the fundamental constraint that everyone has limited time and attention.[20] The essential idea is to give citizens the freedom to choose when and how much to participate, based on their limited attention, while delegating their voice on matters beyond their capacity or interests to others they trust to represent them. In essence, all participants receive an individual choice between direct and representative democracy, on an issue-by-issue or vote-by-vote basis.

There have been many experiments in implementing and deploying liquid democracy throughout the world over the past two decades, with promising but mixed results.[21] The most prominent and large-scale experiment in liquid democracy so far was the German Pirate Party's adoption of the idea for online intraparty discussion via its LiquidFeedback platform.[22] Liquid democracy presents many concerns and potential risks, however.

One important concern with liquid democracy is that different delegates, freely chosen by proxy voters, will necessarily exercise different amounts of voting power in the deliberative process.[23] Concern for such effects seems to be supported by the German Pirate Party's experience of one delegate accumulating (apparently by accident) an outsize share of voting power.[24] Many of these risks may be mere artifacts of immature implementations of liquid democracy, however, with weaknesses that are important but easily fixed. The concentration of power the Pirate Party experienced, for example, may be attributable to the LiquidFeedback software's allowing voters to choose only one person to delegate their entire vote to, artificially creating a "winner-take-all" scenario in which almost-but-not-quite-as-popular delegates lose out completely. Other formulations of liquid democracy allow voters to split their voting power among multiple delegates,[25] enabling delegated power to spread among all the delegates each voter trusts instead of concentrating on a few global winners. Other recent work introduces multiple-delegation mechanisms with provisions specifically designed to limit the inequality of delegated voting power.[26] It is not yet clear whether such provisions are strictly necessary, however, or what the attendant costs and trade-offs might be.

Other concerns that are less fundamental but equally critical in practice center on the immature technology implementations of current online deliberation and liquid democracy platforms, almost all of which rely on a single centralized server, whose compromise could undetectably corrupt the entire democratic process. The experience of Italy's "Five Star" movement, widely suspected to embody more of a techno-autocracy than a democracy facilitated by the software platform designed and run by a father-son duo, illustrates the risks inherent in centralized platforms.[27] There is growing interest in building liquid democracy systems on decentralized blockchain and smart contract platforms,[28] but these experiments and the platforms they build on are still immature, and subject to the same critical security, privacy, and voting fraud risks that apply to e-voting systems.

Information Selection, Reputation, Bias, and Polarization

Voters cannot make informed decisions without access to good information, together with the time and motivation to digest it—the key prerequisite to effective democracy that Robert Dahl terms "enlightened understanding."[29] Throughout most of human history, information was scarce and precious. The digital world stands this problem on its head, creating the equally serious but opposite problem of too much information, with only inadequate, insecure, and essentially undemocratic mechanisms for users to filter, select, and mentally process that information.

The Usenet was the first global, decentralized public forum online allowing anyone to read and post messages and discuss practically any topic.[30] In its time, Usenet was intensely exciting and empowering to many, and was an early entrant in the long line of digital technologies frequently referred to, rightfully or not, as "democratizing."[31] The Usenet is now largely forgotten, not because it stopped working, but because it worked too well, reliably broadcasting signal and noise together and rendering both nearly uncensorable. Spam—both the term and the online practice—were invented on Usenet, and precipitated its effective downfall as uncontrolled spam and trolling finally sent most "netizens" scurrying away to more protected forums on centralized platforms.[32]

But trading the uncontrolled chaos of the decentralized Usenet, for the protection of professional moderators and opaque filtering algorithms owned by profit-motivated technology companies, may in hindsight have been a Faustian social bargain. Centralized technology platforms like Facebook and Twitter did give people freedom to communicate among their friends with greater protection from spam and trolls. These platforms had their own heyday of being called "democratizing"—especially around the time of the Arab Spring.[33] But spammers and trolls learned to adapt and abuse these platforms, leading to the online forum governance and exclusion problems detailed elsewhere in this volume.[34] Further, the effective concentration of information-filtering power into opaque and unaccountable algorithms, designed and run by a few profit-motivated technology giants, represents a crucial threat to democracy in its own right.

Despite the public's retreat to proprietary platforms, technology researchers never lost interest in finding decentralized solutions to the noise and abuse problems that defeated Usenet. The proof-of-work algorithm underlying Bitcoin, for example, was originally proposed as a mechanism to combat spam, by requiring an e-mail's sender to prove to have spent considerable computational effort preparing it.[35] Other clever decentralized algorithms could in principle efficiently pick a set of *guides*, who find and recommend content compatible with the tastes of a given user, out of an ocean of bad content and fake accounts.[36]

The encryption tool PGP (Pretty Good Privacy) popularized the idea of decentralized social trust networks in its "web of trust" model.[37] Many decentralized content governance and filtering algorithms subsequently built on the idea of trust networks.[38] However, actually building trust networks with PGP or other decentralized tools never caught on among the public or even the tech-savvy. Even if decentralized social networks had caught on widely, it is doubtful whether the social networks constructed by the popular centralized platforms actually have the critical properties of trust networks required by decentralized content filtering algorithms.[39]

Another fundamental problem with the social or trust network approach lies in the basic premise that it is desirable for voters to perceive the world through a lens filtered by their immediate social relationships, a practice widely suspected (though not conclusively proved) to create an "echo chamber" effect and contribute to social polarization and tribalism.[40] People who mostly rely on—and most trust—information filtered through their social network may also be more inclined to perceive bias in information sources *not* filtered by their social tribe.[41] Without discounting the popularity and appeals of social communication, it seems clear that the digital ecosystem is missing an objective, unbiased, and usable source of "big picture" information and perspective.

One promising idea is to employ the sampling methods discussed here to the problem of selecting and filtering information.[42] For example, we might try to design news feeds whose topics and viewpoints are chosen through some deliberative information selection process, by members of a representative sample population, to ensure diversity and objectively avoid bias. Although this approach seems worth exploring, it presents further challenges.

A small representative group might conceivably be effective at choosing among and selecting information on topics already of widespread interest. A sample population is much less likely to be effective, however, at identifying rare topics of not-yet-widely-recognized importance or at finding valuable but obscure information about such topics. This is an instance of the perennial "needle in a haystack" problem or of the "rare event" problem in statistics.

Here again, liquid democracy ideas may be useful in synergy with sortition methods. In an online forum dedicated to gathering and selecting information, suppose we initially give voting power only to the members of a small sample population. However, we allow these sampled voters to delegate their voice selectively—in whole or in fractions—to others outside the representative group whom they deem trustworthy and knowledgeable on particular topics. This delegation could enable the small original sample population to spread and multiply their information-gathering power, effectively recruiting a much larger crowd of assistants and advisers to their aid, while preserving the sampling-based diversity and democratic representativeness of the group's composition and perspectives.

Any approach to information filtering and selection runs into the fundamental problem of accounting (or not) for expertise. We generally expect information from domain experts to be more reliable and trustworthy precisely because experts are supposed to know more about the domain. Being able to identify and utilize domain expertise increases in importance as topics and policy questions become more complex and deeply technical. Experts may also bring domain-related biases, however. An obvious example of such bias is the tendency of technology developers to perceive the positive uses of their systems and algorithms far more readily than the negative risks their designs carry.

Further, there is the fundamental question of who decides who is an expert, on what grounds, and whether that expert-selection process can be called "democratic" in any sense. Neither ordinary citizens nor professional politicians without domain knowledge are necessarily good at distinguishing experts from smooth-talking charlatans.[43] But professional organizations and certification systems, in which yesterday's experts vet and choose tomorrow's experts, are subject to narrow groupthink, gradual inbreeding, and cultural ossification.[44] Organizations that vet, fund, or reward experts for their work may become disconnected from and unaccountable to the broader public.[45]

Can we find more democratic and accountable ways to recognize and vet experts and the critical role they play in both producing and evaluating information serving the broader public on deeply technical topics? One observation that may be useful is that although nonexperts may have trouble distinguishing top experts from lesser experts or charlatans who merely speak the language, it might be more feasible to rely on people merely to identify others with *greater expertise than themselves* in a domain. This observation suggests a variation on the delegation ideas explored above: ask members of a community or a representative group to identify a few other people each—inside or outside the original group—who they consider trustworthy and to have more expertise than themselves on some topic. Sort the resulting group by delegated voting weight, eliminate (say) the bottom half, and repeat. The hypothesis, yet to be fully developed and tested, is that each iteration of this process will use progressively better (more expert) information to narrow the population of candidate experts, and ensure that charlatans will be discovered and eliminated at some level at which the genuine experts can reliably distinguish them.[46]

Yet another important question is how we can democratically finance and reward journalism and the production of good information, and manage it once produced.[47] I defer exploration of this topic to the next section.

Access, Inclusion, and Economic Empowerment

Real voters in functioning democracies aren't just disembodied decision-making entities in an academic's theoretical model; they must make decisions and participate (or not) in the context of the real environment they live in. The opportunities and constraints their environment provides—including their education, social networks, money, and free time—has significant practical impact on their effective inclusion or exclusion in political and civic society.[48]

Many practical factors can present exclusionary barriers to the act of voting, such as to voters who have no identity card, no home address, or past criminal convictions.[49] Timing and other logistical factors may affect voter turnout, although in complex and often-unclear ways.[50]

Excluding citizens from voting, however, is merely the most blunt and crude way to compromise Dahl's democratic criterion of inclusiveness.[51] For effective democracy, citizens also need good information, the time to digest and discuss it and form their preferences, and the time and opportunity to participate in controlling the agenda—whether by attending town-hall meetings, joining political demonstrations, calling their representatives, or other activities. People also need the requisite education and political culture to have a basic sense of what democracy is.[52] Those struggling to survive while juggling three precarious part-time jobs may reasonably consider voting, let alone taking the time required for informed voting and active participation, to be a luxury they cannot afford.[53]

Given the evidence that political participation is linked with economic inequality, which has been growing uncontrollably, we may justifiably consider the economic equality and well-being of voters to be as essential to effective democracy as voting itself.[54] This raises the question of whether the economic foundations of today's democracies are adequately "democratic"—and if not, to what extent the proper use of technology could improve that situation.

In the developing world experiencing the highest global inequality, mobile phones have become surprisingly ubiquitous.[55] There is evidence this penetration has helped mitigate inequality and stimulate financial development.[56] This ubiquity of mobile devices could potentially offer a technological foundation for further projects to improve financial inclusion, as exemplified by the M-Pesa project in Kenya.[57]

Bitcoin and its many derivative cryptocurrencies represent another class of technologies often loosely called "democratizing"—this time usually in the sense of enabling people to perform cashlike transactions electronically without relying on trusted parties such as banks.[58] While Bitcoin may in principle be usable by anyone without banks, however, to use it one must either buy Bitcoin from someone, or mine it oneself by competing to solve cryptographic puzzles. Because the nature of Bitcoin mining confers huge advantages on those with access to cheap energy and the latest specialized hardware, however, mining is no longer economically viable to ordinary users—or to anyone but a few large entrenched specialists, in fact.[59] Thus, to use Bitcoin, the have-nots must buy or borrow it from the haves. Bitcoin and most cryptocurrencies thus merely replicate and digitally automate the inequality-increasing status quo and cannot justifiably be described as "democratic" at least in a sense of equality of participation or inclusiveness.

An increasingly popular idea is to replace, or augment, traditional social "safety net" programs with a universal basic income, or UBI—a regular income that citizens of some jurisdiction receive automatically regardless of whether or how much they work.[60] UBI is an intriguing idea that has seen limited experiments.[61] There is at least one important downside of the usually proposed approach of implementing UBI in a political jurisdiction such as a town, state, or country, however: it would create an incentive for anyone outside the relevant jurisdiction to move in—or try to gain residency status fraudulently—thereby exacerbating already-inflamed xenophobic and protectionist tendencies.[62]

An interesting alternative approach would be to build a cryptocurrency with a built-in UBI.[63] Like Bitcoin, such a crypto-UBI currency would be usable by anyone "across borders" and not tied with existing geopolitical jurisdictions or currencies. Several cryptocurrency start-ups are already attempting such projects, in fact.[64] Such crypto-UBI currencies could also conceivably be designed to offer collectively financed economic rewards to the producers of information that the user community finds useful. This possibility suggests new ways to fund news, media, and open-source technologies more democratically rather than via traditional profit-motivated or philanthropic channels.[65]

Many social, economic, and technical issues remain to determine whether and in what form the crypto-UBI idea is viable, however. And like many of the other potentially democratizing technology ideas discussed above, no crypto-UBI scheme can operate fairly or improve equality unless it can identify real people and distinguish them from fake identities of fraudsters, the fundamental challenge we focus on next.

Identity, Personhood, and Digital Citizenship

When humans interact in the real world, we use multiple senses tuned over millions of years of evolution to detect and distinguish other humans, creatures, inanimate objects, and unknown potential threats. We therefore take it for granted that we can easily and reliably distinguish people from nonpeople. Professions such as computer graphics and robotics that try to simulate human forms have discovered just how difficult it is to fool our senses, due to the widely observed *uncanny valley* effect, in which almost-but-not-quite-perfect simulations can unintentionally trigger deep emotional reactions.[66] Our physical-world identity challenges thus tend to focus on classifying and differentiating between people: known or unknown, friend or enemy, attractive or unattractive, insider or outsider, member or nonmember, citizen or foreigner.

But our intuitive assumption that distinguishing between real and fake people is easy completely fails to translate into the digital world—in part because our electronic devices do not have human senses with their millions of years of evolutionary tuning. By default, digital technologies know a "person" only as an electronic record or account someone entered claiming, correctly or incorrectly, to represent a person. This inability to recognize personhood underlies one of the most fundamental unsolved challenges in our technology ecosystem: preventing abusers from creating several (or many) fake identities—whether for fun, for profit, or to undermine democracy. In distributed systems, this problem is termed the Sybil attack, after a famous psychiatric case of multiple-personality disorder.[67]

The reason the Sybil attack is so important is that it renders most of our common and intuitive defenses against abuse ineffective. Blocking a spammer's e-mail address is useless because the spammer will just create many more fake identities automatically, leaving the spam problem unsolved after decades of attempted solutions.[68] As technology companies try to employ more sophisticated automated algorithms such as machine learning to detect fake identities, professional spammers and trolls adapt the same automation technologies to create ever-more-convincing fake identities,[69] with increasingly serious consequences.[70] Automated Turing tests such as CAPTCHAs fail to stem fake accounts in part because machine-learning algorithms are getting better than real humans at solving such tests.[71] Sybil attacks allow trolls to amplify their voices in collaborative forums by creating sock puppets supporting their cause.[72] Because the internet cannot distinguish between real and fake people, and ideologically, politically, or profit-motivated users can exploit this fundamental vulnerability at increasingly massive scales without significant risk, our digital ecosystem is evolving into one in which a large fraction of the "people"—and their online "discourse"—is fake.[73]

This increasingly-correct perception that so much of the internet is fake, including a large portion of its supposed inhabitants, marginalizes *real* people online and fuels the growing technology backlash.[74] Further, the fact that such a high percentage of likes, upvotes, reviews, or any other online artifacts purportedly representing the opinions of "people" are well known to be fake or readily forgeable, undermines any presumption that anything about the internet can be justifiably or legitimately called "democratic."

The scope, generality, and global consequences of the fake identity problem demand a correspondingly robust and general solution, but all of the currently popular proposed solutions have significant flaws. Bitcoin's attempt to address Sybil attacks via proof-of-work failed to ensure either equality or inclusiveness in participation, but also created an environmentally-disastrous runaway competition to waste energy.[75] The cryptocurrency community has explored many variations such as memory-hard proof-of-work, proof-of-space, and proof-of-stake.[76] All these variations reward participants in proportion to some form of investment, however, whether in terms of computation, memory, or purchasing and "staking" existing currency. All these investment-centric mechanisms, therefore, can be expected to retain "rich get richer" tendencies toward inequality in the power and influence of participants, and thus cannot hope to offer person-centric fairness or ensure equality of participation in a truly democratic sense.

The most obvious solution to identifying people is simply to transplant traditional identity documentation and verification processes online. Today's "know your customer" (KYC) regulations for anti-money-laundering (AML) in banking have made checking identification a critical step in online finance businesses and have created a booming market in identity-verification start-ups.[77] These verification processes typically involve asking users to present a physical photo ID over a video-chat session and use machine learning for automation and liveness detection techniques such as eyeball tracking in attempt to detect spoofing attacks.[78]

Besides being privacy-invasive, however, this approach is not particularly secure either. Computer-generated imagery technology has already traversed the "uncanny valley" to produce deep fakes, or video-simulated people that look convincing to real people.[79] Because abusers can use this ever-improving simulation technology to counter advances in detection, this approach offers at best an endless arms race that will likely end in deep fakes eventually becoming more reliably convincing than real people to detection algorithms.

A second problem with ID verification processes is that the physical IDs that they verify are not difficult or costly to fake. Digital passport scans sell for $15 on the black market, for example, with forged passports sufficient for online ID verification but not to cross borders selling for $1,000, and genuine passports usable to cross borders available for around $15,000.[80] Because these are retail prices of black market IDs sold individually, the vendors and their corrupt sources can doubtless perform wholesale ID forgery much more cheaply. In effect, ID verification appears inevitably destined to become little more than security theater and a legal compliance checkbox while offering no real protection against determined identity forgery.[81]

Another approach to Sybil attack protection utilizes automated graph analysis of social trust networks. These algorithms typically rely on the assumption that a Sybil attacker can easily and cheaply create nodes (fake identities) in the social graph but has a harder time creating edges (trust relationships) connecting them to real people.[82] As mentioned, however, it is doubtful that popular online social networks actually constitute trust networks satisfying this assumed property.[83] For example, many Twitter users intentionally engage in *link farming*, or following a large number of other accounts on the (well-placed) hope that a significant fraction will reciprocate.[84] The presence of a significant number of link-promiscuous real users makes it easy for Sybil accounts to hide in that group and defeat graph algorithms that assume attackers are link limited. In more sophisticated infiltration attacks, social bots interact with other users by forwarding or synthesizing content.[85]

Even if real users could build a well-disciplined trust network, these graph algorithms would detect only egregious Sybil attacks, such as the case of one attacker creating a large number of fake identities. Graph algorithms could not and would not prevent many users from each cheating a little bit, by coordinating with their friends to create a few fake identities for example. Finally, Sybil resistance via trust networks would also be exclusionary against real people or groups who are in fact poorly connected socially, and who would likely be falsely eliminated as Sybil identities.

Biometrics present another approach to identity and Sybil attack protection, as exemplified in India's Aadhaar digital identity project, which has issued biometric identities to more than a billion people.[86] While attractive in terms of usability, the use of biometrics for identification is problematic in numerous ways. First, protecting against Sybil attacks and ensuring that each person registers only once requires de-duplication checks during enrollment, or comparing the new enrollee's biometrics against all existing ones, that is, over a billion in the case of Aadhaar.[87] This de-duplication requires all users' biometrics to be collected in a massive searchable database, creating huge privacy and surveillance concerns, in part because biometrics are effectively "passwords you can't change."[88] Second, since biometric matching is inherently imprecise, it can both falsely accept duplicate enrollments and falsely reject legitimate new users. The Aadhaar program estimated a 0.035 percent false accept rate in 2017, but a different method produced an estimate of 0.1 percent the following year, implying that hundreds of thousands of Aadhaar records might be duplicates.[89] There are signs that false rejections may be an increasing problem, as well, leading to another form of digital exclusion.[90] Biometric exclusion threatens not just the participation opportunities, but even the very lives of unlucky or marginalized people who fall through the inevitable gaps left by biometric technologies.[91]

Having exhausted the commonly-proposed but uniformly-flawed solutions to distinguishing real people from fake Sybil accounts, what else is left? One idea is to create digital "proof-of-personhood" tokens via in-person ceremonies called *pseudonym parties*.[92] This idea builds on a back-to-basics security foundation, by relying on a person's physical presence at some time and place. For now, real people still have only one body each, and thus can be in only one place at a time. Expendable clones are still science fiction, and robots have yet to follow Hollywood across the uncanny valley.[93]

Leveraging this property, a few times per year we might organize concurrent pseudonym parties at various locations, wherever a suitable group of organizers is available to run one. Before a certain critical moment, synchronized across a set of coordinated events, anyone is allowed to enter an enclosed or cordoned-off space. After the critical moment, people may only leave, getting one anonymous digital credential scanned on the way out, such as a QR code displayed on a smartphone or printed on paper. If properly run, witnessed, and recorded for public transparency, such a process could ensure that any participant can get one and only one "verified real person" credential valid for a given time period. Because pseudonym parties rely only on physical presence for their security, they avoid requiring any privacy-invasive identity checks or biometrics, or problematic security assumptions about social trust networks.

There is ample precedent for people participating in events requiring physical presence. The billions of members of the world's largest religious traditions often attend in-person ceremonies in churches or temples several times a year, once a week, or more. Two Swiss cantons, Glarus and Appenzell Interior, have used open-air assemblies, or *Landsgemeinden*, for direct democracy for hundreds of years.[94] Political protests play a regular role in many democracies despite producing only rough media estimates of numbers present, and to uncertain and nonobvious concrete political effect.[95] Scheduling and organizing such events to double as pseudonym parties could provide both the organizers and the public more precise statistics on attendance and could give the attendees themselves verifiably Sybil-resistant anonymous credentials that might eventually become useful for many purposes.

After a pseudonym party, for example, attendees could later use their digital credentials merely to "prove they were there," or to form anonymous but abuse-resistant online forums for follow-up discussion or deliberation with attendance restricted to the in-person attendees. More broadly, attendees could use personhood badges to obtain "verified real person" status similar to verified account ("blue checkmark") status on websites and social networks. Attendees could use their digital credentials as voting tokens in online polls or deliberative forums. They could use their digital credentials to represent a one-per-person notion of stake to build proof-of-personhood decentralized blockchains and crypto-UBI currencies.[96]

Regular attendance of pseudonym parties could eventually become part of a social contract that offers a kind of Sybil-resistant formal digital citizenship, with various rights and abilities unlocked by digital proof-of-personhood credentials. These rights include secure, private, and democratically equitable online participation, together with the necessary protection from abuse, trolling, and ballot stuffing by fake identities. Because proof-of-personhood tokens have limited value and validity period, they are inherently "renewable" simply by showing up at a future pseudonym party anywhere. Digital citizenship rights attached to such time-limited but renewable tokens may therefore prove both more democratically equitable (fair) and more inalienable (inclusive) than offline or online identity-based approaches can achieve. The main cost to citizens imposed by this social contract is simply to show up and "prove personhood" periodically.

While promising, many social, process, and implementation challenges remain to develop and test the viability of proof-of-personhood. Addressing these challenges remains an ongoing research project.[97]

An Architecture for Digital Democracy

This chapter has explored several levels of societal functionality that appear to be critical to effective democracy: deliberation and choice, information selection, inclusion and economic empowerment, personhood and digital citizenship. It has also explored ways technology attempts to address these levels of functionality, ways it fails to do so, and potential ways we might improve our technology to address some of those flaws.

I now attempt to stitch these functionality levels together and look at them as a sketch for a potential architecture for digital democracy. This architectural perspective is directly inspired by classic layered network architectures such as the OSI model,[98] which attempts to decompose functionality into layers so that higher-level layers providing more sophisticated functionality depend only on the simpler services of lower layers. Taking this inspiration, we might arrange the functional layers of digital democracy described in the previous sections as follows:

Democratic Deliberation and Choice
Information Filtering and Selection
Inclusion and Economic Empowerment
Personhood and Digital Citizenship

Although this is only one of no doubt many potential architectural perspectives and is most likely not complete or perfect, we can at least briefly justify this particular layering as follows, from bottom to top.

At the base level we need personhood and digital citizenship—specifically, some technological mechanism to different real people from fake accounts, whether or not that means identifying them in a traditional sense—to enable all the layers above to function securely and provide inclusion and democratic equality. Without a secure personhood foundation, financial inclusion technologies such as cryptocurrencies cannot allocate stake or resources (e.g., crypto-UBI) fairly among real people, information filtering and selection technologies are vulnerable to sock puppetry and content reputation manipulation attacks, and deliberation and choice mechanisms are vulnerable to trolling and ballot stuffing. Online democracy can never be legitimate, either in fact or in public perception, without a legitimate demos comprised of real people.

At the next level up, citizens of democracies need a stable social and economic "floor" to stand on before they can be expected to take time for or prioritize enlightened participation in democracy. This is simply an inevitable result of the "survival first" principle. A UBI or crypto-UBI might or might not be the right economic mechanism to help ensure such an economic floor and the assurance of personal independence of dignity it is intended to provide. However, it seems that every conceivable such mechanism, if democratic, will need to rely on some notion of personhood to allocate resources and services of all kinds equitably, and thus must be built atop some form of personhood and digital citizenship layer.

Given sufficient social and economic freedom to participate in democracy, citizens then need access to good information with which to make decisions, whose provision in whatever form is the function of the information filtering and selection layer. Again, we have explored some potential ways current abuse-ridden social information filtering and reputation systems might be improved and made more democratic, for example by relying on representative sample populations, with or without delegation capability, to prioritize topics, to evaluate and filter information, and to choose experts in a democratically egalitarian fashion. While we have not much discussed models for the funding and compensation of news and information, leaving that topic to other chapters in this volume,[99] we might envision such funding and reward mechanisms building on the economic empowerment mechanisms of the layer below, such as cryptocurrencies supporting collective rewards or micropayments for information content. Regardless, because almost all realistic filtering and selection mechanisms become vulnerable if abusers can use Sybil attacks to inject fake upvotes or downvotes or reviews, this layer depends like the others on the personhood and citizenship foundation.

Finally, at the top level, we feel ready to envision more solid digital mechanisms for democratic participation, deliberation, and choice, building on all the functionality of the lower layers. It may not be too far off the mark to consider this layer the "mind" of the democratic digital collective: the decentralized organ at which the deliberative body hopefully achieves awareness and well-informed collective decision-making capability. We can hope for this collective "mind" to make truly democratic decisions, reflecting the interests of the entire population, only if it has the critical lower architectural layers to build on: layers ensuring that people have good information with which to make decisions, that guarantee a universal baseline of access to the time and economic resources to do so, and that protect participants' rights as people from both individual exclusion and collective manipulation through digital fakery.

Again, we offer this perspective only as a likely incomplete and imperfect sketch of a potential reference model fitting together a few of the critical support functions for digital democracy. The hope is merely that it provide be a useful starting point to think about and build on.

Conclusion

Inspired by Robert Dahl's analysis of critical elements of effective democracy,[100] I have attempted a high-level exploration of key areas of functionality where digital technologies seem relevant to the mechanisms of democracy but are currently failing to fill these roles reliably or securely. In this exploration, I have attempted to fit together these functionality areas into a layered architectural perspective designed around the principle of ensuring that higher layers depend only on lower layers. Higher layers derive from lower layers all the functional services they need to operate in a reliable, secure, abuse resistant, and democratically egalitarian fashion.

All elements of digital democratic functionality seem to depend fundamentally on a currently missing personhood or digital citizenship foundation to distinguish real people from fake Sybil accounts. The inclusion and economic empowerment layer depends on the personhood layer to build a "floor" of economic freedom and financial empowerment for all digital citizens to stand on and be able to have the time for real democratic participation. The information filtering and selection layer ensures that citizens have access to good information, depending on the economic layer to fund the production of information and the personhood layer to ensure that information filtering and selection is broad, representative, and objectively unbiased. Finally, the deliberation and choice layer builds on all lower layers—personhood to ensure "one person, one vote" equality in participation, the economic layer to ensure the time and economic freedom to participate, and the information layer to support enlightened understanding.

While only the barest sketch, this architectural perspective might help us break down and think about the complex problems of digital democracy in a more modular, systematic fashion than has been typical, and hopefully will provide a baseline for more detailed future architectural models for digital democracy to build from.

Notes

1. On polarization, see Prior, "Media and Political Polarization"; Iyengar and Westwood, "Fear and Loathing across Party Lines." On tribalism, see Hawkins et al., "Hidden Tribes";

Packer, "A New Report Offers Insights into Tribalism in the Age of Trump." On authoritarianism, see Browning, "The Suffocation of Democracy."

2. Woolley, "Automating Power"; Ferrara et al., "The Rise of Social Bots"; Woolley and Guilbeault, "Computational Propaganda in the United States of America"; Broniatowski et al., "Weaponized Health Communication"; Shao et al., "The Spread of Low-Credibility Content by Social Bots."

3. Smith, "Political Donations Corrupt Democracy in Ways You Might Not Realise"; Gilens and Page, "Testing Theories of American Politics"; Cost, *A Republic No More*; Flavin, "Campaign Finance Laws, Policy Outcomes, and Political Equality in the American States"; Kalla and Broockman, "Campaign Contributions Facilitate Access to Congressional Officials"; Samuel, "Rigging the Vote"; Tisdall, "American Democracy Is in Crisis, and Not Just Because of Trump."

4. Shlapentokh and Woods, *Feudal America.*

5. Keller and Kelly, "Partisan Politics, Financial Deregulation, and the New Gilded Age"; Piketty, *Capital in the Twenty-First Century.*

6. Weitzer, "Human Trafficking and Contemporary Slavery"; Kara, *Modern Slavery.*

7. Dahl, *Who Governs?.*

8. Zinn, *A People's History of the United States.*

9. Berger, "Bot vs. Bot"; Read, "How Much of the Internet Is Fake?"

10. Day and Zimmermann, "The OSI Reference Model."

11. Heinlein, *The Moon Is a Harsh Mistress*; Tullock, *Toward a Mathematics of Politics*; Miller, "A Program for Direct and Proxy Voting in the Legislative Process."

12. Moynihan, "Building Secure Elections"; Alvarez, Hall, and Trechsel, "Internet Voting in Comparative Perspective"; Germann and Serdült, "Internet Voting and Turnout."

13. Luechinger, Rosinger, and Stutzer, "The Impact of Postal Voting on Participation." See also Gerlach and Gasser, "Three Case Studies from Switzerland"; Serdült et al., "Fifteen Years of Internet Voting in Switzerland"; Mendez and Serdült, "What Drives Fidelity to Internet Voting?"; Germann and Serdült, "Internet Voting and Turnout."

14. Blinder, "New Election Ordered in North Carolina Race at Center of Fraud Inquiry"; Ford, "The Remote Voting Minefield."

15. Schryen and Rich, "Security in Large-Scale Internet Elections"; Zetter, "Experts Find Serious Problems with Switzerland's Online Voting System before Public Penetration Test Even Begins."

16. Daian et al., "On-Chain Vote Buying and the Rise of Dark DAOs"; Puddu et al., "TEEvil: Identity Lease via Trusted Execution Environments."

17. Iyengar, Luskin, and Fishkin, "Facilitating Informed Public Opinion"; Grönlund, Strandberg, and Himmelroos, "The Challenge of Deliberative Democracy Online"; Esau, Friess, and Eilders, "Design Matters!"

18. Fishkin, *Democracy and Deliberation*; Iyengar, Luskin, and Fishkin, "Facilitating Informed Public Opinion"; Landemore, this volume.

19. Luu et al., "A Secure Sharding Protocol for Open Blockchains"; Kokoris-Kogias et al., "OmniLedger."

20. Ford, "Delegative Democracy"; Sayke, "Liquid Democracy"; Litvinenko, "Social Media and Perspectives of Liquid Democracy"; Green-Armytage, "Direct Voting and Proxy Voting"; Blum and Zuber, "Liquid Democracy"; Landemore, this volume; Ford, "A Liquid Perspective on Democratic Choice."

21. Litvinenko, "Social Media and Perspectives of Liquid Democracy"; Ford, "Delegative

Democracy Revisited"; Hardt and Lopes, "Google Votes"; Ford, "A Liquid Perspective on Democratic Choice."

22. Swierczek, "5 Years of Liquid Democracy in Germany"; Litvinenko, "Social Media and Perspectives of Liquid Democracy"; Behrens, "The Evolution of Proportional Representation in LiquidFeedback."

23. Blum and Zuber, "Liquid Democracy."

24. Becker, "Liquid Democracy."

25. Ford, "Delegative Democracy"; Boldi et al., "Viscous Democracy for Social Networks"; Ford, "A Liquid Perspective on Democratic Choice."

26. Gölz et al., "The Fluid Mechanics of Liquid Democracy."

27. Horowitz, "The Mystery Man Who Runs Italy's 'Five Star' from the Shadows"; Loucaides, "What Happens When Techno-Utopians Actually Run a Country."

28. Agarwal, "On-Chain Liquid Democracy"; Crichton, "Liquid Democracy Uses Blockchain to Fix Politics, and Now You Can Vote For It"; Zhang and Zhou, "Statement Voting."

29. Dahl, *Democracy and Its Critics.*

30. Hauben and Hauben, *Netizens*; Templeton, "I Remember USENET."

31. Hill and Hughes, "Is the Internet an Instrument of Global Democratization?"; Blumler and Gurevitch, "The New Media and Our Political Communication Discontents."

32. Templeton, "Origin of the Term 'Spam' to Mean Net Abuse"; Templeton, "Reflections on the 25th Anniversary of Spam." See also Templeton, "I Remember USENET."

33. Comninos, "Twitter Revolutions and Cyber Crackdowns"; Howard and Hussain, *Democracy's Fourth Wave?*.

34. See chapters in this volume by Caplan, Gangadharan, Farrell and Schwartzberg, and Cohen and Fung.

35. Nakamoto, "Bitcoin." See also Dwork and Naor, "Pricing via Processing or Combatting Junk Mail."

36. Yu et al., "DSybil."

37. Stallings, "The PGP Web of Trust"; Abdul-Rahman, "The PGP Trust Model."

38. Kamvar, Schlosser, and Garcia-Molina, "The EigenTrust Algorithm for Reputation Management in P2P Networks"; Mislove et al., "Ostra"; Yu et al., "SybilLimit"; Tran et al., "Sybil-Resilient Online Content Voting"; Viswanath et al., "Canal."

39. Mislove et al., "Measurement and Analysis of Online Social Networks"; Viswanath and Post, "An Analysis of Social Network-Based Sybil Defenses"; Ghosh et al., "Understanding and Combating Link Farming in the Twitter Social Network"; Messias et al., "You Followed My Bot!"

40. On echo chambers, see Barberá et al., "Tweeting from Left to Right: Is Online Political Communication More Than an Echo Chamber?"; Dubois and Blank, "The Echo Chamber Is Overstated." On social polarization, see Iyengar and Westwood, "Fear and Loathing across Party Lines." On tribalism, see Hawkins et al., "Hidden Tribes"; Packer, "A New Report Offers Insights into Tribalism in the Age of Trump."

41. Saez-Trumper, Castillo, and Lalmas, "Social Media News Communities"; Eberl, Boomgaarden, and Wagner, "One Bias Fits All?"; Kaye and Johnson, "Across the Great Divide"; Budak, Goel, and Rao, *Fair and Balanced?*; Ribeiro et al., "Media Bias Monitor"; Farrell and Schwartzberg, this volume.

42. Iyengar, Luskin, and Fishkin, "Facilitating Informed Public Opinion"; Landemore, this volume.

43. Edens et al., "'Hired Guns,' 'Charlatans,' and Their 'Voodoo Psychobabble'"; Gemberling and Cramer, "Expert Testimony on Sensitive Myth-Ridden Topics."

44. Collins and Evans, "The Third Wave of Science Studies"; "'Democratising' Expertise, 'Expertising' Democracy"; Kotzee, "Expertise, Fluency and Social Realism about Professional Knowledge."

45. Brewster, *Unaccountable*; Reich, *Just Giving Why Philanthropy Is Failing Democracy and How It Can Do Better.*

46. Ford, "Experts and Charlatans, Breakthroughs and Bandwagons."

47. See the chapters in this volume by Cagé; Lee, Levi, and Brown; Farrell and Schwartzberg; and Bernholz.

48. See the chapters in this volume by Landemore, Gangadharan, and Ananny.

49. Hicks et al., "A Principle or a Strategy?"; Hajnal, Lajevardi, and Nielson, "Voter Identification Laws and the Suppression of Minority Votes"; Highton, "Voter Identification Laws and Turnout in the United States." See also Feldman, *Citizens without Shelter*; Manza and Uggen, *Locked Out.*

50. Quinlan, "Facilitating the Electorate."

51. Dahl, *Democracy and Its Critics.*

52. Cho, "How Well Are Global Citizenries Informed about Democracy?"

53. Standing, *The Precariat.*

54. Armingeon and Schädel, "Social Inequality in Political Participation"; Filetti, "Participating Unequally?" See also Piketty, *Capital in the Twenty-First Century.*

55. Aker and Mbiti, "Mobile Phones and Economic Development in Africa."

56. Asongu, "The Impact of Mobile Phone Penetration on African Inequality"; Asongu, "How Has Mobile Phone Penetration Stimulated Financial Development in Africa?"

57. Mbiti and Weil, "Mobile Banking."

58. Nakamoto, "Bitcoin."

59. Vorick, "The State of Cryptocurrency Mining."

60. Parijs, *Basic Income*; Standing, *Basic Income*; Jackson and Victor, "Confronting Inequality in a Post-Growth World"; Bidadanure, "The Political Theory of Universal Basic Income."

61. Forget, "The Town with No Poverty"; Koistinen and Perkiö, "Good and Bad Times of Social Innovations."

62. Wagner, "The Swiss Universal Basic Income Vote 2016."

63. Ford, "Democratic Value and Money for Decentralized Digital Society."

64. eternalgloom, "Overview of Universal Basic Income Crypto Projects."

65. Cagé, this volume.

66. Mori, "The Uncanny Valley"; MacDorman, "Subjective Ratings of Robot Video Clips for Human Likeness, Familiarity, and Eeriness."

67. Douceur, "The Sybil Attack"; Schreiber, *Sybil.*

68. Dwork and Naor, "Pricing via Processing or Combatting Junk Mail"; Cranor and LaMacchia, "Spam!"; Koprowski, "Spam Filtering and the Plague of False Positives"; Templeton, "Reflections on the 25th Anniversary of Spam"; Shaw, "Avoid the Spam Filter"; Chellapilla et al., "Computers Beat Humans at Single Character Recognition in Reading Based Human Interaction Proofs (HIPs)"; Ramachandran, Dagon, and Feamster, "Can DNS-Based Blacklists Keep up with Bots?"; Mislove et al., "Ostra."

69. Ferrara et al., "The Rise of Social Bots."

70. Bessi and Ferrara, "Social Bots Distort the 2016 US Presidential Election Online Discussion"; Broniatowski et al., "Weaponized Health Communication."

71. Ahn et al., "CAPTCHA." See also Chellapilla et al., "Computers Beat Humans at Single Character Recognition in Reading Based Human Interaction Proofs (HIPs)"; May, "Inaccessibility of CAPTCHA."

72. Bu, Xia, and Wang, "A Sock Puppet Detection Algorithm on Virtual Spaces"; Solorio, Hasan, and Mizan, "Sockpuppet Detection in Wikipedia"; Liu et al., "Sockpuppet Gang Detection on Social Media Sites"; Yamak, Saunier, and Vercouter, "Detection of Multiple Identity Manipulation in Collaborative Projects."

73. Berger, "Bot vs. Bot"; Read, "How Much of the Internet Is Fake?"

74. Doward, "The Big Tech Backlash."

75. Dwork and Naor, "Pricing via Processing or Combatting Junk Mail." See also Vorick, "The State of Cryptocurrency Mining." See de Vries, "Bitcoin's Growing Energy Problem"; Digiconomist, "Bitcoin Energy Consumption Index."

76. Boneh, Corrigan-Gibbs, and Schechter, "Balloon Hashing"; Park et al., "SpaceMint"; Kiayias et al., "Ouroboros"; Gilad et al., "Algorand."

77. nanalyze, "6 Digital Identity Verification Startups to Check Out"; Abhishek and Mandal, "Digital ID Verification."

78. Doughty, "Know Your Customer"; nanalyze, "6 Digital Identity Verification Startups to Check Out"; Pan, Wu, and Sun, "Liveness Detection for Face Recognition"; Bao et al., "A Liveness Detection Method for Face Recognition Based on Optical Flow Field"; Wen and Jain, "Face Spoof Detection with Image Distortion Analysis."

79. Chesney and Citron, "Deep Fakes"; Mack, "This PSA about Fake News from Barack Obama Is Not What It Appears."

80. Durden, "From $1,300 Tiger Penis to $800K Snipers"; Bischoff, "Passports on the Dark Web"; Havocscope, "Fake ID Cards, Driver Licenses, and Stolen Passports."

81. Kline, "Security Theater and Database-Driven Information Markets"; Sethi, Kantardzic, and Ryu, "'Security Theater.'"

82. Mislove et al., "Ostra"; Yu et al., "SybilLimit"; Tran et al., "Sybil-Resilient Online Content Voting"; Viswanath et al., "Canal."

83. Mislove et al., "Measurement and Analysis of Online Social Networks"; Viswanath and Post, "An Analysis of Social Network-Based Sybil Defenses."

84. Ghosh et al., "Understanding and Combating Link Farming in the Twitter Social Network"; Messias et al., "You Followed My Bot!"

85. Freitas et al., "Reverse Engineering Socialbot Infiltration Strategies in Twitter"; Ferrara et al., "The Rise of Social Bots"; Bessi and Ferrara, "Social Bots Distort the 2016 US Presidential Election Online Discussion"; Broniatowski et al., "Weaponized Health Communication."

86. Bhatia and Bhabha, "India's Aadhaar Scheme and the Promise of Inclusive Social Protection"; Chaudhuri and König, "The Aadhaar Scheme."

87. Abraham et al., "State of Aadhaar Report 2016–17."

88. Dixon, "A Failure to 'Do No Harm'"; Srinivasan et al., "The Poverty of Privacy"; Schneier, "Tigers Use Scent, Birds Use Calls"; Chanthadavong, "Biometrics."

89. Abraham et al., "State of Aadhaar Report 2016–17"; Abraham et al., "State of Aadhaar Report 2017–18."

90. Venkatanarayanan, "Enrolment Rejections Are Accelerating"; Mathews, "Flaws in the UIDAI Process."

91. Ratcliffe, "How a Glitch in India's Biometric Welfare System Can Be Lethal."

92. Ford and Strauss, "An Offline Foundation for Online Accountable Pseudonyms"; Ford, "Let's Verify Real People, Not Real Names"; Borge et al., "Proof-of-Personhood."

93. Brin, *Kiln People*; Scott, "5 Lifelike Robots That Take You Straight into the Uncanny Valley."

94. Dürst, "The 'Landsgemeinde'"; Reinisch, "Swiss *Landsgemeinden*"; Schaub, "Maximising Direct Democracy."

95. Madestam et al., "Do Political Protests Matter?"; Acemoglu, Hassan, and Tahoun, "The Power of the Street"; Enikolopov, Makarin, and Petrova, "Social Media and Protest Participation."

96. Borge et al., "Proof-of-Personhood."

97. Ford, "Privacy-Preserving Foundation for Online Personal Identity."

98. Day and Zimmermann, "The OSI Reference Model."

99. Cagé, this volume.

100. Dahl, *Who Governs?*; Dahl, *Democracy and Its Critics.*

References

Abdul-Rahman, Alfarez. "The PGP Trust Model." *Journal of Electronic Commerce* 10, no. 3 (April 1997): 27–31.

Abhishek, Kumar, and Diwakar Mandal. "Digital ID Verification: Competitive Analysis of Key Players." MEDICI, October 2017. https://gomedici.com/diving-deep-into-id-verification-market-comprehensive-evaluation-of-competitive-landscape/.

Abraham, Ronald, Elizabeth S. Bennett, Rajesh Bhusal, Shreya Dubey, Qian (Sindy) Li, Akash Pattanayak, and Neil Buddy Shah. "State of Aadhaar Report 2017–18." IDinsight, May 2018. https://www.idinsight.org/state-of-aadhaar.

Abraham, Ronald, Elizabeth S. Bennett, Noopur Sen, and Neil Buddy Shah. "State of Aadhaar Report 2016–17." IDinsight, May 2017. https://www.idinsight.org/state-of-aadhaar.

Acemoglu, Daron, Tarek A. Hassan, and Ahmed Tahoun. "The Power of the Street: Evidence from Egypt's Arab Spring." *Review of Financial Studies* 31, no. 1 (January 2018): 1–42. https://doi.org/10.1093/rfs/hhx086.

Agarwal, Arpit. "On-Chain Liquid Democracy." *Medium*, November 5, 2018. https://medium.com/coinmonks/on-chain-liquid-democracy-c08ed8c07f6e.

Ahn, Luis von, Manuel Blum, Nicholas J. Hopper, and John Langford. "CAPTCHA: Using Hard AI Problems for Security." In *Advances in Cryptology—Eurocrypt*, International Association for Cryptologic Research (IACR), Warsaw, May 2003, 294–311. https://doi.org/10.1007/3-540-39200-9_18.

Aker, Jenny C., and Isaac M. Mbiti. "Mobile Phones and Economic Development in Africa." *Journal of Economic Perspectives* 24, no. 3 (September 2010): 207–32. https://doi.org/10.1257/jep.24.3.207.

Alvarez, R. Michael, Thad E. Hall, and Alexander H. Trechsel. "Internet Voting in Comparative Perspective: The Case of Estonia." *Political Science & Politics* 42, no. 3 (June 26, 2009): 497–505. https://doi.org/10.1017/S1049096509090787.

Armingeon, Klaus, and Lisa Schädel. "Social Inequality in Political Participation: The Dark Sides of Individualisation." *West European Politics* 38, no. 3 (January 2015): 1–27. https://doi.org/10.1080/01402382.2014.929341.

Asongu, Simplice A. "How Has Mobile Phone Penetration Stimulated Financial Development in Africa?" *Journal of African Business* 14, no. 1 (April 1, 2013): 7–18. https://doi.org/10.1080/15228916.2013.765309.

———. "The Impact of Mobile Phone Penetration on African Inequality." *International Journal of Social Economics* 42, no. 8 (August 10, 2015): 706–16. https://doi.org/10.1108/IJSE-11-2012-0228.

Bao, Wei, Hong Li, Nan Li, and Wei Jiang. "A Liveness Detection Method for Face Recognition Based on Optical Flow Field." In *International Conference on Image Analysis and Signal Processing*, IEEE, Taizhou, China, April 2009. https://doi.org/10.1109/IASP.2009.5054589.

Barberá, Pablo, John T. Jost, Jonathan Nagler, Joshua A. Tucker, and Richard Bonneau. "Tweeting from Left to Right: Is Online Political Communication More Than an Echo Chamber?" *Psychological Science* 26, no. 10 (October 1, 2015): 1531–42. https://doi.org/10.1177/0956797615594620.

Becker, Sven. "Liquid Democracy: Web Platform Makes Professor Most Powerful Pirate." *Der Spiegel Online*, March 2, 2012. https://www.spiegel.de/international/germany/liquid-democracy-web-platform-makes-professor-most-powerful-pirate-a-818683.html.

Behrens, Jan. "The Evolution of Proportional Representation in LiquidFeedback." *Liquid Democracy Journal* 1 (March 2014): 32–41. https://liquid-democracy-journal.org/issue/1/The_Liquid_Democracy_Journal-Issue001-04-The_evolution_of_proportional_representation_in_LiquidFeedback.html.

Berger, Andreas. "Bot vs. Bot: Will the Internet Soon Be a Place without Humans?" Singularity Hub, July 7, 2018. https://singularityhub.com/2018/07/07/bot-vs-bot-will-the-internet-soon-be-a-place-without-humans/#sm.0001ny1wyu1iuco7q081h41v7wnyr.

Bessi, Alessandro, and Emilio Ferrara. "Social Bots Distort the 2016 US Presidential Election Online Discussion." *First Monday* 21, no. 11 (November 7, 2016). https://doi.org/10.5210/fm.v21i11.7090.

Bhatia, Amiya, and Jacqueline Bhabha. "India's Aadhaar Scheme and the Promise of Inclusive Social Protection." *Oxford Development Studies* 45, no. 1 (January 2017): 64–79. https://doi.org/10.1080/13600818.2016.1263726.

Bidadanure, Juliana Uhuru. "The Political Theory of Universal Basic Income." *Annual Review of Political Science* 22 (May 2019): 481–501. https://doi.org/10.1146/annurev-polisci-050317-070954.

Bischoff, Paul. "Passports on the Dark Web: How Much Is Yours Worth?" *Comparitech*, October 4, 2018. https://www.comparitech.com/blog/vpn-privacy/passports-on-the-dark-web-how-much-is-yours-worth/.

Blinder, Alan. "New Election Ordered in North Carolina Race at Center of Fraud Inquiry." *New York Times*, February 21, 2019. https://www.nytimes.com/2019/02/21/us/mark-harris-nc-voter-fraud.html.

Blum, Christian, and Christina Isabel Zuber. "Liquid Democracy: Potentials, Problems, and Perspectives." *Journal of Political Philosophy* 24, no. 2 (June 2016): 162–82. https://doi.org/10.1111/jopp.12065.

Blumler, Jay G., and Michael Gurevitch. "The New Media and Our Political Communication Discontents: Democratizing Cyberspace" *Information, Communication & Society* 4, no. 1 (2001): 1–13. https://doi.org/10.1080/713768514.

Boldi, Paolo, Francesco Bonchi, Carlos Castillo, and Sebastiano Vigna. "Viscous Democracy for Social Networks." *Communications of the ACM* 54, no. 6 (June 2011): 129–37. https://doi.org/10.1145/1953122.1953154.

Boneh, Dan, Henry Corrigan-Gibbs, and Stuart Schechter. "Balloon Hashing: A Memory-Hard Function Providing Provable Protection against Sequential Attacks." In *Advances in Cryptology—Asiacrypt*, International Association for Cryptologic Research (IACR), Hanoi, Vietnam, December 2016, 220–48. https://doi.org/10.1007/978-3-662-53887-6_8.

Borge, Maria, Eleftherios Kokoris-Kogias, Philipp Jovanovic, Nicolas Gailly, Linus Gasser, and Bryan Ford. "Proof-of-Personhood: Redemocratizing Permissionless Cryptocurrencies." In *1st IEEE Security and Privacy on the Blockchain*, IEEE, Paris, April 29, 2017. https://doi.org/10.1109/EuroSPW.2017.46.

Brewster, Mike. *Unaccountable: How the Accounting Profession Forfeited a Public Trust.* Hoboken, NJ: Wiley, 2003. https://www.wiley.com/en-us/Unaccountable:+How+the+Accounting+Profession+Forfeited+a+Public+Trust-p-9780471423621.

Brin, David. *Kiln People.* New York: Tor Books, 2002. https://us.macmillan.com/books/9780765342614

Broniatowski, David A., Amelia M. Jamison, SiHua Qi, Lulwah AlKulaib, Tao Chen, Adrian Benton, and Sandra C. Quinn, and Mark Dredze. "Weaponized Health Communication: Twitter Bots and Russian Trolls Amplify the Vaccine Debate." *American Journal of Public Health* 108, no. 10 (October 1, 2018): 1378–84. https://doi.org/10.2105/AJPH.2018.304567.

Browning, Christopher R. "The Suffocation of Democracy." *New York Review of Books*, October 25, 2018. https://www.nybooks.com/articles/2018/10/25/suffocation-of-democracy/.

Bu, Zhan, Zhengyou Xia, and Jiandong Wang. "A Sock Puppet Detection Algorithm on Virtual Spaces." *Knowledge-Based Systems* 37 (January 2013): 366–77. https://doi.org/10.1016/j.knosys.2012.08.016.

Budak, Ceren, Sharad Goel, and Justin M. Rao. "Fair and Balanced? Quantifying Media Bias through Crowdsourced Content Analysis." *Public Opinion Quarterly* 80, no. S1 (January 2016): 250–71. https://doi.org/10.1093/poq/nfw007.

Chanthadavong, Aimee. "Biometrics: The Password You Cannot Change." *ZDNet*, August 26, 2015. https://www.zdnet.com/article/biometrics-the-password-you-cannot-change/.

Chaudhuri, Bidisha, and Lion König. "The Aadhaar Scheme: A Cornerstone of a New Citizenship Regime in India?" *Contemporary South Asia* 26, no. 2 (September 2017): 127–42. https://doi.org/10.1080/09584935.2017.1369934.

Chellapilla, Kumar, Kevin Larson, Patrice Simard, and Mary Czerwinski. "Computers Beat Humans at Single Character Recognition in Reading Based Human Interaction Proofs (HIPs)." In *Second Conference on E-Mail and Anti-Spam (CEAS 2005), IACR and IEEE. Stanford University*, July 2005. https://web.archive.org/web/20051210090512/http://www.ceas.cc/papers-2005/160.pdf.

Chesney, Robert, and Danielle Keats Citron. "Deep Fakes: A Looming Challenge for Privacy, Democracy, and National Security." *California Law Review* 107, no. 6 (July 2018): 1753–820. http://dx.doi.org/10.15779/Z38RV0D15J

Cho, Youngho. "How Well Are Global Citizenries Informed about Democracy? Ascertaining the Breadth and Distribution of Their Democratic Enlightenment and Its Sources." *Political Studies* 63, no. 1 (March 2015): 240–58. https://doi.org/10.1111/1467-9248.12088.

Collins, H. M., and Robert Evans. "The Third Wave of Science Studies: Studies of Expertise and Experience" *Social Studies of Science* 32, no. 2 (April 2002): 235–96. https://doi.org/10.1177/0306312702032002003.

Comninos, Alex. "Twitter Revolutions and Cyber Crackdowns: User-Generated Content and Social Networking in the Arab Spring and Beyond." Association for Progressive Communications, June 2011. https://www.apc.org/sites/default/files/AlexComninos_MobileInternet.pdf.

Cost, Jay. *A Republic No More: Big Government and the Rise of American Political Corruption.* New York: Encounter Books, 2015. https://www.encounterbooks.com/books/a-republic-no-more-big-government-and-the-rise-of-american-political-corruption-paperback/.

Cranor, Lorrie Faith, and Brian A. LaMacchia. "Spam!" *Communications of the ACM* 41, no. 8 (August 1998): 74–83. https://doi.org/10.1145/280324.280336.

Crichton, Danny. "Liquid Democracy Uses Blockchain to Fix Politics, and Now You Can Vote for It." *TechCrunch*, February 24, 2018. https://techcrunch.com/2018/02/24/liquid-democracy-uses-blockchain/.

Dahl, Robert A. *Democracy and Its Critics.* New Haven, CT: Yale University Press, 1989. https://yalebooks.yale.edu/book/9780300049381/democracy-and-its-critics.

———. *Who Governs? Democracy and Power in an American City.* New Haven, CT: Yale University Press, 1961. https://yalebooks.yale.edu/book/9780300103922/who-governs.

Daian, Philip, Tyler Kell, Ian Miers, and Ari Juels. "On-Chain Vote Buying and the Rise of Dark DAOs," *Hacking, Distributed*, July 2, 2018. https://hackingdistributed.com/2018/07/02/on-chain-vote-buying/.

Day, John D., and Hubert Zimmermann. "The OSI Reference Model." In *Proceedings of the IEEE* 71, no. 12 (December 1983): 1334–40. https://doi.org/10.1109/PROC.1983.12775.

"'Democratising' Expertise, 'Expertising' Democracy: What Does This Mean, and Why Bother?" *Science and Public Policy* 30, no. 3 (June 2003): 146–50. https://doi.org/10.3152/147154303781780551.

de Vries, Alex. "Bitcoin's Growing Energy Problem." *Joule* 2, no. 5 (May 2018): 801–5. https://doi.org/10.1016/j.joule.2018.04.016.

Digiconomist. "Bitcoin Energy Consumption Index," 2020. https://digiconomist.net/bitcoin-energy-consumption.

Dixon, Pam. "A Failure to 'Do No Harm'—India's Aadhaar Biometric ID Program and Its Inability to Protect Privacy in Relation to Measures in Europe and the US." *Health and Technology* 7, no. 4 (December 2017): 539–67. https://doi.org/10.1007/s12553-017-0202-6.

Douceur, John R. "The Sybil Attack." In *First International Workshop on Peer-to-Peer Systems* (IPTPS), Cambridge, MA, February 2002: 251–260. http://research.microsoft.com/pubs/74220/IPTPS2002.pdf.

Doughty, Caroline. "Know Your Customer: Automation Is Key to Comply with Legislation." *Business Information Review* 22, no. 4 (December 2005): 248–52. https://doi.org/10.1177/0266382105060603.

Doward, Jamie. "The Big Tech Backlash: Tech Giants Are Drawing Political Fire over Fake News and Russian Meddling." *The Guardian*, January 28, 2018. https://www.theguardian.com/technology/2018/jan/28/tech-backlash-facebook-google-fake-news-business-monopoly-regulation.

Dubois, Elizabeth, and Grant Blank. "The Echo Chamber Is Overstated: The Moderating Effect of Political Interest and Diverse Media." *Information, Communication & Society* 21, no. 5 (January 2018): 729–45. https://doi.org/10.1080/1369118X.2018.1428656.

Durden, Tyler. "From $1,300 Tiger Penis to $800K Snipers: The Complete Black Market Price Guide." *ZeroHedge*, August 13, 2015. https://www.zerohedge.com/news/2015-08-13/1300-tiger-penis-800k-snipers-complete-black-market-price-guide.

Dürst, Hansjörg. "The 'Landsgemeinde': The Cantonal Assembly of Glarus (Switzerland),

History, Present and Future." In *IX Congreso internacional del CLAD sobre la reforma del Estado y de la administración pública*, Centro Latinoamericano de Administración para el Desarrollo (CLAD), Madrid, November 2004. https://cladista.clad.org/bitstream/handle/123456789/2992/0049818.pdf.

Dwork, Cynthia, and Moni Naor. "Pricing via Processing or Combatting Junk Mail." In *Advances in Cryptology—Crypto*, International Association for Cryptologic Research (IACR), Santa Barbara, CA, August 1992. https://doi.org/10.1007/3-540-48071-4_10.

Eberl, Jakob-Moritz, Hajo G. Boomgaarden, and Markus Wagner. "One Bias Fits All? Three Types of Media Bias and Their Effects on Party Preferences." *Communication Research* 44, no. 8 (December 2017): 1125–48. https://doi.org/10.1177/0093650215614364.

Edens, John F., Shannon Toney Smith, Melissa S. Magyar, Kacy Mullen, Amy Pitta, and John Petrila. "'Hired Guns,' 'Charlatans,' and Their 'Voodoo Psychobabble': Case Law References to Various Forms of Perceived Bias among Mental Health Expert Witnesses." *Psychological Services* 9, no. 3 (August 2012): 259–71. https://doi.org/10.1037/a0028264.

Enikolopov, Ruben, Alexey Makarin, and Maria Petrova. "Social Media and Protest Participation: Evidence from Russia." *Econometrica*, November 15, 2019. https://dx.doi.org/10.2139/ssrn.2696236.

Esau, Katharina, Dennis Friess, and Christiane Eilders. "Design Matters! An Empirical Analysis of Online Deliberation on Different News Platforms." *Policy and Internet* 9, no. 3 (September 2017): 321–42. https://doi.org/10.1002/poi3.154.

eternalgloom. "Overview of Universal Basic Income Crypto Projects." Bitcoin Forum, April 2, 2018. https://bitcointalk.org/index.php?topic=3242065.0.

Feldman, Leonard C. *Citizens without Shelter: Homelessness, Democracy, and Political Exclusion.* Ithaca, NY: Cornell University Press, 2006. https://www.cornellpress.cornell.edu/book/9780801441240/citizens-without-shelter/.

Ferrara, Emilio, Onur Varol, Clayton Davis, Filippo Menczer, and Alessandro Flammini. "The Rise of Social Bots." *Communications of the ACM* 59, no. 7 (July 2016): 96–104. https://cacm.acm.org/magazines/2016/7/204021-the-rise-of-social-bots/.

Filetti, Andrea. "Participating Unequally? Assessing the Macro-Micro Relationship Between Income Inequality and Political Engagement in Europe." *Partecipazione e Conflitto* 9, no. 1 (2016): 72–100. http://siba-ese.unisalento.it/index.php/paco/article/view/15893.

Fishkin, James S. *Democracy and Deliberation: New Directions for Democratic Reform.* New Haven, CT: Yale University Press, September 1993. https://yalebooks.yale.edu/book/9780300051636/democracy-and-deliberation.

Flavin, Patrick. "Campaign Finance Laws, Policy Outcomes, and Political Equality in the American States." *Political Research Quarterly* 68, no. 1 (March 2015): 77–88. https://www.jstor.org/stable/24371973.

Ford, Bryan. "A Liquid Perspective on Democratic Choice," November 2018. https://bford.info/book/.

———. "Delegative Democracy," May 15, 2002. https://bford.info/deleg/deleg.pdf.

———. "Delegative Democracy Revisited," November 16, 2014. https://bford.info/2014/11/16/deleg.html.

———. "Democratic Value and Money for Decentralized Digital Society," June 2018. https://bford.info/book/.

———. "Experts and Charlatans, Breakthroughs and Bandwagons: Collectively Distinguishing Signal from Noise under Attack," 2020. https://bford.info/book/.

———. "Let's Verify Real People, Not Real Names." October 7, 2015. https://bford.info/2015/10/07/names.html.

———. "Privacy-Preserving Foundation for Online Personal Identity," US Office of Naval Research Grant No. N000141912361, May 2019.

———. "The Remote Voting Minefield: From North Carolina to Switzerland," February 22, 2019. https://bford.info/2019/02/22/voting/.

Ford, Bryan, and Jacob Strauss. "An Offline Foundation for Online Accountable Pseudonyms." In *1st Workshop on Social Network Systems (SocialNets)*, Association for Computing Machinery (ACM), Glasgow, April 2008, 31–36. http://bford.info/pub/net/sybil.pdf.

Forget, Evelyn L. "The Town with No Poverty: The Health Effects of a Canadian Guaranteed Annual Income Field Experiment." *Canadian Public Policy* 37, no. 3 (September 2011): 283–305. https://doi.org/10.3138/cpp.37.3.283

Freitas, Carlos A., Fabrício Benevenuto, Saptarshi Ghosh, and Adriano Veloso. "Reverse Engineering Socialbot Infiltration Strategies in Twitter." In *International Conference on Advances in Social Networks Analysis and Mining* (ASONAM), IEEE/ACM, Paris, August 2015, 25–32. https://doi.org/10.1145/2808797.2809292.

Gemberling, Tess M., and Robert J. Cramer. "Expert Testimony on Sensitive Myth-Ridden Topics: Ethics and Recommendations for Psychological Professionals." *Professional Psychology: Research and Practice* 45, no. 2 (April 2014): 120–27. https://psycnet.apa.org/doi/10.1037/a0036184

Gerlach, Jan, and Urs Gasser. "Three Case Studies from Switzerland: E-Voting." Research Publication No. 2009-03, Berkman Center for Internet and Society at Harvard University, Cambridge, MA, March 2009. https://cyber.harvard.edu/sites/cyber.harvard.edu/files/Gasser_SwissCases_ExecutiveSummary.pdf.

Germann, Micha, and Uwe Serdült. "Internet Voting and Turnout: Evidence from Switzerland." *Electoral Studies* 47 (June 2017): 1–12. https://www.zora.uzh.ch/id/eprint/136119/.

Ghosh, Saptarshi, Bimal Viswanath, Farshad Kooti, Naveen Kumar Sharma, Gautam Korlam, Fabrício Benevenuto, Niloy Ganguly, and Krishna Phani Gummadi. "Understanding and Combating Link Farming in the Twitter Social Network." In *21st International Conference on World Wide Web* (WWW), Association for Computing Machinery (ACM), Lyon, France, April 2012, 61–70. https://doi.org/10.1145/2187836.2187846

Gilad, Yossi, Rotem Hemo, Silvio Micali, Georgios Vlachos, and Nickolai Zeldovich. "Algorand: Scaling Byzantine Agreements for Cryptocurrencies." In *26th Symposium on Operating Systems Principles* (SOSP), Association for Computing Machinery (ACM), Shanghai, October 2017, 51–68. https://dl.acm.org/authorize?N47148.

Gilens, Martin, and Benjamin I. Page. "Testing Theories of American Politics: Elites, Interest Groups, and Average Citizens." *Perspectives on Politics* 12, no. 3 (September 2014): 564–81. https://doi.org/10.1017/S1537592714001595.

Gölz, Paul, Anson Kahng, Simon Mackenzie, and Ariel D. Procaccia. "The Fluid Mechanics of Liquid Democracy." In *International Conference on Web and Internet Economics*, 188–202. New York: Springer, December 2019. https://paulgoelz.de/papers/fluid.pdf.

Green-Armytage, James. "Direct Voting and Proxy Voting." *Constitutional Political Economy* 26, no. 2 (June 2015): 190–220. https://jamesgreenarmytage.com/proxy.pdf.

Grönlund, Kimmo, Kim Strandberg, and Staffan Himmelroos. "The Challenge of Deliberative Democracy Online—A Comparison of Face-to-Face and Virtual Experiments in Citizen

Deliberation." *Information Polity* 14, no. 3 (August 2009): 187–201. https://dl.acm.org/doi/abs/10.5555/1735346.1735352.

Hajnal, Zoltan, Nazita Lajevardi, and Lindsay Nielson. "Voter Identification Laws and the Suppression of Minority Votes." *Journal of Politics* 79, no. 2 (April 2017): 363–79. https://doi.org/10.1086/688343.

Hardt, Steve, and Lia C. R. Lopes. "Google Votes: A Liquid Democracy Experiment on a Corporate Social Network." Technical Disclosure Commons, June 5, 2015. https://www.tdcommons.org/dpubs_series/79.

Hauben, Michael, and Ronda Hauben. *Netizens: On the History and Impact of Usenet and the Internet.* Hoboken, NJ: Wiley-IEEE Computer Society Press, May 1997.

Havocscope. "Fake ID Cards, Driver Licenses, and Stolen Passports," December 2, 2019. https://web.archive.org/web/20191202092629/https://www.havocscope.com/fake-id/.

Hawkins, Stephen, Daniel Yudkin, Míriam Juan-Torres, and Tim Dixon. "Hidden Tribes: A Study of America's Polarized Landscape." London: More in Common, October 2018. https://hiddentribes.us.

Heinlein, Robert A. *The Moon Is a Harsh Mistress.* New York: G. P. Putnam's Sons, 1966.

Hicks, William D., Seth C. McKee, Mitchell D. Sellers, and Daniel A. Smith. "A Principle or a Strategy? Voter Identification Laws and Partisan Competition in the American States." *Political Research Quarterly* 68, no. 1 (March 2015): 18–33. https://doi.org/10.1177/1065912914554039

Highton, Benjamin. "Voter Identification Laws and Turnout in the United States" *Annual Review of Political Science* 20 (May 2017): 149–67. https://doi.org/10.1146/annurev-polisci-051215-022822.

Hill, Kevin A., and John E. Hughes. "Is the Internet an Instrument of Global Democratization?" *Democratization* 6, no. 2 (September 2007): 99–127. https://doi.org/10.1080/13510349908403613.

Horowitz, Jason. "The Mystery Man Who Runs Italy's 'Five Star' from the Shadows." *New York Times,* February 28, 2018. https://www.nytimes.com/2018/02/28/world/europe/italy-election-davide-casaleggio-five-star.html.

Howard, Philip N., and Muzammil M. Hussain. *Democracy's Fourth Wave? Digital Media and the Arab Spring.* New York: Oxford University Press, 2013. https://doi.org/10.1093/acprof:oso/9780199936953.001.0001.

Iyengar, Shanto, Robert C. Luskin, and James S. Fishkin. "Facilitating Informed Public Opinion: Evidence from Face-to-face and Online Deliberative Polls." In *Annual Meeting of the American Political Science Association,* Philadelphia, August 27, 2003. https://pcl.stanford.edu/common/docs/research/iyengar/2003/facilitating.pdf.

Iyengar, Shanto, and Sean J. Westwood. "Fear and Loathing across Party Lines: New Evidence on Group Polarization." *American Journal of Political Science* 59, no. 3 (July 2015): 690–707. https://doi.org/10.1111/ajps.12152

Jackson, Tim, and Peter Victor. "Confronting Inequality in a Post-Growth World—Basic Income, Factor Substitution and the Future of Work." *CUSP Working Paper Series* No. 11. Guildford, UK: Centre for the Understanding of Sustainable Prosperity, University of Surrey, April 2018. https://www.cusp.ac.uk/themes/s2/wp11/.

Kalla, Joshua L., and David E. Broockman. "Campaign Contributions Facilitate Access to Congressional Officials: A Randomized Field Experiment." *American Journal of Political Science* 60, no. 3 (July 2016): 545–58. https://doi.org/10.1111/ajps.12180.

Kamvar, Sepandar D., Mario T. Schlosser, and Hector Garcia-Molina. "The EigenTrust Algorithm for Reputation Management in P2P Networks," In *The Twelfth International World Wide Web Conference* (WWW 2003), Association for Computing Machinery (ACM). Budapest, May 2003, 640–51. https://doi.org/10.1145/775152.775242.

Kara, Siddharth. *Modern Slavery: A Global Perspective.* New York: Columbia University Press, October 2017. https://cup.columbia.edu/book/modern-slavery/9780231158466.

Kaye, Barbara K., and Thomas J. Johnson. "Across the Great Divide: How Partisanship and Perceptions of Media Bias Influence Changes in Time Spent with Media." *Journal of Broadcasting & Electronic Media* 60, no. 4 (November 2016): 604–23. https://doi.org/10.1080/08838151.2016.1234477.

Keller, Eric, and Nathan J. Kelly. "Partisan Politics, Financial Deregulation, and the New Gilded Age." *Political Research Quarterly* 68, no. 3 (September 2015): 428–42. https://doi.org/10.1177/1065912915591218.

Kiayias, Aggelos, Alexander Russell, Bernardo David, and Roman Oliynykov. "Ouroboros: A Provably Secure Proof-of-Stake Blockchain Protocol." *Advances in Cryptology—Crypto*, International Association for Cryptologic Research (IACR). Santa Barbara, CA, August 2017, 357–88. https://doi.org/10.1007/978-3-319-63688-7_12.

Kline, Candice L. "Security Theater and Database-Driven Information Markets: A Case for an Omnibus US Data Privacy Statute." *University of Toledo Law Review* 39, no. 1 (February 2008): 443–95. https://www.utoledo.edu/law/studentlife/lawreview/pdf/v39n2/Kline%20Corr%20Final.pdf

Koistinen, Pertti, and Johanna Perkiö. "Good and Bad Times of Social Innovations: The Case of Universal Basic Income in Finland." *Basic Income Studies* 9, nos. 1–2 (December 2014): 25–57. https://ideas.repec.org/a/bpj/bistud/v9y2014i1-2p25-57n5.html.

Kokoris-Kogias, Eleftherios, Philipp Jovanovic, Linus Gasser, Nicolas Gailly, Ewa Syta, and Bryan Ford. "OmniLedger: A Secure, Scale-Out, Decentralized Ledger via Sharding." In *IEEE Symposium on Security and Privacy* (S&P), IEEE, San Francisco, May 2018, 583–98. https://bford.info/pub/dec/omniledger-abs/.

Koprowski, Gene J. "Spam Filtering and the Plague of False Positives." *TechNewsWorld*, September 30, 2003. http://www.technewsworld.com/story/31703.html.

Kotzee, Ben. "Expertise, Fluency and Social Realism about Professional Knowledge." *Journal of Education and Work* 27, no. 2 (November 2012): 161–78. http://dx.doi.org/10.1080/13639080.2012.738291.

Litvinenko, Anna. "Social Media and Perspectives of Liquid Democracy: The Example of Political Communication in the Pirate Party in Germany." In *12th European Conference on E-Government (ECEG 2012), European Commission*, Barcelona, June 2012, 403–7. https://pureportal.spbu.ru/en/publications/social-media-and-perspectives-of-liquid-democracy-the-example-of—2.

Liu, Dong, Quanyuan Wu, Weihong Han, and Bin Zhou. "Sock Puppet Gang Detection on Social Media Sites." *Frontiers of Computer Science* 10, no. 1 (February 2016): 124–35. https://doi.org/10.1007/s11704-015-4287-7

Loucaides, Darren. "What Happens When Techno-Utopians Actually Run a Country." *Wired*, February 14, 2019. https://www.wired.com/story/italy-five-star-movement-techno-utopians/.

Luechinger, Simon, Myra Rosinger, and Alois Stutzer. "The Impact of Postal Voting on Participation: Evidence for Switzerland." *Swiss Political Science Review* 13, no. 2 (Summer 2007): 167–202. https://doi.org/10.1002/j.1662-6370.2007.tb00075.x.

Luu, Loi, Viswesh Narayanan, Chaodong Zheng, Kunal Baweja, Seth Gilbert, and Prateek Saxena. "A Secure Sharding Protocol for Open Blockchains." In *ACM SIGSAC Conference on Computer and Communications Security* (CCS 2016), Association for Computing Machinery (ACM). Vienna, Austria, October 2016: 17–30. https://doi.org/10.1145/2976749.2978389.

MacDorman, Karl F. "Subjective Ratings of Robot Video Clips for Human Likeness, Familiarity, and Eeriness: An Exploration of the Uncanny Valley." In *5th International Conference of the Cognitive Science* (ICCS 2006), Cognitive Science Society. Vancouver, BC, July 2006, 48–51. http://www.macdorman.com/kfm/writings/pubs/MacDorman2006SubjectiveRatings.pdf.

Mack, David. "This PSA about Fake News from Barack Obama Is Not What It Appears." *BuzzFeed News*, April 17, 2018. https://www.buzzfeednews.com/article/davidmack/obama-fake-news-jordan-peele-psa-video-buzzfeed.

Madestam, Andreas, Daniel Shoag, Stan Veuger, and David Yanagizawa-Drott. "Do Political Protests Matter? Evidence from the Tea Party Movement." *Quarterly Journal of Economics* 128, no. 4 (November 2013): 1633–85. https://doi.org/10.1093/qje/qjt021.

Manza, Jeff, and Christopher Uggen. *Locked Out: Felon Disenfranchisement and American Democracy*. New York: Oxford University Press, 2008. https://doi.org/10.1093/acprof:oso/9780195149326.001.0001.

Mathews, Hans Verghese. "Flaws in the UIDAI Process." *Economic & Political Weekly* 51, no. 9 (February 2016). https://www.epw.in/journal/2016/9/special-articles/flaws-uidai-process.html.

May, Matt. "Inaccessibility of CAPTCHA: Alternatives to Visual Turing Tests on the Web," *W3C Working Group Note*, World Wide Web Consortium (W3C), December 9, 2019. https://www.w3.org/TR/turingtest/.

Mbiti, Isaac, and David N. Weil. "Mobile Banking: The Impact of M-Pesa in Kenya." In *African Successes*, vol. 3, *Modernization and Development*, edited by Sebastian Edwards, Simon Johnson, and David N. Weil, 247–93. Chicago: University of Chicago Press, 2016. https://www.nber.org/books/afri14-3.

Mendez, Fernando, and Uwe Serdült. "What Drives Fidelity to Internet Voting? Evidence from the Roll-Out of Internet Voting in Switzerland." *Government Information Quarterly* 34, no. 3 (September 2017): 511–23. https://doi.org/10.1016/j.giq.2017.05.005

Messias, Johnnatan, Lucas Schmidt, Ricardo Oliveira, and Fabrício Benevenuto. "You Followed My Bot! Transforming Robots into Influential Users in Twitter" *First Monday* 18, no. 7 (July 1, 2013). https://doi.org/10.5210/fm.v18i7.4217.

Miller, James C., III. "A Program for Direct and Proxy Voting in the Legislative Process." *Public Choice* 7 (1969): 107–13. https://doi.org/10.1007/BF01718736.

Mislove, Alan, Massimiliano Marcon, Krishna P. Gummadi, Peter Druschel, and Bobby Bhattacharjee. "Measurement and Analysis of Online Social Networks." In *Internet Measurement Conference*, ACM/USENIX. San Diego, CA, October 2007, 29–42. https://doi.org/10.1145/1298306.1298311.

Mislove, Alan, Ansley Post, Peter Druschel, and Krishna P. Gummadi. "Ostra: Leveraging Trust to Thwart Unwanted Communication." In 5th *USENIX Symposium on Networked Systems Design and Implementation* (NSDI '08), USENIX Association, San Francisco, April 2008, 15–30. https://www.usenix.org/legacy/events/nsdi08/tech/mislove.html.

Mori, Masahiro. "The Uncanny Valley [From the Field]." *IEEE Robotics & Automation Magazine* 19, no. 2, June 2012: 98–100. https://doi.org/10.1109/MRA.2012.2192811.

Moynihan, Donald P. "Building Secure Elections: E-Voting, Security, and Systems Theory."

Public Administration Review 64, no. 5 (September 2004): 515–28. https://doi.org/10.1111/j.1540-6210.2004.00400.x.

Nakamoto, Satoshi. "Bitcoin: A Peer-to-Peer Electronic Cash System." 2008. https://bitcoin.org/bitcoin.pdf.

nanalyze. "6 Digital Identity Verification Startups to Check Out," September 5, 2017. https://www.nanalyze.com/2017/09/6-digital-identity-verification-startups/.

Packer, George. "A New Report Offers Insights into Tribalism in the Age of Trump." *New Yorker*, October 13, 2018. https://www.newyorker.com/news/daily-comment/a-new-report-offers-insights-into-tribalism-in-the-age-of-trump.

Pan, Gang, Zhaohui Wu, and Lin Sun. "Liveness Detection for Face Recognition." In *Recent Advances in Face Recognition*, edited by Kresimir Delac, Mislav Grgic, and Marian Stewart Bartlett, 109–24. Rijeka, Croatia: InTech, December 1, 2008. https://doi.org/10.5772/6397.

Parijs, Philippe Van. *Basic Income: A Radical Proposal for a Free Society and a Sane Economy*. Cambridge, MA: Harvard University Press, 2017. https://www.hup.harvard.edu/catalog.php?isbn=9780674052284.

Park, Sunoo, Albert Kwon, Georg Fuchsbauer, Peter Gaž, Joël Alwen, and Krzysztof Pietrzak. "SpaceMint: A Cryptocurrency Based on Proofs of Space." In *Financial Cryptography and Data Security* (FC '18), International Financial Cryptography Association, Curaçao, February 2018, 480–99. https://fc18.ifca.ai/preproceedings/78.pdf.

Piketty, Thomas. *Capital in the Twenty-First Century*. Cambridge, MA: Belknap Press of Harvard University Press, April 15, 2014. https://www.hup.harvard.edu/catalog.php?isbn=9780674430006.

Prior, Markus. "Media and Political Polarization." *Annual Review of Political Science* 16 (May 2013): 101–27. https://doi.org/10.1146/annurev-polisci-100711-135242.

Puddu, Ivan, Daniele Lain, Moritz Schneider, Elizaveta Tretiakova, Sinisa Matetic, and Srdjan apkun. "TEEvil: Identity Lease via Trusted Execution Environments," May 9, 2019. https://arxiv.org/pdf/1903.00449.pdf.

Quinlan, Stephen. "Facilitating the Electorate: A Multilevel Analysis of Election Timing, Registration Procedures, and Turnout." *Irish Political Studies* 30, no. 4 (November 2015): 482–509. https://doi.org/10.1080/07907184.2015.1099041.

Ramachandran, Anirudh, David Dagon, and Nick Feamster. "Can DNS-Based Blacklists Keep up with Bots?" In *3rd Conference on Email and Anti-Spam* (CEAS 2006). Mountain View, CA, USA, July 2006. https://web.archive.org/web/20070625203520fw_/http://www.ceas.cc/2006/listabs.html#14.pdf.

Ratcliffe, Rebecca. "How a Glitch in India's Biometric Welfare System Can Be Lethal." *The Guardian*, October 16, 2019. https://www.theguardian.com/technology/2019/oct/16/glitch-india-biometric-welfare-system-starvation.

Read, Max. "How Much of the Internet Is Fake? Turns Out, a Lot of It, Actually." *New York Magazine*, December 26, 2018. https://nymag.com/intelligencer/2018/12/how-much-of-the-internet-is-fake.html.

Reich, Rob. *Just Giving: Why Philanthropy Is Failing Democracy and How It Can Do Better*. Princeton, NJ: Princeton University Press, November 20, 2018. https://press.princeton.edu/books/hardcover/9780691183497/just-giving.

Reinisch, Charlotte. "Swiss *Landsgemeinden*: A Deliberative Democratic Evaluation of Two Outdoor Parliaments." In *ECPR Joint Sessions*, European Consortium for Political Research. Helsinki, Finland, May 2007. https://ecpr.eu/Events/PaperDetails.aspx?PaperID=12373

Ribeiro, Filipe N., Lucas Henriqueo, Fabrício Benevenuto, Abhijnan Chakraborty, Juhi Kulshrestha, Mahmoudreza Babaei, and Krishna P. Gummadi. "Media Bias Monitor: Quantifying Biases of Social Media News Outlets at Large-Scale." In *Twelfth International AAAI Conference on Web and Social Media*, Association for the Advancement of Artificial Intelligence (AAAI), Stanford, CA, June 2018, 290–99.

Saez-Trumper, Diego, Carlos Castillo, and Mounia Lalmas. "Social Media News Communities: Gatekeeping, Coverage, and Statement Bias." In *22nd International Conference on Information & Knowledge Management* (CIKM '13), Association for Computing Machinery, San Francisco, October 2013, 1679–84. https://doi.org/10.1145/2505515.2505623.

Samuel, Ian. "Rigging the Vote: How the American Right Is on the Way to Permanent Minority Rule." *The Guardian*, November 4, 2018. https://www.theguardian.com/commentisfree/2018/nov/04/america-minority-rule-voter-suppression-gerrymandering-supreme-court.

Sayke. "Liquid Democracy." 2003. https://web.archive.org/web/20040726071737/http://twistedmatrix.com/wiki/python/LiquidDemocracy.

Schaub, Hans-Peter. "Maximising Direct Democracy—By Popular Assemblies or by Ballot Votes?" *Swiss Political Science Review* 18, no. 3 (August 2012): 305–31. https://doi.org/10.1111/j.1662-6370.2012.02075.x.

Schneier, Bruce. "Tigers Use Scent, Birds Use Calls—Biometrics Are Just Animal Instinct." *The Guardian*, January 8, 2009. https://www.theguardian.com/technology/2009/jan/08/identity-fraud-security-biometrics-schneier-id.

Schreiber, Flora Rheta. *Sybil: The True Story of a Woman Possessed by Sixteen Separate Personalities*. New York: Warner Books, 1973. https://www.grandcentralpublishing.com/titles/flora-rheta-schreiber/sybil/9780446550123/

Schryen, Guido, and Eliot Rich. "Security in Large-Scale Internet Elections: A Retrospective Analysis of Elections in Estonia, The Netherlands, and Switzerland." *IEEE Transactions on Information Forensics and Security* 4, no. 4 (December 2009): 729–44. https://doi.org/10.1109/TIFS.2009.2033230.

Scott, Grace Lisa. "5 Lifelike Robots That Take You Straight into the Uncanny Valley." *Inverse*, September 24, 2017. https://www.inverse.com/article/36745-5-lifelike-robots-that-take-you-straight-into-the-uncanny-valley

Serdült, Uwe, Micha Germann, Fernando Mendez, Alicia Portenier, and Christoph Wellig. "Fifteen Years of Internet Voting in Switzerland: History, Governance and Use." In *Second International Conference on eDemocracy and eGovernment* (ICEDEG 2015), IEEE, Quito, April 2015, 126–32. https://doi.org/10.1109/ICEDEG.2015.7114482.

Sethi, Tegjyot Singh, Mehmed Kantardzic, and Joung Woo Ryu. "'Security Theater': On the Vulnerability of Classifiers to Exploratory Attacks." In *12th Pacific Asia Workshop on Intelligence and Security Informatics* (PAISI 2017), Jeju Island, South Korea, May 2017, 49–63. https://doi.org/10.1007/978-3-319-57463-9_4.

Shao, Chengcheng, Giovanni Luca Ciampaglia, Onur Varol, Kai-Cheng Yang, Alessandro Flammini, and Filippo Menczer. "The Spread of Low-Credibility Content by Social Bots." *Nature Communications* 9 (November 2018). https://doi.org/10.1038/s41467-018-06930-7.

Shaw, Russell. "Avoid the Spam Filter." *iMedia Connection*, June 18, 2004. http://www.imediaconnection.com/content/3649.asp.

Shlapentokh, Vladimir, and Joshua Woods. *Feudal America: Elements of the Middle Ages in Contemporary Society*. University Park, PA: Penn State University Press, 2011. http://www.psupress.org/books/titles/978-0-271-03781-3.html

Smith, Warwick. "Political Donations Corrupt Democracy in Ways You Might Not Realise." *The Guardian*, September 11, 2014. https://www.theguardian.com/commentisfree/2014/sep/11/political-donations-corrupt-democracy-in-ways-you-might-not-realise

Solorio, Thamar, Ragib Hasan, and Mainul Mizan. "Sockpuppet Detection in Wikipedia: A Corpus of Real-World Deceptive Writing for Linking Identities." In *Ninth International Conference on Language Resources and Evaluation* (LREC 2014), European Language Resources Association (ELRA), Reykjavik, May 2014, 1355–58. http://www.lrec-conf.org/proceedings/lrec2014/index.html.

Srinivasan, Janaki, Savita Bailur, Emrys Schoemaker, and Sarita Seshagiri. "The Poverty of Privacy: Understanding Privacy Trade-Offs From Identity Infrastructure Users in India." *International Journal of Communication* 12 (March 2018): 1228–47. https://ijoc.org/index.php/ijoc/article/view/7046/2296

Stallings, William. "The PGP Web of Trust." *BYTE Magazine* 20, no. 2 (February 1995): 161–62. https://archive.org/details/eu_BYTE-1995-02_OCR/page/n210/mode/1up.

Standing, Guy. *Basic Income: A Guide for the Open-Minded.* New Haven, CT: Yale University Press, 2017. https://yalebooks.yale.edu/book/9780300230840/basic-income.

———. *The Precariat: The New Dangerous Class.* London: Bloomsbury Academic, 2011. https://www.bloomsbury.com/uk/the-precariat-9781849664561/.

Swierczek, Björn. "Five Years of Liquid Democracy in Germany." *Liquid Democracy Journal* 1 (March 2014): 8–19. https://liquid-democracy-journal.org/issue/1/The_Liquid_Democracy_Journal-Issue001-02-Five_years_of_Liquid_Democracy_in_Germany.html.

Templeton, Brad. "I Remember USENET." O'Reilly Network, December 21, 2001. https://web.archive.org/web/20050824032345/http://www.oreillynet.com/pub/a/network/2001/12/21/usenet.html?page=1.

———. "Origin of the Term 'Spam' to Mean Net Abuse." O'Reilly Network, December 2001. https://web.archive.org/web/20050825120621/http://www.templetons.com/brad/spamterm.html.

———. "Reflections on the 25th Anniversary of Spam." O'Reilly Network, May 2003. https://web.archive.org/web/20050826005609/http://www.templetons.com/brad/spam/spam25.html

Tisdall, Simon. "American Democracy Is in Crisis, and Not Just Because of Trump." *The Guardian*, August 7, 2018. https://www.theguardian.com/commentisfree/2018/aug/07/american-democracy-crisis-trump-supreme-court.

Tran, Nguyen, Bonan Min, Jinyang Li, and Lakshminarayanan Submaranian. "Sybil-Resilient Online Content Voting." In *6th USENIX Symposium on Networked Systems Design and Implementation* (NSDI), USENIX Association, Boston, April 2009, 15–28. https://www.usenix.org/legacy/events/nsdi09/tech/full_papers/tran/tran.pdf.

Tullock, Gordon. *Toward a Mathematics of Politics.* Ann Arbor: University of Michigan Press, 1967.

Venkatanarayanan, Anand. "Enrolment Rejections Are Accelerating." *Medium*, November 22, 2017. https://medium.com/karana/aadhaar-enrollment-rejections-are-accelerating-5aa76191d9a9.

Viswanath, Bimal, Mainack Mondal, Krishna P. Gummadi, Alan Mislove, and Ansley Post. "Canal: Scaling Social Network-Based Sybil Tolerance Schemes." In *EuroSys 2012*, Association for Computing Machinery (ACM), Bern, Switzerland, April 2012, 309–22. https://doi.org/10.1145/2168836.2168867.

Viswanath, Bimal, and Ansley Post. "An Analysis of Social Network-Based Sybil Defenses." ACM SIGCOMM Computer Communication Review 40, no. 4 (October 2010): 363–74. https://doi.org/10.1145/1851182.1851226.

Vorick, David. "The State of Cryptocurrency Mining," *Sia Blog*, May 13, 2018. https://blog.sia.tech/the-state-of-cryptocurrency-mining-538004a37f9b.

Wagner, Che. "The Swiss Universal Basic Income Vote 2016: What's Next?" Economic Security Project, February 7, 2017. https://medium.com/economicsecproj/the-swiss-universal-basic-income-vote-2016-d74ae9beafea.

Weitzer, Ronald. "Human Trafficking and Contemporary Slavery." *Annual Review of Sociology* 41 (August 2015): 223–42. https://doi.org/10.1146/annurev-soc-073014-112506.

Wen, Di, Hu Han, and Anil K. Jain. "Face Spoof Detection with Image Distortion Analysis." *IEEE Transactions on Information Forensics and Security* 10, no. 4 (April 2015): 746–761. https://doi.org/10.1109/TIFS.2015.2400395.

Woolley, Samuel C. "Automating Power: Social Bot Interference in Global Politics." *First Monday* 21, no. 4 (April 4, 2016). https://doi.org/10.5210/fm.v21i4.6161.

Woolley, Samuel C., and Douglas R. Guilbeault. "Computational Propaganda in the United States of America: Manufacturing Consensus Online." Working Paper No. 2017.5, Computational Propaganda Research Project, Oxford Internet Institute, Oxford, UK, June 19, 2017. https://comprop.oii.ox.ac.uk/research/working-papers/computational-propaganda-in-the-united-states-of-america-manufacturing-consensus-online/.

Yamak, Zaher, Julien Saunier, and Laurent Vercouter. "Detection of Multiple Identity Manipulation in Collaborative Projects." In *25th International World Wide Web Conference* (WWW 2016), World Wide Web Consortium (W3C), Montreal, April 2016, 955–60. https://doi.org/10.1145/2872518.2890586.

Yu, Haifeng, Phillip B. Gibbons, Michael Kaminsky, and Feng Xiao. "SybilLimit: A Near-Optimal Social Network Defense against Sybil Attacks." *IEEE/ACM Transactions on Networking* 18, no. 3 (June 2010): 885–98. https://doi.org/10.1109/TNET.2009.2034047.

Yu, Haifeng, Chenwei Shi, Michael Kaminsky, Phillip B. Gibbons, and Feng Xiao. "DSybil: Optimal Sybil-Resistance for Recommendation Systems." In *30th IEEE Symposium on Security & Privacy*, IEEE, Oakland, CA, May 2009, 283–98. https://doi.org/10.1109/SP.2009.26.

Zetter, Kim. "Experts Find Serious Problems with Switzerland's Online Voting System before Public Penetration Test Even Begins." *Motherboard*, February 21, 2019. https://www.vice.com/en_us/article/vbwz94/experts-find-serious-problems-with-switzerlands-online-voting-system-before-public-penetration-test-even-begins.

Zhang, Bingsheng, and Hong-Sheng Zhou. "Statement Voting." In *Financial Cryptography and Data Security* (FC '19), International Financial Cryptography Association. St. Kitts, February 2019. https://fc19.ifca.ai/preproceedings/97-preproceedings.pdf.

Zinn, Howard. *A People's History of the United States.* New York: Harper Perennial Modern Classics, November 17, 2005. https://www.harpercollins.com/9780062397348/a-peoples-history-of-the-united-states/.

Acknowledgments

Writing a book like this is an exercise in time travel. It began with gathering the chapter authors, each of whom has a different time horizon in mind for their writing. It moved into the three workshops that led to these chapters, during which the authors and editors debated issues of timeliness and time-boundedness. Each of us struggled with the pendulum swings of technological triumphalism and dystopia. We strove to include breaking insights and current topics, from digital disinformation to concerns about election hacking, while trying to write beyond any specific moment. And we went to press in early 2020, in the midst of the COVID-19 pandemic, during which the capacity of democracies to respond effectively was called into question and major technology companies were simultaneously battling misinformation on their platforms and rushing to assist public authorities with health surveillance. And of course, we sent the book off knowing that it would reach you, the reader, after the 2020 US presidential election, for which none of us dared predict the process, let alone the outcome. One thing was a certainty: the rise of digital technologies has profound consequences for democratic institutions, even, and perhaps especially, as those institutions are confronted with historic challenges.

Our greatest gratitude goes to the chapter authors who contributed to the book. They came from different methodological perspectives; brought with them a willingness to engage, learn, and disagree; and pushed one another to produce chapters that do the same.

An undertaking like this also depends on the work of many whose names do not appear as authors. The support of the Stanford Center on Philanthropy and Civil Society (PACS), particularly executive director Kim Meredith, was critical. Of all PACS's contributions, the most notable may be the support it provided for Hélène to join Rob and Lucy at Stanford for six weeks in 2017, during which time the idea for this editorial collaboration came to fruition. We are grateful to Heather Robinson and Laura Seaman, of PACS's Digital Civil Society Lab, who handled all the logistics of this three-year project with humor and exactitude. Margaret Levi and the Center of Advanced Study in the Behavioral Science provided space, food, and research assistance in Federica Carugati. We'd also like to thank Hilary Cohen and Gabriel Karger for their help taking notes at the author workshops. Finally, Elizabeth Branch Dyson and her team at University of Chicago Press took on the book and the process of expediting it for publication. We are grateful to Elizabeth and her team specifically and appreciative of the role that independent university presses fill in today's landscape of ideas.

Each of the people named herein spent countless hours on this volume. The choice to do so requires time spent away from home; in some cases this time was spent continents away. We are grateful to the authors' families for making space and time for this work.

Finally, some personal acknowledgments.

Lucy Bernholz: To my father, Peter Bernholz, who let me help "pull the levers" in those old election booths. Those were my first opportunities to ask questions about technology and democracy. Decades later, I'm still asking.

Hélène Landemore: To my husband, Darko Jelaca, who likes to say he builds unnecessary electronics and to whom I owe some serendipitous time spent in California and much of my interest in questions of democracy and technology.

Rob Reich: To my parents, Nancy Reich and Bob Reich, whose enthusiasm for my enthusiasms, bullishness for my bookishness, and rock-solid love made it seem possible that I might one day make a life out of ideas.

Index